# Family are the Friends You Choose

Marthe Kiley-Worthington

TSL Publications

This edition published in Great Britain in 2019 & 2021
By TSL Publications, Rickmansworth

Copyright © 2019 & 2021 Marthe Kiley-Worthington

Revised edition
ISBN: 978-1-914245-18-3

The right of Marthe Kiley-Worthington to be identified as the author of this work has been asserted by the author in accordance with the UK Copyright, Designs and Patents Act 1988.

All rights reserved. No part of this publication may be reproduced, stored in a retrieval system or transmitted, in any form or by any means without the prior written permission of the publisher, nor be otherwise circulated in any form of binding or cover other than that in which it is published and without a similar condition being imposed on the subsequent buyer.

# Contents

| | |
|---|---|
| Acknowledgements | 5 |
| The Irish setter, Fairy & Jackob. Setting the scene | 7 |
| Konyok & Mosquito helping me grow up in Kenya. A Colonial Kenya childhood | 19 |
| The Mau Mau & the Belgian Congo | 39 |
| Riding, farming, cows, dogs & horses as a teenager | 55 |
| University & back to Africa | 71 |
| The Katangese War, wedding & other African adventures | 98 |
| Animal communication, animal welfare & eco-farming | 113 |
| South Africa, people, politics, eland & blesbok | 133 |
| Milton Court Farm, horses, cattle & animal welfare | 147 |
| Ecological agriculture, horses', cattle & humans' welfare & behaviour: Can a multi-species community work? | 168 |
| Druimghigha, Isle of Mull | 190 |
| Circuses & zoos: animals & people | 208 |
| To the West Country, teaching & measuring quality of life in animals | 227 |
| Volunteers, students, multi-species living & teaching elephants, buffalo & black rhino | 243 |
| Cattle, elephants, India, Australia & France | 271 |
| La Combe, Drome, France | 295 |
| Similarities between gender & animal issues | 317 |
| Animal stories. What they can teach us | 332 |
| Beyond the Anthropocene | 369 |
| MKW Publications referenced in text | 410 |
| Select Index | 413 |

# Acknowledgements

There are so many sentient beings, experiences, landscapes and plants to thank, I don't know where to start. My Mum and Dad, I suppose, for giving me life is a good beginning, then the wisteria which flowered by my first window and encouraged my nascent being to take account of its beauty and smell, landscapes and bubbling brooks, deserts and forests, mountains and plains, snow fields and seas, often accompanied by a host of friends and family, high days, holidays, low days and difficulties, all in the mix of life. This book is a testament to them all. But more directly, those who have helped directly with this book are Angela Kingston for reading the first bits, and encouragement, Anne Samson of TSL who took the risk to publish the full article and converted my erratic English into a readable version. Kate Rawls, the outdoor philosopher and bike rider, and Christine Nicol, a thoughtful open minded welfare scientist, two of my long-time associates are deeply thanked for their critical reading of the first version of the manuscript and particularly for their help in pointing out my gross bloomers or lack of clarity. There have been many mentors, both bipedal and quadrupedal, in my life who have kept me encouraged and motivated to report it all. The list is so long that I will rely on the book to point out their donations. Finally, my long-term partner and friend, Chris Rendle for his help in all life enhancing directions, and today my black boy Shatish and spotty Moto for their lifelong loyalties and the constant surprises of the mental aptitudes they display, now I have learnt a little more how to read them. There is more to learn, but perhaps this book will help those who follow to jump over the first hurdle or two so they can advance further in our understanding of all mammals.

# The Irish Setter, Fairy & Jackob.
## Setting the scene

It was 1939 and my pretty dark-haired mum was very pregnant for the fourth or fifth time, she already had produced two girls and then had a couple of miscarriages. The upper middle-class family she had married into very badly needed a boy to carry on the family name and do all the other things that only boys were allowed to do. It was considered by her mother-in-law and the aunts of both sides of her husband's family that she should have the baby in London. So, at a time when everything was heating up for the beginning of the Second World War and the bombing of London, she was packed off to 1 St John's Wood in London, where one of the female relatives resided, so she could have the baby under the strict eye of the best doctors in London.

May 16th, 1939 arrived. According to Google, Ravensbruck concentration camp in Germany was opened on that day, and my mum went into labour late at night. The family waited, holding their breath and probably counting worry beads, or whatever they did for wishes to come true. All the signs were that it would be a boy, surely, for the Worthington line to continue. Much anxiety as my black-haired scruffy head appeared followed by the rest, and alas, despite everyone's efforts it was another girl ... what a disappointment. Instead of celebrations, my mum and I were quickly packed off back to the Lake District where my father was running a research station, finding out about the life of eels, trout, perch and all the other living things in the numerous small tarns dotted about in the moors.

So that was that, another girl. But, my earliest memories are not about what eventually was considered a "fiasco". They are rather the pleasure of lying in my cot next to the window of the family's beautiful Georgian mansion just outside Hawkshead (where Wordsworth lived) in the Lake District with the dangling blue wisteria flowers pouring scent and visual delight right through me. It was an unusually early summer for the Lake District, and the delicate scent drifted over this semi-conscious baby hour after hour; perhaps it was this that imprinted on her the glories of the natural world. Even today I smile if I smell wisteria in flower. For that instant at least, the sun is shining, flowers blooming, and all is well in the world.

My mum suffered deeply from having disappointed the family in the sex of her last offspring, although why she was to blame and not equally my father, was a mystery. Some of this disappointment was reflected on me as I grew up but, for the first few years, unaware of any such chagrin, from all accounts I was a noisy, boisterous, cheerful, energetic kid, always in the way. The youngest by a long way, five years is a long time when you are young, and the most incapable – which was frequently pointed out to me.

Another very early memory was lying in my cot under the enormous old beech trees at the top of the drive. Occasionally, Mr B, the tenant farmer of the home farm, would peer down at me with his old eyes all clenched up and say things like, "'es right ugly this one", followed by reels of North country swearwords. According to my father, my first words were swear words and I have to confess I have had a propensity to use them all my life, to the horror of many.

Not far from us in Barrow-in-Furness on the coast, they were building submarines. The Blitz had started in London and soon Barrow attracted the attention of the German bombers. As I began to scurry around, I remember having the background noise of bombs drop and the earth shudder. But my job was to swing in hammocks or run down to my favourite place: a hidden pond under a red Japanese maple where the moss dripped into the transparent still pond. How could it be that among such natural beauty, humans could beat each other up?

Fifteen refugee relatives descended on us. My mum did her best, she had a couple of sows to add some pork to the ration every now and then, hens for eggs, and she cultivated a vegetable garden. I can't really remember any of the relatives helping cook, clean, dig the garden or anything else, although they had little else to do. They were, after all, "towns people" who spent their time arguing with each other or about clothes and did a great deal of dressing and undressing, all of which to me seemed a bit of a bore, even at the age of three or four.

My mum made me "frocks" from bits and pieces of left over cotton, and insisted that I had to have matching underpants. For some reason I really loathed this. Worse was to come. An out of work portrait painter was hired to paint each of us. My sisters looking grown up, proper and in control. But, I sat on a chair in one of those awful frocks with matching underpants feeling awkward and bored. To keep me still, someone read *Moorland Mousey*, a very sad story about a Dartmoor pony, so I wept and wept, which was not helpful for the portrait painter, but I did love the story. We graduated to Hans Andersen and his horror stories of people cutting each other up which at least stopped me weeping! But, the frock and matching underpants were there for perpetuity when the painting was finished.

Sometimes, my father would take us on a walk to fish in one of the tarns not far up in the moors. The rest of the family were older, longer legged and walked a lot faster. I was always being left behind on these moorland walks, however hard I tried to keep up. My sisters who I greatly admired, would constantly turn around with a sigh, saying with long faces, "Oh Martha, do keep up, what is wrong with you?" ... the human rot began to set in. Clearly, I was not up to much. I also remember having to sit still in a very wobbly dinghy in the rain where we had to wear those peculiar large yellow so'wester hats where the water dripped regularly down your neck and in your eyes, and the smell of fish pungently oozed through the rain as my sister Grizelda and father gleefully gutted the trout they had caught to see what they had been eating. All of this, they said was designed to "make a man of me"; at least it might have done, if I had been the right sex!

I began to wonder if I would ever make it as a human. I seemed to be accepted more by our Irish setter who was kind but kept me in my place, and our two fell ponies, black, long-haired, ragged chaps whom my sisters rode and finally, one day, I was allowed to sit on Fairy. I could feel her warm, soft, living, breathing, feeling self. Should the humans be in a bad mood with me, which was normal, I began to have friends who accepted me for who I was, not who I ought to be and they tolerated me and gently ensured that I learnt their rules of behaviour. With brown, gentle, shaggy Fairy, I could keep up on expeditions, even outstrip the others ... escape the rocking, smelly, cold, wet boats. It was a new beginning.

Those first five formative years are still deeply felt and remembered. My sisters rarely allowed me to join in their work or games, but on one occasion they allowed me to muck out the drains in the Victorian stables. I was so proud that I had been included, even though they made me do it again and again and I finally emerged covered in everything that drains contain in stables, much to their glee. But by being allowed to join them, I was deeply flattered and soon realized that this must be how a neglected dog feels when finally allowed to take part in some human activity. We were not a church going family, but I somehow discovered the parish priest, or he discovered me. His rectory was just the other side of a beautiful Capability Brown designed open parkland with large oaks and grazing brown and white Ayrshire cows where I spent much time chatting to the cows and not requiring any answers, but noting their tolerant looks; they were treating me as an equal. The vicar was an old man who also seemed to need some company. We would sit and admire the bubbling stream together or wander slowly around his garden, filled mostly with evergreen laurel if I remember. He made me a set of stables out of oak for a birthday present, and for years they were the best toy I ever had. I accumulated Dinky metal models of horses and carts to live in them. They are dilapidated now, but I still have them ... I wanted to give them to one of my grandchildren, but none are interested!

There were some memories of the war. All the various radio broadcasts were avidly listened to by all members of the household, except me. I do remember my father joined the Home Guard and told us stories which resembled closely *Dad's Army* ... they rushed around carrying guns (usually with no bullets). They spent a lot of time crawling around on the moors, chatting and drinking homemade beer. There was no petrol, so no motor car could be used. This pleased me. Mum, who always had a plan B, put one of the ponies into a trap she had acquired, and we drove off to Ambleside in the rain to claim the rations from the ration cards of all the relatives, and maybe buy the odd treat of a bun.

Once or twice she took me to visit Beatrix Potter who lived in a neighbouring village and who had become a friend of hers. My mum hankered after conversation with literary or musical people I think, but the most I remember at Beatrix Potter's stone cottage in the moors was looking at her books of pictures of the animals around her, even if they were somewhat humanised. The cakes were gorgeous too, they were probably plain madeira, this was after all "the war", but they were very special as we never had cakes at home. We would trot back in the trap as it was getting dark, usually in the rain and always getting freezing cold. As usual one had to get out on the hills as Fairy was unable, or unwilling (I suspect the latter), to pull us both and the two-wheeled tipping trap; still it did warm you up.

When I was about five, I was sent to nursery school in Ambleside in order to learn to read and write. We would trot in with Fairy and often after a somewhat boring day, Fairy and Mum would turn up again to take me home. I only remember obtaining a prize for "tidiness" because my books were neatly piled up ... not because of my appearance. Never again was I to get such a prize!

My father was not often at home, what he did during the war and why being of military age he was not drafted into the army, remained a mystery. People would say he was exempt because he was doing "useful work understanding eels, so we could all eat eels and increase our protein intake". But in 2016 I received a package through the post (from relatives of his second wife who had died). In this were a host of diaries and photos of his early life. The diaries of most interest are those that were written of his trips to the Middle East between 1942-45. Apparently, as a biologist, he was hired by the Foreign Office to travel to all the countries in the Middle East administered by Britain at the time. This was to investigate their natural resources and how they could be better used sustainably to feed, clothe and supply building materials and so on for the local populations, and how they could be "developed" to create industries. Scientific research on diseases of animals and humans and how to prevent or cure them was included. There is somewhere in the bowels of some Foreign Office library the original scientific reports, but his diaries included much of the science and also his personal impressions and feelings about how and what was going on in these countries. As a special envoy of the British Government, he was grandly entertained and hosted by governors, the leading scientists and all those with influence. He describes large meals and plenty of drinks, sitting on verandas in the sun discussing this and that. At the time, his family was not exactly starving, but we had a restricted and boring diet of porridge, bread, potatoes and cabbages, with the odd egg thrown in and two ounces of butter a week. But I am sure my dad earned his high living!

The most interesting part of these diaries are his visits to Palestine before Israel had even been proposed. In the early 1940s most of the Middle East was administered by Britain, and Palestine is drawn by him as an interesting, exciting country with many educated and thinking people. Both Muslims and Jews worked alongside each other trying to understand and combat diseases of animals, plants and humans, and doing research and development in agriculture, forestry, and fisheries. At that time, Britain was still respecting the deal that Lawrence of Arabia had organised between the Arabs and Britain. What a shame that this did not continue, and the treaty was broken. It might have been a better world if such a state had remained.

Father did the colonial things of the time, visiting the Gazira Club in Egypt, and watching polo played on the edge of the desert, but it is fair to say that I never saw any sign of racism in his dealings with people. He believed strongly in a meritocracy and not the "old school tie" or familiarity network. He had some great mentors in this. One was my godfather whom I never met until he was well into his eighties and in a wheelchair: Lord Hailey, one of the last viceroys of India. They wrote a book together encouraging cooperation in science throughout Africa: *Science in Africa*. Another was Bad Uncle John. My father and John Corner had been at preparatory and public school together, and at every opportunity had sneaked away to collect, measure and study bugs and beetles, butterflies and birds, mosses and liverworts. They went to Cambridge together, he to study zoology, and John botany. During the war, John Corner was director of the Singapore Botanic Gardens, the most comprehensive collection of tropical plants in the world at the time. After the Japanese invasion, instead of being captured and sent to a camp something like *On the River Kwai*, or choosing to blow up all the plants, he negotiated with the Japanese to keep the plants, and eventually the emperor, who was a keen botanist, gave orders to ensure the safety of the botanic gardens. Forever after Uncle John was Bad, it was treason to negotiate with the Japanese to save plants from extinction. Luckily, he could bury his head in his flowers, and came up with some of the most original and exciting ideas on the evolution of flowering plants that were quoted with reverence when I studied botany later. He was one of the early deep ecologists who recognised the intrinsic value of all living things, and was not prepared to give that up for a human squabble, and he was great fun. I remember him visiting and games, cavorting around our large house and garden together. At the end of their lives, my father and Uncle John fell out, I never did discover why, whether it was because he dropped his pretty American wife in favour of the Czechoslovakian housekeeper, or what. Whatever it was, there was a feud of such power that it broke a friendship and their common interests which had spanned some seventy-five years, it seemed a bit silly! Another godfather I never knew was a professor of zoology at Cambridge. I imagine he had been very influential in my father's education. His name was George Carter and he wrote numerous textbooks which I ploughed through when I grew up. I do not remember having any godmothers. Godparents are supposed to be available to discuss, comfort, reassure and maybe even put a different point of view from the parents to the godchild. They should not just be famous, important people, but my father was ambitious, and my mum did not seem to have had a say, so godparents did not play a role in my life.

When not travelling around the Middle East, my father was the director of the Fresh Water Research Centre on Lake Windermere, situated in Wray Castle, a pseudo-castle built by some rich guy with pretend turrets, moats and draw bridges on the edge of the lake. There was a bevy of young women scientists who were very bright and fun, researching the life of this or that fish or plant in the lake. Fresh water ecology was the first type of scientific ecology that developed, probably because there were clear limits to that ecosystem: the shoreline. As a result, it had attracted many young women who had recently been allowed to become ecologists. The universities in the UK had only been able to grant women degrees for the previous twenty years or so and some of the biologists had found an exciting opening and were hired to do research at Wray Castle. About five of those I vaguely knew became very well-known internationally, pushing back the barriers of ignorance about a range of freshwater habitats worldwide, and influential in starting the Open University and other academic pursuits. Occasionally, as a small girl I visited Wray Castle but all I can remember are dark, dank passages with small fish-smelling laboratories opening off them, and swimming in the very cold deep lake: "never dive in after a meal, you will not come up again," they would shout at me.

Perhaps because of his success getting along with people and writing good reports, at the end of the war, the colonial government offered my father a post doing more or less the same thing he had done in the Middle East in East Africa: Kenya, Uganda and Tanzania. He was to be based in Nairobi and apart from paper pushing, the idea was that he would travel around and work out what was best to do with the forests, agriculture, veterinary and medical research and of course fisheries, in order to benefit firstly, the people who lived there, and secondly "development" for the colonial government.

In 1946, we were all packed up and ready to catch the *Lancaster Castle*, an enormous cruise ship commandeered by the military to take soldiers around. It took us through the Suez Canal, down the east coast of Africa and to Mombasa. There were many soldiers aboard. The colonial servants and returning settlers were squashed onto one deck. The journey took three weeks and for us kids, it was full-on fun. There were things organised daily: deck quoits, races, fancy dress competitions, treasure hunts and so on. On top of that it was hot and although there was no swimming pool, there were plenty of hose pipes and shower heads around to play with. I made particular friends of the soldiers. I suppose because I was very young and bumptious, they took me to their hearts, giving me sweets and biscuits, taking me down to their common rooms and the boatswain made a special skirt out of sisal rope for me to wear in the fancy dress competition, which I duly won. My mother kept threatening me that I would be thrown overboard if I continued to visit the soldiers, but how was she to know? My solemn sisters read books and lay in the sun and did all the things that teenage girls need to do.

We stopped at both Port Suez and Port Said at the beginning and the end of the Suez Canal. It was a riot: hundreds of small boats came storming out to our enormous ship with every conceivable type of consumable. We were not allowed off the boat, but things were taken on and off with ropes and pullies, including the Gili Gili men, Arabs in large nighties who had baby chicks hidden all over themselves and did tricks producing them. I was transfixed and spent many hours watching, until of course they were discovered and roughly removed to dangle down on ropes to their dinghies, and I didn't see a single chick fall in the water!

Another ten days or so in the Indian Ocean chasing flying fish. These are incredible fish whose pectoral fins open up into wings and they can fly and fly along the bows of the ship in hundreds. I watched the clear, clear water to try and find a mermaid or monster and then we had a "crossing the equator" celebration. I swear, there were real mermaids and monsters on deck, at least briefly, followed by orange juice and cakes for tea, what luxury. I spent the evening watching the water go down the plughole because on the other side of the equator, it is supposed to circle down the other way around. Difficult to tell of course, as I could not remember which way it had gone before.

We glided quietly into Mombasa harbour, and witnessed the milling multitudes of all colours, races and genders on the pier awaiting the ship's docking. The diversity of humans was astounding, white colonial servants with big topies (white helmets they wore on all occasions ... if they forgot to put them on they died of sun stroke, so my mum told me), off-white shirts and shorts that were very baggy, and long white socks with their muscley legs showing through (I was about leg height so legs became important to me). Then there were Swahilis with their brightly coloured kangas and kikoyes wound around them, the females around their tops and carrying their heads high with enormous brightly coloured turbans and usually a plate of mangos or avocados on top, and the men looking urbane and colourful, strutting about in kangas wound around their waists with very rude messages written on them in Swahili (someone translated one or two for me). Their legs if visible were deep brown and not so muscley. Severe looking Arabs with flowing gowns, enormous beards and pieces of cloth over their heads held on with pieces of wire, and Arab women all in black, gliding about like witches. No visible legs either.

After several hours of hassling and waiting in the sun, Mum had managed to get us onto the train that would take us to Nairobi. The train line had been constructed at enormous cost of money and lives by the "coolies" who had come from India. It wound up 5,000 feet in around 350 miles and took a day and a night as the wood-powered engine puffed its way along. We went right through Tsavo where the man-eating lions had eaten many of the workers, and then passed herd after herd of antelopes: Grants and Thomson's gazelles, occasional bush buck and waterbuck near the rivers with crocodiles and hippopotamuses blowing air, buffalo, zebra and elephants wandering around smashing trees and generally being at home. At that time, wild animals were everywhere throughout Kenya, it was their day. What a contrast to today where people are everywhere, and the animals confined to what are really large zoos.

We passed Mount Kilimanjaro with snow on its summit and chugged into Nairobi station in the early morning where Father met us and took us out to Dagoretti Corner where he had rented a house. At that time, this was a sort of open, rather upper crust suburb where each house had a large garden and a field or two in case of need. After the compulsory unpacking, there was exploring, and I found that less than half a mile away were the plains of Africa with their occasional flat-topped acacia trees, *totos* (children) minding cattle, sheep and goats, somewhat fleeting herds of gazelles and the odd warthog rushing off with its tail in the air. The garden was also exciting, with all sorts of unknown plants and trees and just around the corner was this enormous valley where a car could reach what seemed like enormous speeds on the descent and then use this mobility to gradually get up the other side. We had a very smart Dodge saloon car and the speedometer changed colour as you went faster or slower. One of the treats was to get Father to drive it up and down the valley so we could watch the speedometer and ooh and aah at its purples and reds, blues and yellows.

Then it was school time again, this time a smart school for the suburb dwellers just down the road. I seem to remember that I had to wear green jerseys and green-checked shirts, while my sisters went off to Kenya Girls High School, a school set up primarily for the daughters of colonial servants but with a considerable academic reputation, and they had blue shirts and grey skirts.

Much of everything was different, but very luxurious with servants cooking and cleaning, and masses of fruit, pineapples, pawpaws, avocados, mangos, bananas, the list was never-ending. I had never seen such a wealth of food and fruit before. We went to the markets and passed shops selling such cream cakes you would not believe, piled high with jams and creams, all colours and types … just standing and staring at them was enough. There was no rationing on clothing, so my mum visited many Indian shops to buy a whole variety of beautifully coloured cottons for curtains, tablecloths, napkins and anything else we needed ... and of course the dreaded frocks ... she never got to the stage of buying "ready-made" clothes as everyone does today. Father's office was on the fourth floor, and there was a grumbly old lift. What a treat, I went up and down many times; I had never seen a lift before.

We soon had a puppy belonging to my sister and finally I managed to persuade my mother that I had to have a pony as well as my sisters, so a dumpy little Somali chap, Jacob, joined us. He taught me a great deal about what not to do at least, but he was never very keen on doing anything, which in itself is a good lesson: how to enthuse him ... it did not often work for me then. Our neighbours were retirees and Ma Gillies would help me make fires and cook revolting mixes in her garden, so we became close friends. I really don't know what they had been doing all their lives, probably care-taking farms and houses for people, but eventually they joined us with other friends, and three families bought five acres next to the sea at Diani, north of Mombasa where we used to go together for the annual or biannual holiday to look at the coral reefs and splash about keeping well away from the "chicken fish", beautiful fish with multi-coloured fins dangling and swaying everywhere. If you touched one of them, you would die within a couple of hours, so admiring them but keeping clear was important. We would roll around on the white beach in the hot, hot sun or under the coconut palm trees. Every now and then a coconut would drop, you hoped not on your head, and its top would be cut off quickly for the delicious cool sweet *madafu* milk to be greedily slurped down. We slept in *banda*s made of banana leaves under a mosquito net and fruit and vegetables came on the heads of ladies dressed in multi-coloured swaying cloaks or severe Arab men in nighties with funny black hats with whom we bargained and bantered for hours ... They were idyllic holidays.

It took a couple of years for Mum to persuade my father that she needed a small farm in the country not too far away from Nairobi so he could continue to go to the "office" and she to play music with her friends when not farming. They finally bought twenty acres on the edge of the escarpment in the rift valley, followed by another twenty being sold off by an old Kenya settler, Jack Dyer, who had started farming there some forty years previously and now that his paunch and his sons were growing, he was downsizing. It was there at Liloni that my real multi-species life began.

What I had learnt from animals during my short life was that they were more accepting, reliable, less grumpy and they did not make wars and blatantly cause others of any species to suffer. They had rules in their societies and if one knew and obeyed them, they were more tolerant and accepting. I had had more security and joy in their company than with humans thus far.

## Konyok & Mosquito helping me grow up in Kenya. A Colonial Kenya childhood

It all really began with Konyok Gabonyok, a lively golden Cocker Spaniel puppy, who was about the same mental age as myself. We both had a lot to learn, from each other and from our elders and betters. He grew rather faster than I, but we became friends and partners in fun and crime: we fought, we laughed, we played, we had "early supper" together in order to go to bed earlier than the others. Although I was the same species as the elders and betters, I was younger, self-willed, incompetent and frequently "got things wrong". I was "a bit of a bore" as my elder sisters would say. They took a similar line to both of us: sweet but silly, mindless, but jolly fellows who understood nothing.

But even knowing this, both Konyok and I loved attention from our elders and betters. He would sidle up to one of them and push his cold nose into their hands, lick and rub his body against them and jump onto their lap, if he thought he could without being scolded. Another trick was to wag his tail so hard that precious breakables smashed to the ground. I would wrap up things I liked and present them to my parents and sisters at breakfast. As they unwrapped a sticky red ribbon or a broken plate with the drawing of half a horse, there would be indulgent sighs and smiles. But, although we understood their reactions, it never put either of us off trying to please them.

Konyok had been taken from his family at eight weeks old, but I was being raised by mine. The similar attitude that the elders held towards both of us was, in part, the result of me having been a disappointment from birth. I was the wrong sex and the last hope. My father who was an Edwardian began his family with strong ideas concerning gender differences, even if these eventually became blurred after twenty years working with some very bright female biologists and also living for many years in a capable all-female family.

My mother had struggled for girls' rights within her own natal family of five girls, mother and parson father. She had excelled at St Paul's School in academic work, sports and music, but was forbidden to go to university by her father. She ran away from home because of this; goodness knows how, as at that time no girl was even given the money for a bus fare by her parents. Eventually, with the aid of the ferocious ex-suffragette head mistress of Rodean School (founded to give equal education to girls at the beginning of the twentieth century) who, I am glad to say, terrified my grandfather, she won a place at Newnham College, Cambridge University. But after a year, she met my father who was just off to northern Kenya to study Lake Turkana. She volunteered to go too and make maps since she was reading geography, but this was not allowed unless they married, so they got married, although I don't think my mother was ever really considered the "right sort" by his family, but they tried to get on at least.

It was the early 1920s and my mum was the first white female to visit this remote part of northern Kenya. One of her greatest achievements was to record by ear the music of the El Molo tribe (which is now extinct). She could do this since she had perfect pitch and could write down exactly the note that they sang as a music score. The problem was, she told me, that they often sang in quarter tones, which was difficult to record in western music. On returning to Cambridge a couple of years later however, she was not permitted to continue her studies; they did not allow married women to study for degrees! She suffered all her life, I believe, from this unfair treatment as she helped three daughters grow up, go to university and have professions; it must have galled her greatly.

Konyok and I aged six, were in awe of "the family", but knew that we would never come "up to scratch", although my mum was at least indulgent. Soon we were three, joined by Mosquito, a Somali pony crossed with an Arab stallion. He had trekked for months across the deserts of northern Kenya with the Somali tribesmen to be sold to white settlers in Kenya. He was a small, vivacious, unreliable, and energetic pony with strong opinions and fitted well into our erratic, quarrelsome, fun-loving group.

Mosquito had his own equine associates, but enjoyed our threesome outings into the African bush, larking around, and squabbling in particular. One game, I am ashamed to say, was to frighten the tribal Kikuyu women working in their maize fields by galloping past them. Our adventures were invented by any one of us, willingly entered into by all three and had unpredictable rules. But, like all play, there was one unchangeable rule: we must not hurt each other intentionally ... or it was no longer a game. In the process of chasing, running, hiding, leaping and dare games, we learnt a lot about each others' skills and limits. I could never leap as high or run as fast as the others, but I could shout loudly and frequently dreamt up new games. We could never hide from Konyok, he always found us by smell and could hide in small places where he was impossible to find. Mosquito would gallop around in circles and terrify us by coming up silently behind to leap over us. The others knew that I hurt easily and then wailed, so they learnt to be less rough with me than they were with each other.

My somewhat avant-garde family of musician/farmer mother and remote biologist father and two teenage sisters were all seriously engaged in their own lives, so I ran a little wild. The sensible solution was to send me off to boarding school in the highlands of Kenya where I could join a group of other somewhat wild settler kids. But, bossy Martha found that they were easier to organize than my friends of different species, so quickly organised a gang of four. For the next four years, we had an encyclopaedia of adventures in the beautiful Kenya highlands. As the leader of the gang of what were really gentle pleasant girls, I often received "six of the best" from "Pa's" (the headmaster's) taki (his tennis shoe). My mother received mildly disapproving reports of my behaviour and inability to conform; I was leading others astray. But she usually ignored these and in her weekly letters, instead of lectures on behaviour, I was sent news of my playmates, and other animals on the farm, how the puppy had peed in Father's shoe, the cow eaten the cook's trousers, and Mosquito had knocked Johanna (the groom/syce) over by nuzzling him too hard.

In the multi-species holiday play-group there was one unbreakable law which has been reinforced throughout my life by all kinds of four-legged pals. This is that our emotional attachments were unarguable; for better or worse, wherever, whenever, forever. Even after all these years living with, teaching and making friends with humans, I am shocked by how often loyalty is overlooked among humans, including those one had regarded as "lifelong friends". Never, once friendship has become established with an individual of another species, is loyalty ever questioned, as long as the friendship is not threatened by bad behaviour to the friend. Elephants, buffalo, lions, dogs, horses, cows, bulls and many other individuals of different species have all been my loyal, reliable friends.

It was therefore aged six that I first asked the questions that have occupied me all my life. What made my friends of different species "tick"? What were they feeling or thinking? How did they see the world? Humans were easier, after all I was one. It was clear to me that my friends of different species felt similar emotions to me, and displayed them in similar ways: pleasure, joy, anger, irritation, confusion, frustration, uncertainty, for example, but yet, they could also see, hear, taste and smell things that I could not. I had a complicated *verbal context independent* language; the same word meant the same thing whatever the context. But the meaning of their messages was often context dependent, that is the meaning depended on what was happening around them, not the word, gesture or movement itself. We liked some similar things but also some different, so how did other species' minds compare to my thinking, learning, dreaming, problem solving, imagining, pretending, believing and feeling? What did we have in common and what was different?

Another puzzle was why my non-human friends cooperated with and often liked humans, even though humans could be brusquer, crueller and more unreasonable with them than any other species with its own. Why do they still willingly do things with us, even when we cannot physically force them to? What is in it for them? What compensates for humans' bad behaviour towards them? These questions dictated how the rest of my life panned out.

At that time, zoologists and many philosophers argued that animals did not have minds. In the 1950s when I was considering my future options, Nikolaas Tinbergen at Oxford and Konrad Lorenz in Germany had started studying "instinctive behaviour". This is behaviour which is inflexible and somehow preprogrammed in the brain. If behaviour was dominated by "instincts" then there was no need for a mind to make choices and decisions, to change behaviours or to consider things. If a non-human had feelings, these were therefore "automatic reactions". Only sentimentalists they said, considered that non-human mammals felt feelings, had thoughts, were conscious and could work out problems ... or even had experiences. Scientists, philosophers, zoologists and everyday people in the street were believed to be the only rational thinkers. They admitted that some species could learn, but this, with the help of Pavlov and Skinner (a psychologist of the time), was (and still is by some) believed to be a curious automatic "conditioning process" where they had no need for a "mind" ... and if they had any feelings, well ... these were best ignored since they could not be measured. Even thinking people apparently did not notice that if an individual learnt something, even in some automatic way, he still had to have a motivation; a "want" or need to do this, that is a feeling. Even as a six year old, it was clear that my friends learnt things like I did, although sometimes quicker or slower. Like me, sometimes they did things they had learnt to do, and sometimes they did not, that is, they made choices and decisions. This is why they were my playmates after all. We all have bodies able to do a variety of athletic actions and minds full of similar and different mental abilities ... yes, us mammals are all alive, body/mind persons.

I found it difficult to understand how feelings, something felt, could be just a "reaction" and not experienced ... surely a feeling is felt? I knew that Konyok and Mosquito liked me, and I liked them. This was a feeling that we experienced: we felt it. But we were doing nothing very different to everyone else who had a close relationship with their animals. For example, my scientific father took it for granted when training his gun dog to retrieve, that she had desires, needs, feelings and made choices; she demonstrated this every day. He could not teach her if he did not take these mental attributes into account. How come then that he, like the vast majority of people, argued that the dog did not have a mind? Fifty years later it turned out that others also found this confusing, but at the time no one would discuss animal minds with me, they just smiled and patted my head ... a bit like the reaction I sometimes get today from some people, some sixty-five years later! Then it was being a young female and silly, now it is being an old female, opinionated and silly!

When I was told in secondary school that it was not "scientific" to think that non-humans had a mind and thought about things, I decided that science was not for me; maybe the army might do. It looked like a life of adventure and fun; they had mules, horses and dogs, maybe I could be in the cavalry? No, no, not at this time sister, although they did indeed still have a mounted division who had fun playing polo and trekking about in the mountains, I was the wrong sex ... although I could be in an office, wear a tight khaki skirt and make the tea they told me.

So what about becoming a game warden and living and learning from all the wild animals that were around in East Africa at that time? With luck, I could spend my life with other species wandering around in the bush and learning from them, there were genuinely wild animals wandering around as far as the eye could see in huge crowds. No, again, it was no good ... I was again the wrong sex.

The only route left open to me (and that had only been opened by some tough women thirty years before) was academic zoology where I could at least study what was known about the bodies of other mammals, even though there was no academic teaching about behaviour or animal minds. So it was to academia, only too aware of its limits that I migrated. Eventually I learnt that by pestering and nagging others with questions concerning the hows and whys of their beliefs, I began to have a grounding in examining my own. It helped me fill in some of the gaps and rethink some conclusions. But of course, not everyone wanted to talk about their strongly held beliefs, whether a scientist, or anyone else. The most usual response was "because what would happen if we all did/believe that", or "because that is right and good" ... but what are "right" and "good" when they are at home? Or "an animal does not really feel or think like you, so where is the harm in doing this or that to it?"

My father sometimes went on safari to assess possible areas for wildlife reserves (Game Reserves they were called), and sometimes he took me with him. I remember vividly a trip to the Serengeti plains to assess whether or not it should become a "Game Reserve", long before it became a National Park. We were to spend a week finding and counting the lavatories of Thompson gazelle (a small common antelope). "Tommies" defecate in chosen places and from the number of their lavatories, it is possible to estimate their population. At that time, there were few other methodologies available; air photography, computer population models, drones or even roads were not available, so no one had a clear idea about the populations of different species in different areas. They were either Abundant, Common, Frequent, Rare or Absent. The first study on the ecology and population of Thomson's gazelles was by one of my father's game-wardens and was based on our lavatory counts. It could be said that I took part in the first ever scientific study of an African animal, in a small way! Jane Goodall (and the other heroes and heroines of wildlife behaviour) were still in their cots, at school or doing other things, but not in Africa. Jane Goodall did not join Louis Leakey (a friend of my father) to be helped into her chimp study until a decade or more later.

After a couple of days' driving to the Serengeti: slow, dusty, painful, jilting, sick-making travel over murram (earthy gravel), we left roads behind and lurched over the plains, falling into ant eater holes and knocking termite mounds here and there. This was known by those who suffered voyages as "MMBA" (Miles and Miles of Bloody Africa). Eventually as evening fell, my father selected a site under a group of acacia trees to camp. However, a solitary male buffalo had taken up residence under it too. I was surprised to see how frightened everyone was of him. It turned out that buffaloes are considered the most dangerous and consequently are the most feared of the African animals. It is quite likely that a lone male such as him, had had bad experiences with humans and might indeed charge first and ask questions afterwards, so everyone panicked and ran about scared which seemed to me a recipe to scare him enough to charge.

The brigade withdrew for a while, and watched, finally the buffalo ambled off. Then it was all hands to the camping equipment which consisted of lugging huge old-fashioned bed rolls containing mattress and blankets wrapped up in heavy canvas and tied with leather belts. Each needed two people to carry, and even with the canvas cover, they always seemed to be wet and musty inside. Then, the complicated camp beds with wooden legs had to be erected by the scouts and askaris and placed where one wanted them. Luckily, we did not need tents, or we would have had to have another truck. Apart from the ten or so people with us (cooks, houseboys, askaris, scouts, kitchen *totos* or "boys", etc.), there was not a sign of any human presence. The lions roared all night and we had to make sure we did not tread on acacia thorns or puff adder snakes when we went to pee. It was second nature to empty one's shoes of scorpions before putting them on, and to keep an eye above for mambas (a very poisonous snake) in the acacia trees. A prickly pear bush separated me from my father, and everyone else was camped around a fire about 200 metres away. It was a cool and starry night with the constant noises of the African bush lulling one to sleep.

When I opened my eyes next morning, I realised I had human company, there was an audience of around fifteen Masai, women with babies on their backs, bangled girls and boys, young and old men all silently staring at this curious white girl with straight brown hair lying on a bed in the bush. The men were as usual standing on one leg and leaning on their spears, their elongated ear lobes (they enlarge them so that they dangle down their necks), and their long ocra dreadknots bouncing as they puffed and chirped as they chatted. When they saw me open my eyes, there was a shout, and everyone creased themselves up with laughter. Well, I am not a particularly shy dresser, but that time it was a challenge, every movement was very carefully observed, commented on and giggled over by the Masai. Later Mturi (the old Kikuyu "house boy" who usually sorted me out) ticked me off strongly for wearing my pyjamas under my shorts because I was not showing off the correct *mzungu* (white person) dress to the Masai ... he was strong on dress code, and did not like the Masai much, as all Kikuyus. After the dress show, the Masai drifted off chatting and cackling to meet up with their dozing cudding cattle relaxing under another group of acacia trees about a mile away. The women and children ambled off to the *manyata* (temporary huts surrounded by thorn hedges) screeching with laughter and mocking the antics of the white campers and their retinue.

We spent a couple of weeks counting Tommies' lavatories, camping around and being at home in real Africa. The plains were never dull, they were dappled with antelope and other large mammals of many types, Tommies and the slightly bigger Grants gazelles who would lift their heads, stare and then spronk off taking enormous leaps or rush away so fast it was impossible to follow them even with your eyes. Gnus socializing in their large herds, looking down their long black noses almost with disdain as they cantered about on their longer front than back legs, chestnut hartebeest standing staring in their own special territorial places. Waterbuck gliding out of the thick bush near the rivers to watch and occasional glimpses of bush buck as they darted among the acacia trees. Even little duiker and diki diki were occasionally seen flitting around at sun rise and sun set. Crocodiles lying with their mouths open looking like logs on the edge of the rivers while hippopotamuses honked and puffed as they floated elegantly with their great fat bulks through the pools. Grand looking eland briefly staring at us before walking off arrogantly. Tail swishing buffalo in large herds socialising with each other and sharing their grazing with hundreds of striped little horses, zebras, each one with their own individual stripe pattern who would alert the buffalo to any serious intrusion and rush off with their bums bouncing and tails swishing.* Lions, cheetahs and even the odd leopard would watch us or slyly invade our camp. Even wild dogs (now almost extinct) rushed through the neighbourhood once, and hyaena cackling and cooing kept me awake at night. Yes, as I wandered about, I could be eaten, stung, or killed in many ways, but that was all part of life or death and no way was anyone going to restrict one's wandering. If one kept one's ears and eyes open, someone of another species would warn me if there was something really to avoid.

Then it was back to school and the farm. I was the world's worst car traveller, just the sight of the motor car being prepared for one of these endless journeys was enough to make me begin to feel sick ... but what partly saved me, was to imagine I was driving the fifteen or twenty horses that were the "horse-power" of our motor car. We raced along bumping and crashing as I held the reins, took the corners wide so that all the horses could gallop around. I would shout instructions to Blackie or Ludo the leaders and helped turn with the imaginary reins, all under my breath so no one knew what I was doing. Car sickness and driving the cars'

---

* H Prins. *Ecology and behaviour of the African buffalo: social inequality and decision making*. 1996, Springer Science & Business Medi

horses has stayed with me all my life. Today, though, it is more usual to imagine galloping along on one of my horse friends next to the road as I sit in a car and predict the leaps, the stops and goes, the fences to jump, the forests to negotiate, the rivers to swim ... it does pass the travel time well, with added imagined adventures along the way.

Some of my fondest memories were travelling on the very slow wood-powered train with my friends Mosquito and Konyok (and various other girls and their ponies and dogs). There were never any boys, I have no idea what they were doing, they hardly featured in my childhood. We would have to travel days and nights to take part in competitions in other parts of Kenya at a distance of 300 miles or so. The train was so slow that the dogs and girls would jump off and run beside it at times. The ponies were shut in their boxes on the train, but they could put their heads out of the open doors and keep an eye on us running alongside. We slept with the horses, lying in the mown dry grass that had been collected by the syces (grooms), sucking Nestlés thick sweet condensed milk from tins, eating lumps of bread, all curled up next to our hay-chewing ponies as we chuntered and lurched slowly through the night.

The Pony Club was run by a Major Shepherd along military lines. We had to be "spick and span" as did our horses, but there was no going out to buy the "right tack", there was nowhere to buy it. We made do with scrubbed, polished, oiled equipment and the clothes that we had. It was military but a little wild ... perhaps a bit like the Foreign Legion, but without the fighting or the sex! There were races and adventures around the African countryside, galloping, falling, leaping, laughing, crying, avoiding jackal holes, snakes, thorn trees, and leaping whatever came in our way with little regard for safety. It was what I later discovered was training in the "British Hunting" mentality. The idea is to take very considerable risks riding your horse at high speed, either you survive to laugh and try again, or you don't. In fact, I don't remember anyone having a particularly serious accident, we broke arms and legs occasionally but that was just life, at least we did not have to ride side saddle like those stoic hunting female Brits did!

In Africa there is a different attitude to life and safety; life is a bit iffy. We were all quite likely to die or be paralysed by cerebral malaria, poliomyelitis, get sick with bilharzia or tick fever, be eaten by a lion, killed by a buffalo, or die quickly from a snake bite. It was not infrequent that friends died in these ways. Carol Jones hopped about with paralyzed legs from poliomyelitis and Jenny Car never came back from her holiday at the coast as she had died of cerebral malaria. African life, at that time for all, might be short, but it was not necessarily brutish or bloody as Hobbes believed. It was fully lived, free and fun ... not the sort of life that risk-conscious technological American or European children generally embrace today.

We had individual and team events, gymkhanas, polo playing, racing, jumping, cross country and even a form of hunting. This was galloping after a drag since there were no foxes, and jackals were not out and about during the day. A drag is when the hounds follow an evil smelling rag that is dragged along the ground. The idea is that when the dragger wants to have a breather, he lifts the drag and the hounds are meant to stop and search around for the continuation of the trail. But, hounds are not fools and usually disregard this rule by following the scent of the dragger, who is either running or riding. This means there are no stops or "checks" as they are called. The hounds just run flat out with everyone galloping along behind until they come to the end or catch up with the dragger.

On one occasion, the person laying the scent happened to be an English vet on secondment who was keen on long distance running. He found he could not run as fast as the hounds and was exhausted after about five miles. When he was caught up by them bounding and barking, he leaped up a nearby tree in fear and trembling. We arrived flat out on our horses to see the hounds baying at him and trying to climb the tree. The added twist was that there was a large police presence. At least ten askaris (African policemen) wearing their traditional fez top hats and shorts were up the surrounding trees. It turned out that since this man had been running and was being chased by the hounds, the police thought he must be the burglar they were looking for who had broken into someone's house locally. The quick arrival of the hounds, however, led to a stalemate with suspected burglar and police all up trees. The hounds were eventually pacified, and we all returned to the nearby farm to drink a toast with the revolting local beer made from motor car batteries, maize, water and anything else anyone wanted to throw in.

Mosquito and I took part in all sorts of events, winning some, losing some. He rushed around, leaping this way and that, in control of both himself and the small two-legged leech attached to his back. My eldest sister Shelagh's pony was a pure-bred Arab of whom we were very proud. He was called Paris and his parents had been imported from England. He was grey, and we all thought him very beautiful and superior, just like my sister. My other sister, Grizelda, who was generally considered better at riding (and everything else) than either of us, had a pony called Gazelle who was a sweet tolerant Arab cross Somali mare who had obviously had some good teaching before she turned up at our farm. She was cooperative and relatively easy, a bit like my sister. But Mosquito, a tough Somali pony was a rugged, ugly tyke but always fun. When he had had bellyfuls of food, to which he was not accustomed, he was more or less unmanageable, I suppose a bit like me.

There was a team hunter trial event at Naro Moru, a village on the slopes of Mount Kenya, which we attended yearly. A two-day trip in the train and then a long hack up the mountain in the sun through the forests and eventually to the open grassy uplands which welcomed us with glorious views of Mount Kenya's snow-covered peaks. Johannah, our syce, a wizen old man with a stoop from the Wakamba tribe whom I admired greatly and who was one of my mentors, said that the witches lived in the snow which we could see from the hot sunny plains. They would swoop down on little girls straying off the path and fly off with them to eat ... I still regard the snow-capped peaks of Mount Kenya with enormous dread.

I was usually too young to take part in the major competitions.

But the last Naro Moru hunter trail we attended, I was old enough, and we were to have the Worthington Team for the first time ... riding with my sisters, what an honour. Paris, Gazelle, my sisters, and Mosquito and me. Unfortunately, someone had told my figure conscious elder sister, Shelagh, that she was too fat. She had won the lead part of St Joan in the school play and felt that fatness would not do. As a result, she slimed and developed what would now be called anorexia. This was not a psychological disease at the time, so she was bullied to eat more. She was wobbly and weak and not going to be much good in the team that Mosquito and I had been so looking forward to riding in. Luckily a substitute was found, my best human friend Angie was to ride Paris. I can't remember if we won, but I do remember galloping far too fast over all sorts of obstacles and finishing all abreast, quite out of control of course!

Shortly after this, sister Shelagh left home to go to London and become an actress or an artist.[1] After an iffy start failing to get into RADA, she spent her life as a sculptress and became widely known internationally with what I always found her rather way-out art ... now, of course, it is all the vogue: an insect trail through gold dust on a table, rain splashing into a puddle, or a pile of rubbish against a fence. She won a whole batch of awards. From my artistically illiterate position, the most intriguing part of her art was that she could talk about it and explain to me what she was trying to make people see in the sculpture ... something I have found rare among artists.

Back to school, jelly on Sundays for tea, and foot washing in the communal bathroom daily (something to do with jiggers: eggs laid in your toes which hatch into little caterpillars that itch and itch). Then no jelly one Sunday and we heard that King George VI had died. Princess Elizabeth and her new husband were on holiday at Tree Tops, the new and possibly the first tourist hotel in the trees which was not far from Mount Kenya. So, she was to be queen which we were all thrilled about, a pretty young woman who was highly respected. (Barak Obama, USA President, 21 January 2009

---

[1] Shelagh Wakely, *Installations and commissions*, 1991-2009

to 20 January 2017, later joined the hallowed company of people from Kenya who sort of rule the world, well, have influence at least.)

My father thought I should be a ballet dancer since he loved ballet and I was small and quite athletic. We had a White Russian lady who had escaped from Russia during the revolution who taught us. She had been with the Bolshoi and we all called her Madame ... small, athletic, and enthusiastic, she badgered and bullied us with exercises at the bar, pliés and arabesques and foot positions 1 to 10. She gave me a solo to dance on my points: Red Riding Hood, I can still remember the somewhat brief dance (since of course Red Riding Hood was quickly eaten by a wolf!) It might have been fun to go to Saddler's Wells and dance away with those queer dresses made of mosquito net which stick out sideways ... but it was the fresh spring-like mornings, galloping across the high mountain grasslands at over 9,000 feet that tore me away. We built tunnels in the woods and stored illicit food for midnight feasts, sneaking and creeping silently out of Bible Story on Sunday afternoons where the whole school sat on the grass in the sun and dreamed. But soon I was eleven and it was time to leave.

A secondary school modelled on the British private girls' school tradition of Roedean and the suffragettes, called Limuru Girls School, is where I spent the next three years. This was at the top of the Rift Valley, no more than twenty miles from home, but except for the two weekends "exeat" a term, we all had to stay at school. Several of us had moved on together, even one of my gang. Rumour has it that the head mistress "Miss Fisher" was a cousin of the queen, or was it the Archbishop of Canterbury? Someone very important, so everyone said, although it was a very long way away from our African lives, like the biology we had to learn, all about ash and oak trees which we had never seen!

I do not remember much about the teaching, except for a Miss Lake who was a small, spherical woman who rolled about on the teacher's desk and threw chalk. But she did try and answer my questions, and finally told me that despite my incompliant behaviour I would probably end up as a nun. Well she was wrong about that, but her geography and history lessons were certainly enjoyable and even challenging.

The dormitories where we spent probably the majority of our time, had elaborate routines and an extraordinarily long list of rules which I constantly forgot. As a result, I was frequently awarded "Order Marks": bad marks. I wanted to know why I should not run in the corridor, wear my jersey before breakfast, not talk when waiting in a line and so on. The only reason I was given was, "what would happen if everyone did it?" The answer to this is of course, "but they don't, and anyway would it be so awful if they did?" This was considered heresy. The other girls would quickly withdraw their breath and put their hands over their mouths as I replied, and the matron would glower, sniff and glide away with her nose in the air. Such behaviour did not encourage me to obey the silly rules, and anyway I had not meant to disobey, I had just forgotten!

My Order Marks accumulated, and the occasional Conduct Mark joined them (these were awarded for terrible things, like eating passion fruit in the dormitory, and they were equivalent to four Order Marks). I found myself frequently bemused, hauled up in front of the high priestesses of authority (the prefects) and receiving all sorts of punishments which seemed to lack rhyme or reason: cutting a lawn with nail scissors or writing lines about 200 times, rather than learning something that might have been of use, a piece of Shakespeare or a Yeats poem or two, perhaps. I did, however, grandiosely take part in the "girdle winning". This was for standing up straight, first a red girdle and then if you were very straight all the time, you were awarded a green one, the green girdle became my most treasured award from that school. At least I was not disqualified from that, but I always counted the hours to the holidays, lying in bed and studying the peculiar fat or thin, spotty, straight or bent legs of my neighbours as they undressed.

Humans were deeply bemusing and confusing. What put another nail in the casket was that my best human friends did not "own up" to "crimes" that we had all committed together. "Owning up" is very closely allied to loyalty. It means you admit that it was you who did the bad deed or at least it was your idea, and then because loyalty to the group and honesty are important, all the other members of the group also "own up". This gives the misdemeanour a sort of justification and moral weight since one is honest, even when knowing that you have broken the social code and it will lead to personal disadvantage. If one suffers, everyone does. An "all for one and one for all" sort of approach. I had learnt therefore that there was a moral obligation to "own up". It is clearly the case with dogs, if one is ticked off, the others will cringe and try their hardest to say "sorry" too, sometimes even if they have not done the bad deed. But my human friends did not "own up", so I took all the blame; what was their moral code then?

The straw that broke the camel's back was when several of us picked some passion fruit (granadilla) as we walked down to the stables for our weekly ride. Instead of eating them immediately, which would never have been found out, we saved some for a "midnight feast" in the dormitory. But this was classified as "stealing", although no one else was going to pick or eat them. The matron came into the dormitory one night pronouncing: "I smell granadilla, who has them?" I owned up, and this time was expelled ... the other girls, far from "owning up", threw knowing cynical smiles in my direction as I packed my things.

At that school, I spent much of my time with sweaty hands and fast beating heart waiting to be pulled up in front of the prefects and torn down "a peg or two". Like most other persons of any mammal species, I wished to socially integrate with others, even if they were irrational and behaved inexplicably. But the whole experience was profoundly confusing. Obeying rules for their own sake remains a mystery to me today; perhaps I had absorbed too many of my pup's and pony's mental attitudes to be a successful human!

My mother shook her head and tut tutted, but I don't think she really thought it was so bad. There was no particular rebel in me (at that time anyway). Like my dog and my pony, all I wanted was to get on with life, enjoy it and work and play hard, with the odd adventure thrown in perhaps ... yet I constantly found I was unable to find out what the right thing to do was supposed to be.

The next change was a trip to Europe, called "home leave", something that all colonial servants had to take every few years. This trip was to "educate" daughters into European culture. For me, a resistant ten-year-old, it consisted of being dragged around numerous art galleries in the freezing cold in Venice and Florence. It was winter and not long after the war, there was no heating and very few gondolas to amuse small girls. But I did find a glorious penknife made with what looked like mother of pearl and with a pair of scissors on it. Whenever I could escape from the art galleries, I would stare at it in its show case. My mum bought it for me before we left and for the next fifteen years I wore it around my neck ... but I was put off art galleries for life. The highlight of this trip was a few weeks learning to ski in a veritable picture postcard snowy remote alpine village, the type that no longer exists, with enormous hot-tiled stoves you could rest your back against, checked tablecloths and gallons of hot glühwein, enormous beds with fat duvets, snow-encrusted windows with icicles, bells on the sleigh pulled by a Contoise chestnut mare taking us to the one horse station, all just like the alpine stories I had read. I tumbled and fell but tried and tried and eventually, I could more or less ski ... enough anyway to make it something that I have enjoyed through my life, gliding through deep snow down a mountain, usually totally out of control, then with luck still standing up, through a pine woodland in the snowy silence, perhaps even catching sight of a chamois or white hare darting away.

We must have caught the train from Switzerland to England, and when we arrived in England, I kept asking when the town would end, and the country start. All the way from Dover to London was built up, I was used to the grand African spaces of that time. Even then in Britain there were only occasional small precious pieces of woodland, or Downland, national park, and occasional farms all preserved from human development only by special planning orders from central government. No, the town never did end, what a mind blower!

Our next stop was in what was considered rural Sussex, and my father's mother's family home: Standen, a would-be landed gents Victorian house, garden and farm.* This enormous pile had been built by my great grandfather, a "nouveau riche" money-making lawyer who wished he had been born into the landed gentry, so bought into it. William Morris had been hired to do the internal decorations and there were Morris wall papers with queer flowers, bright colours and enormous heavy curtains with similar designs which one could roll up and hide in. The gardens were designed by Lutyens or one of his school, and with the backdrop of the South Downs, consisted of ten acres of lawns for tennis and croquet, hidden terraces and Ha Ha's, masses of rhododendrons and azaleas and gently swaying oaks and ash trees dotted about in the rolling meadows grazed by red and white Shorthorn cows. It was spring which I had never consciously experienced before, all the shrubs were out and the grass growing almost as you watched. Rhododendrons in every colour, and best of all, the secret water garden with mossy paths, waterfalls, little groups of miniature daffodils and crocuses. I had a passion for roller skates at the time and rolled around everywhere much to everyone's annoyance, particularly in the Orangery (which had seen better days). There were no oranges, or other plants ("not enough staff"), but the glorious sun beating through the glass made it warm enough to be Africa again.

We stayed in the main house with my great Aunt Helen. Aunt Helen Beale was the last generation where one daughter had to remain unmarried to look after aging parents. My grandmother, her elder sister, had married my grandfather, and the next sister Aunt Maggie was considered an artist and consequently too occupied to look after parents. Helen, the youngest, had drawn the short straw and spent most of her adult life looking after her parents, originally with a vast staff. But, in her youth she had escaped to nurse at Gallipoli or somewhere like that, and in addition, she had been allowed to accompany her brothers to be one of the first women downhill skiers, careering down the slopes in

---

* https://www.nationaltrust.org.uk/standen-house-and-garden

deep snow in her Edwardian frocks and enormous flower bedecked hats, one leg in front of the other trying to do "Kandahar" turns with skis with no edges. Then spending hours walking up the mountain with seal skins fixed to the skis, only to run down again. It was the very early days of skiing as a sport which, curiously, was taken up first by the Brits and there were of course no ski lifts. She had also, I believe, had contact with the suffragettes in her youth, although this was never spoken of, it was definitely "hush, hush", not something to be proud of, associating with those "awful women who broke the law."

By the time we came to stay in 1950, Aunt Helen's parents were long dead, and the staff gone, so she ran a large cold house organised in military style. She was long, thin and always had a drip on the end of her purple nose. Rules and traditions were to be obeyed by all. When my father, aged fifty-five, used the wrong knife to cut his bread, he was sent out to the kitchen (much to my glee). The kitchen was at least 200 yards down dark passages, so the food was usually cold on arrival. We had to trolley everything from it to the "dining room". One thing I did learn which has been useful ever since: when pushing something with moveable wheels always "keep the moveable wheels behind Martha" ... as Aunt Helen shouted at me in her sonorous nasal tones every time I crashed the trolley into vases and other works of art. There were no dogs, but the home farm had a dairy herd with whom I spent much of my time when not gallivanting around the gardens.

I remember an outing to East Grinstead to a shop that supplied "Equestrian Garments". I was trying on jodhpurs and hats in order to look more the part in Kenya, when my mum picked up a newspaper on the counter with the headlines: "War in Korea declared". A silent stillness hit the shop ... was this to prove another world war? Britain was still short of food and we still had to have ration cards (although I subsequently discovered this was not the case in Germany whom we had beaten in the war; seemed a bit unfair). After two world wars both of which many of the older members of the population had been involved in, it was clear that any war could turn into a world war. The Korean war did not turn into a world war, but it very easily could have, and British troops again found themselves fighting abroad.

On balance, whatever else Europe might be, it was a very built up, slightly scary place, which was cold, very well organised and inhabited by severe people who did not laugh much. But the natural world never disappoints, and if you could find it in the UK, there were some beautiful and interesting semi-wild places with exciting and different plants and animals. Whatever else it was, it was very un-African, even at age ten, my cultural history was Afro-Anglo, so England was a foreign place to me.

At the end of this chapter of my life I had learnt that my animal friends were unconditionally loyal and that different species as well as individuals have different, but not necessarily inferior or superior mental skills, which today "Science is just discovering"![1]

I had also found out that fun and joy seemed to be more a part of my non-human friends' lives than the humans I met. Human lives were confusing and seemed to be unnecessarily complicated, often without any particular reason. Finally, I had discovered that the living world in all its aspects is profoundly rich and varied, but humans can, and do, cause horrible changes to it and destroy it.[2] However, if they think about it, they can, by using their manipulative skills, at least keep it around here and there.

## The Mau Mau & the Belgian Congo

We lived on a relatively isolated farm at the top of the Rift Valley where the weather was almost perpetually spring. On a clear day we could see the snow on Mount Kenya in the north and that on Mount Kilimanjaro to the south. Our house was a sprawling bungalow designed by my parents and built, with endless delays and price re-negotiations, by a Sikh builder and his large clan of

---

[1] A Chapman, A Colin & Michael A Huffman, "Why do we want to think humans are different?" in *Animal Sentience*, 2018, 23:1

[2] https://www.nytimes.com/2018/10/31/world/australia/australia-wilderness-environment-gone.html

Kikuyu and Luo helpers.

It was built out of local stone with a tin roof and was surrounded by wattle (Mimoza trees), which had been planted to harvest the bark for leather tanning. Our water came from a curiously designed swimming pool which saved the roof water and which we swam in, although at the end of the dry season, there was only a muddy pool. There was a tank high up on a wooden tower to which the water was pumped by hand daily, and then ran by gravity to the house. My father was a high diver, so whenever we went swimming, which was everywhere we went if possible, from a rocky cliffy sea shore, to a muddy pool in the savannah occupied by hippopotamus, there was always a challenge to dive off the highest, and often most dangerous rock, cliff or tree that hung over the water. I spent many an hour at home trying to amass courage to dive from the five metre high water tank into the pool. Jumping was easier, not head first, but not "proper"; head first diving was another matter entirely.

My mum had terraced the hill and planted a garden with both tropical and temperate plants, she was a particular wizard with sweet peas and shrubs ... the smell of sweet peas still reminds me of those high trellises of sweet peas she grew, picking them with long stems (short stems were no good for selling or putting in vases) and every now and then peeking through them to look at the white snow on Kilimanjaro. Both parents were keen on grass tennis so there was much discussion on how to make a tennis court, and then hours of rolling and digging, but they eventually had a grass court at the bottom of the garden. It was, of course, an experiment in using the different grasses as grass courts were unknown in Kenya at the time. The parts with Kikuyu grass were spongey and it grew and spread everywhere, a very persistent colonizer. By contrast, Naivasha Star Grass was hard and narrow, and the ball bounced well. If you knew which grass bounced the ball and which did not, it was not too difficult to win, as long as you could hit the ball at all. I went down to the court to do ball practice, but the temptation was to visit my sister's hut just behind the court where she collected and kept all the skulls of animals that she had found. I was not allowed anywhere near of course, but I could not resist spying. It was spooky squatting in the shrubs watching her stumping around arranging, dusting and polishing the queer, threatening, light white skulls of snakes or the enormous hunking half-skull of a hippopotamus.

We had sixty acres and employed people part-time who came in from the surrounding peasant farms. One rather dapper Kikuyu who often helped with fences and told funny stories that amused the rest of the workers was Green Shirt, he always wore a very ancient torn had-been green shirt of which he was very proud. Then there was *Mzee* (old man) who was in charge of the dairy cows. The cows would not let their milk down unless the calf suckled first, I would watch him sighing and doting on the cows when they were letting him take the milk, and swearing and cursing them when they would not, as the calf dribbled milk that Mzee assumed would be his. There were also daily dramas, for example, one day Mpishi (the cook) came charging into the dining room swearing that the cow had eaten his trousers. He had left them on the line to dry, and they had been quietly munched up by Daisy. Mum sorted it out and bought him another pair, but that evening there were peals of laughter from the servants' houses as the story was retold and elaborated by all over their cooking fires.

Mosquito, Konyok and I rode around the neighbourhood and became very familiar with it. There were miles of wattle trees that must have been grown for us to have fun playing chase in and out of. There were *shambas* (gardens/small farms) in other areas where the Kikuyu women would be chatting, singing and jembying (*jemba* – to dig with a big hoe-type tool) within the thorn hedges they had built to keep animals, and other humans, out. There were still a few pieces left of the indigenous multi-layered, multi-species thick forest, still hanging on by its claws ... but long since vanished now in that area. There were open grass glades where we might see bush buck leaping shyly into the surrounding forest if we were lucky, or warthog with their tails up determinedly running across, grunting in front of Mosquito, giving him an excuse to leap sideways and lark about. There were villages of mud huts with thatched roofs with children squeaking and running around, particularly excited when they saw us, with their grandmothers standing and staring, or aunts lumbered with huge piles of sticks tied onto their backs, arriving from their wood searches.

Once or twice, I was allowed to join my elder sisters in cowboy games galloping around and hiding in the forests; these were red letter days; they usually left me to get on with, what they thought, their own much more interesting games.

But, the neighbourhood riding was curtailed at the start of the Mau Mau uprising when I was about twelve. The Mau Mau were a group of Kikuyu men fighting for independence for Kenya. A white officer of the security services came to explain that we must be gated. He wore the colonial uniform, khaki shirt and big baggy khaki shorts with long socks. These baggy shorts were very revealing when the wearer was seated, particularly if one was a small giggling girl!

To avoid the askaris, the security services, the Mau Mau had run off into the forests in various parts of Kenya. Most of the everyday tribal people were probably not greatly interested in Self Determination and Independence, but the emerging middle-class Kikuyus were, and they had some charismatic leaders such as Jomo Kenyatta. Jomo was a large, very black, bearded, domineering man who demanded a lot of respect. He shrewdly traded on the tribemen's superstitions to accumulate a fighting force in the forests out of sight of the whites, and equipped them with guns, *pangas* (a universal tool for cutting, digging, killing or anything else) or anything available. The Kikuyus combined African witch doctor voodoo with guerrilla warfare, and they terrified themselves, or were terrified, into fighting for independence. A lot of them died, but Jomo Kenyatta eventually became an excellent first president of independent Kenya.

During the Mau Mau uprising, all the white men (and some of their black servants), were supposed to go off to meetings one evening a week. This was a sort of *Dad's Army* Home Guard. There, they let off guns, talked, drank beer and marched about in the night, leaving the women and children, somewhat unwisely one would have thought, alone in their houses.

As a small girl, I was not particularly aware of any of this, but had been told exaggerated terrifying, girly stories: how one of my school friends had been attacked, but her dog had jumped at the man and saved her life, or how someone's father had been chopped into small pieces with a panga ... all untrue but very exciting and frightening. Having to go in the dark night to the recycling long-drop toilet, was a teeth-grinding, heart-racing experience. It was a hut seventy-five metres from the house. One thing that I realised quickly was that with a torch, you can be quickly spotted and knifed, shot or bunged down the toilet hole (some of the ducks had fallen down before, and we had had to fish them out). The only way therefore was to glide silently in the dark with heart beating towards the hut taking a wiggly path. It was also wise, I felt, to have Konyok with me for support in case of need.

Mpishi, our cook, was very black, very serious, very terrifying and had a beard and red piercing eyes. He usually carried a knife because he was always cutting things up in the kitchen, so it was not difficult to imagine that he might be a Mau Mau, and Mtoto (the kitchen boy) told me terrible stories in vivid colonial Swahili of how Mpishi had been initiated into the Mau Mau and how he had sworn to kill all white girls.

One evening as my mum, sister Grizelda and I sat with all the curtains drawn (to avoid being seen and shot through the window), and with my mother's revolver on the table (all the women also had to go to shooting practice and carried revolvers around with them), Mpishi walked in in a fury waving a large kitchen knife, I thought that this was the end, and hid behind a chair. Of course, neither Konyok or Hippolata (our boxer dog) were at all concerned as they knew him ... The cause of his fury turned out to be that the bread would not rise. Mum took him back to the kitchen to sort it out as our hearts gradually recovered their usual beat.

Then one night when my father was out with the Home Guard, one of the lads who looked after the cattle came in, very scared indeed, to report that all the cattle had been stolen by the Mau Mau. It was a very dark windy rainy night. We really did think they would return for us. Luckily, they did not, they only wanted the cattle to feed the itinerant Mau Mau, but this was sad, all our lovely pregnant Boran cows going to be slaughtered.

We had had a herd of around fifteen bovines at the time, and my mother had spent a great deal of time and energy selecting the best of the indigenous humped breed (*Bos indicus*): Borans because they were immune to East Coast Fever, a cattle disease that all the imported non-humped more productive cattle (*Bos taurus*) suffered from. She had crossed the Boran with imported bulls and after a few generations had a herd of good beef cattle who matured early, had lots of milk for their calves, and were immune to ECF as it was called. It was a sad day when the Mau Mau stole them to eat.

My long-suffering parents were not confronted with what to do with this erratic girl for long as my father had changed jobs to become the scientific director of the first international organisation to combine biological research efforts throughout Africa south of the Sahara called the International Biological Programme. It was not permitted for the secretary general to have the central administration in a country run by his nationals, but the Belgians had offered to host the offices, so we had to relocate to the Belgian Congo by road.

The cattle had been stolen and then Konyok contracted tick fever (a common and much dreaded disease for dogs for which at that time there was no vaccine). He lingered for a couple of days and then died, his death was certainly a release from his pain and suffering, but it was a very sad time. Hippolata my mother's Boxer bitch was sympathetic, and I tried to gain solace from her company, but she was not him. Then it was threatened that we would have to leave all the horses behind when we moved to the Congo. With constant whining and misery, I finally persuaded my mum that we really could NOT leave Mosquito behind. With much head shaking from my father, and lectures on the risk of him contracting other terrible diseases, he and his friend Menelik were finally to come to the Belgian Congo with us. This meant working out a way in which they could travel on the open lorry packed with all the household furniture, piano, books and so on. A two-horse crate was designed and built so they could enter from the rear and have their heads pointing out over the baggage to the front.

We piled Mother's grand piano, books, furniture, hens, dog (Hip), and all the rest of our possessions into the open five-ton truck. The two horses were persuaded to enter with bribes of special delicious buckets of food. Johanna, who was a Wakamba and the head syce (groom), had decided to come with us and he had with him a young Kikuyu lad, the stable *toto* who was up for the adventure. We three and Hip found places around the horses' heads on top of the baggage and we set off, with many a sad backward stare.

First, we drove down the 4,000-foot escarpment into the great Rift Valley on a beautifully engineered road that had been built by Italian prisoners of war. Some of them had fallen off the rocks and been killed so they had constructed a tiny Catholic chapel at the foot of the valley. From its windows one could view the large herds of zebra, Thomson's gazelles, impala and wildebeest who, at that time, roamed the Rift Valley. We travelled on and on, leaving the country, its lakes, forests and plains that we knew. Eventually entering Uganda protectorate, with its lush greenness, where fruit just fell in your lap. It was considered to be the most successful African state administered by London. It was a protectorate, not a colony, which meant that there were no settlers (whites who come to live and buy land). It had a budding university, secondary schools throughout, and expanding medical care; a shining example of an emerging democracy in a semi-independent African state. Little did we know about Idi Amin and its bloody future.

We stopped in Jinga on the shores of Lake Victoria where my father, who was a limnologist (studied fresh water ecology), had a posse of fish researchers making ground-breaking ecological finds. Then to Western Uganda, driving through what are now national parks which were to be my home for a while when I was older. Finally, to Goma at the top of Lake Kivu. We drove at night when the tsetse fly was inactive so in the dark we creaked, wheezed and rolled over the rutted dusty tracks. During the day, we unloaded the animals. My father had people to see, new research stations to look at and fisheries experts to go fishing with wherever we stopped. My mum and I would sometimes tag along, or she would visit musicians she knew and bash out a few quartets while I would play with our animal pals who were being carefully kept in the dark, away from tsetse fly attack. Sleeping sickness: Trypanosomiasis, which was endemic in most of these areas at the time, was generally fatal for horses.

The last part of the trek was the most memorable. We passed by Nyamuragira, an erupting volcano at the edge of the Congo forest where the last of the pygmies lived. At that time, some still lived as hunter gatherers and slash-and-burn agriculturalists, untouched by the West. Because their population was low, this type of agriculture was sustainable in the forest and had been continued for thousands of years. We visited some of them, and much to my joy, as a badly grown twelve-year-old, I found I was taller than the men and was invited to spend an afternoon sneaking around with them hunting and gathering in the forests and appreciating their freedoms and enjoyment of the living world. They lived very well in my opinion, they had enough to eat, drink, plenty to do, adventures to have and time to lie in the sun chatting with their various dogs. They were familiar with all the other species who lived there, sharing resources as they had to, genuinely "at home" in their environment which they both treasured and used. They were always free to go where they wanted, not tied by money, work, mortgages or a "need" for more of anything. It was perhaps that brief visit which showed me how it was possible for humans to live well, simply and freely in the natural world which seeded the ideas of how it might still be at least partly possible, even in the modern consumer-orientated world.

There was a single-track road hewn out of the mountains that bordered the lake from Goma to Bukavu, our destination at the south end of Lake Kivu. One day the traffic went south, the next day it went north and one hoped that no one broke down in the middle. The scenery was spectacular, a gloriously beautiful silvery lake set among great big spreading trees with creepers and climbers: a lush green equatorial forest. There were jabbering black and white Colobus monkeys peeking at us now and then (now almost extinct) and the cloud-capped mountains behind. Then, suddenly, a recent larva flow, not a plant in sight, just mounds of what looked like giant helpings of black treacle that had solidified on the way down the mountain side. One or two areas were still glowing red as the molten lava slid down on top of the previous layer which had solidified. A moment later, we would be plunged back into the thick green indigenous forest with its colours, flowers, creepers and architectural trees.

Eventually we arrived at an unlikely research station, alpine wood chalets, neat, business-like and not at all African. It was a new Belgian biological research station erected to study the mountain biology, the erupting volcanoes and the colonisation of plants and animals on the larva flows. The director must have dreamed about the Alps and had completed the illusion with geraniums in the windows and all the plants only growing where they should. There were also horses in grandiose European stables, and not a laugh to be heard ... it was a very un-African place, set in a tropical equatorial African forest bordered by a silver lake and an erupting volcano. Black servants scurried about with trays, swept drives and stared at us with the long faces they had learnt to use when with the white Belgians. Had we fallen into a hybrid European/African nightmare? As a twelve/thirteen-year-old, I had goose bumps and bit my lips. Later developments in the Belgian Congo confirmed that we had indeed tumbled into an outlandish dream. My father seemed unaffected as he batted ecological knowledge to and fro with the Belgian scientists, but my mother winced.

After a night or two, we forged on in our lorry to Bukavu on lake Kivu, and to the curious *Centre Equestre* where the horses were to live until we had a place for them. There was an outside manège where the serious long-faced Belgian riders rode around and around with curious equipment. One was a plank fixed behind the rider's back and through his arms, another with pieces of leather tying the horse's mouth together and his head down, so his eyes were popping out. It looked like another nightmare. I never understood why they did these things nor why they never rode outside through the forests and quinine plantations that surrounded the stables. I had misgivings about leaving the horses there, but it had to be, temporarily. They had lost weight after their seven days' and nights' journey, but otherwise seemed less daunted by their new surroundings than I was.

Our new house was a small bungalow on the hillside overlooking Lake Kivu. It had a sloping piece of land in front which my mother immediately took over for her garden, and a piece of grassland behind for our horses. The first thing was to plant the cuttings that Mum had brought with us and build the mud and wattle stables. Johannah, our syce, and the stable *toto* erected their huts next to the stables. It quickly became a multi-species, multi-cultural "ex pat" enclave, entertaining visiting research biologists from all over Africa, as well as locals of all colours and persuasions.

I was sent to a Belgian convent girls' school just outside the town. We were all locked in during the morning, and locked out at night, and taught Latin through the medium of Flemish. The nuns spoke Flemish and wore weird dresses and veils. It was not long before I realised this school was really a no-go area for me and what I had in mind to do. The only language we had in common was Kiswahili which some of the local girls and the servants spoke even worse than me. Trips to the kitchens to chatter to the cooks and kitchen *totos* was more fun than attending class, but soon I was discovered by a nun and banned. It was clear, this schooling was not getting me anywhere, and was also very boring.

After school, there was usually enough light for me to jump on Mosquito and gallop off into the quinine plantations for an hour or two. But after a few months, it seemed silly to be going to such a school at all, so in the morning, I would creep back from the bus stop, jump on Mosquito with Johannah's encouragement, and with sandwiches in my pocket, vanish for the day into the quinine and coffee plantations. Mosquito and I would stop for lunch and a siesta, he would have a patch of grass to munch, while I read a book. Since my mother's French was as bad as mine, she never discovered my school absences. On weekends Johannah, the horses and I would wander down to the town to compete in jumping events against important serious Belgian men, who felt they should win. There were no children or women riders; I had the impression that these freaky serious Belgians did not permit them to ride. We had been winning in Kenya where the competition was much greater, so it was not surprising that we usually won. They gave us enormous cups made of aluminium meant to look like silver, some of which I still have and over the years they have made good vases for plants and flowers in many countries. Our winning performances did not help the curious English family's acceptance into the Belgian community one bit, but because we had plenty of visitors, it was not of much consequence to any of us. My dad was fully occupied organising biological research throughout Africa, so had no time for anything else, my mum fully occupied with her garden and entertaining visitors from other countries, and I was busy with our horses, dogs and helpers from Kenya. Integrating ourselves into the Belgian community was not a priority.

Mother was an excellent hostess and gardener. She entertained the visitors, played the piano and even found the odd violinist and cellist to play trios. Biologists came from all over Africa and Europe and even a few of the first newly graduated Ugandans, Kenyans and Tanzanians came as my father was committed to teach them to do research and run research stations so that biological research would continue after independence in East Africa, and at least some indigenous people could fight for the survival of the natural world in the future.

The Belgians had a strong colour bar. Although I came from colonial Kenya, and had been to all-white schools, my home life had been multi-racial as well as multi-speciesist. Human racism for me was as confusing as speciesism (discriminating between humans and other species). Why did some humans only associate with some other humans and treat others as they treated other mammals, that is as inferior beings?

After a few months, a horrifying event took place which confirmed my belief in the Belgian Congo nightmare. Johannah felt ill, my mother persuaded him to go to the bilharzia clinic, since there was a great deal of bilharzia in the lake. Bilharzia is a debilitating disease caused by a worm that lives in continual copulation in the bladder, it lays millions of eggs a day which are peed out and absorbed by snails where they develop into several stages of worms and nymphs, before again being liberated into the water and burrowing into a passing human. Johannah attended the clinic a couple of times, but seemed to get sicker. After a short period in bed, he died. We visited the clinic and the doctor in charge to find out WHY he should have died of bilharzia which was not a fatal illness. The Belgian doctor shrugged his shoulders and said that, well there were too many people anyway and most of them had bilharzia and we found out that far from curing bilharzia, Johannah had been used for a drug trial. My soul turned to stone. Was this really what medical care and colonialism was about in the Congo? The answer it turned out later, was yes indeed it was. There was only one graduate in the Belgian Congo at independence. The Belgians had successfully destroyed how the tribal societies worked and replaced them with ... nothing. Perhaps they thought they would be asked back to sort it out ... at least that did not happen even though there was mass genocide, total social breakdown and pillage in human societies. This continues today in the natural environment with the pygmies and anyone else who revered and lived with nature long gone, as well as many of the species they lived with.

There were other incidents which made my hair stand on end. One night I was sleeping next to the window with Hippolata at the foot of my bed. I woke up to feel hands sliding up my leg, I half opened an eye and saw a shape in the window, so I leaped up and hit out hard at what appeared to be the head of a human. It crashed on the windowsill and then the creature disappeared. If this had been an unknown thief or a rapist why had Hip not barked, admittedly, she was a deep and noisy sleeper; but surely she would have barked at a stranger coming in my window? The house servant did not turn up for a couple of days and when he did, he had a wound on his head, and it was generally agreed it must have been him. The police were duly called and he was carted off to prison.

Sometimes the prisoners who were working on improving the roads would clatter past our bungalow all chained up together and accompanied by whip-snapping policemen who shouted at any straggler and whipped them ... one morning, there was our ex-house servant in the line being whipped ... it was just like something out of those awful films of the old testament or the American south and their slaves. In my naivety, although I had lived a relatively physically dangerous life, it was difficult to believe that people did this sort of thing to each other, certainly dogs and horses, elephants and antelope do not. My memories of the Belgian colonial presence are not pleasant ones. We have heard a great deal about the outrageous behaviour of white South Africans and Zimbabweans, why has the Congo story never been told? On balance, it was probably even more evil and covered the whole country. How could sentient beings behave like that in such a beautiful natural world, with the shimmering silver lake surrounded by dense forest and chattering monkeys dominating the landscape and its feelings?

There were also some pleasant memories: the day we visited the last hunter-gathering pygmies in their forest, days swimming and picnicking on Lake Kivu, the hot sunny days riding aimlessly on Mosquito through the quinine plantations with a pocket full of sandwiches and a loyal dog. Even the days beating serious Belgian doctors, lawyers and chiefs in the jumping ring, coming first on Menelik, and second on Mosquito, and having all the animals blessed by a Catholic father in a very dirty habit.

Two sister friends of mine from Kenya visited, we galloped about on the horses, swam, laughed, ran, jumped and generally cavorted for a few weeks before they left in one of the first aeroplanes that had made it to Bukavu. After this, at fourteen, I pestered my parents to send me to a school where I would learn something, anxious not to be left illiterate in this academically oriented family and thus confirm my family's belief in my consummate stupidity.

Eventually it worked, and that part of my African life ended. I was sent to what was considered an alternative, advanced boarding school in the UK where my mum hoped I would not get thrown out. I said goodbye to Mosquito and Menelik who stayed in the Congo for the rest of their lives, but with people who did more than trot around the arena. I boarded a small plane that took me to Kampala where I had a trip on a seaplane, one with floats that takes off from the water, arrived in London and began a "European experience".

Boarding school and being farmed out to relations whom I had never met for the holidays was next. Perhaps it was then that I finally realised that "friends were family that you choose". My human family was often a collection of unknowns from a different culture who thought they knew everything about me. It was a bit like when a dog has to join a new human family ... they know everything about him and particularly his mental limits, and he knows nothing about those who pontificate about him. How wrong they are. If there is one thing that dogs born and raised with humans know, it is something about humans, although the humans usually really know next to nothing about the dog, however much they tell each other stories.

My mum hoped that if I went to an ultra-modern girls' school where the girls were at least considered as individuals, they might not throw me out. For me, the great thing about the school was that it was in a beautiful old Georgian house with glorious gardens, a lake and stables. It was less important to me that we had committees, discussed things with the staff, and fewer rules, I had grown used to the peculiar restrictions that humans put on each other, even though I thought and still do, that they are daft. Shortly after the end of the war in the fifties many ex-military men had no jobs but were offered posts teaching in private schools whose women teachers had joined the land girls or gone off to higher things. The result was we had some great teachers who had had war experiences all over the world, totally strange to us naïve, but questioning, girls. We could ask them anything and discuss anything with them. One very thin fellow, Mr Johnson, who had been a prisoner in a Japanese war camp somewhere in Asia, taught geography through discussion, Mr Jenke was German, and I think was interned during the war. He taught me to play the bassoon, carpentry, upholstery, book binding and the guitar. He organised our chamber music and I well remember attempting a Mozart trio, creasing up with mirth when his daughter, who played the clarinet with us, was always making curious breathy noises we called "wind by the sea". Mr Brakenbury taught the classics, and founded our interests in philosophy and the "big questions of life". I have great memories discussing Tolstoy's confessions under the flowering wisterias ... we nearly always had our classes outdoors if the weather was good. There were half a dozen large, heavy, fat British horses who took me around the green and pleasant Dorset countryside winter and summer. There was a Capability Brown designed lake and gardens around which we could run, swim, picnic or just stand and stare, and at the beginning of the autumn term, the dahlias were every colour of the rainbow nodding their heads in the wind.

On the human front, I made a few lifelong friends, but although staying on for two years in the sixth form, I never was a prefect ... probably correctly considered too irresponsible, but I had the distinction of being the only non-prefect in my year.

Our interest in Thomas Hardy was sparked by studying *The Woodlanders* and *Tess of the d'Urbervilles*, and a group of us took up cycling around the Dorset country looking up his haunts and wandering wood-wise. Then we were taken on camping field courses to the Isle of Purbeck to study its geology and the ash woods. We looked for crabs and all the other seaweeds and living things on the seashore, and appreciated the gentle beauty of the Dorset countryside with the scent of the gorse in flower always surrounding us. There were no "health and safety" regulations, just freedom to do and go where you liked, a great foundation for a biological career.

Towards the end of my school career, I was discovered brewing beer on the flat roof of the Georgian mansion, thinking it would be a safe place. But going on the roof was a major sin since the proprietor of the school particularly forbade it in case damage was done. I was discovered (probably reported by one of my classmates who was a "prefect") and sent to the highly respected, terrifying head mistress: Miss Galton who crushed me by telling me I was "immature". This was a blot on the landscape. In retrospect, it is difficult to see the advantages of being "mature". My life has been a lot more fun probably than that of the middle-aged risk conscious twenty-year-olds one meets today, and no more irresponsible than theirs. Ah well, we are creatures of our cultures, good or bad. Anyway, I obtained my A and Scholarship levels and enjoyed the teachers and the subjects and could, as a result, go to university if I wanted to.

This period of my life confirmed me in realising that the animals I had to do with and knew were my family, they were more predictable, often more fun, loyal, less restrictive and cared about things, they were at least tolerant of me, and I began to realise I could learn much from them. It also emphasised the natural world; gardens and farmed fields, and English woodland, are full of wonder and joy. And it confirmed that humans surround themselves with rules, consider themselves superior in every way to non-human mammals and carry a lot of unquestioned mental baggage around. They are also keen to make life difficult for those who do not conform to their peculiarities. I grew up, and became adolescent, but it brought me no nearer to wanting to do what adolescent humans normally do; another reason for exclusion and further confirmation of my fondness for a multi-species family.

## Riding, farming, cows, dogs & horses as a teenager

When I was about fifteen, my parents returned from Africa and we moved into a cottage in Granchester Cambridge ... where the church clock still stands at ten to three. Browning's poems helped me to appreciate the different but very gentle, flat countryside. Hip, the boxer bitch, had come with my parents and after six months in jail (quarantine), she was back to her old bouncing self. My mother, realising that my animal interests were not going away (they usually do with teenage girls), sent me off to study with the first exponent of the Spanish Riding School in the UK. Kongoni (a tall thoroughbred gelding my mum had bought from a horse dealer) and I went by train. It took a whole day from Cambridge to Slough, although probably only fifty miles. It must have been one of the last years that British Rail took horses anywhere. We had a superior apartment with a padded cell for Kongoni, and a groom's bedroom complete with bed for me. We spent most of the day stopped in sidings while Kongoni munched hay, and I read a book. Eventually we arrived at Slough station from which Kongoni jogged, and I bounced about on top for five road miles in the wet fog to arrive at a very British stable yard. I have visited many stable yards all over the world now, but British ones are, or were, in a world of their own, although all who work in them think they are normal!

Robert Hall had been in Vienna with the Spanish Riding School. He was the "Grand Maitre" of the school, but he was rarely there, I have no idea where he went. The fearsome large, severe stable manager Miss Manning was there, though ... rather too often. She fastidiously ensured that every i was dotted and t crossed for the terrified working pupils (girls who do stable work in exchange for lessons). These rules had been established when "riding to hounds" was the be all and end all of contact with horses for the horse establishment called the British Horse Society. At that time Brits had not been involved in the "art" of riding (with one or two exceptions). They had forsaken an intricate communication with their mount for galloping around too fast, taking phenomenal risks, jumping anything in the way and hoping to stay on while hunting foxes. The military colonel who had taught me in Kenya came from this school, but although I had enjoyed this enormously, I had an inkling that there was more to riding than that.

At British stable yards, there was always a bevy of chattering, giggling, rushing around girls who flirted shamelessly with any two-legged male who might stray into the yard. The girls wore tight jodhpurs and t-shirts to show their curves, but they were expected to work very hard ... albeit difficult to see what benefit was gained for either the horse or the girl from much of what they did. I had six enormous horses to look after. Three were to compete at badminton (a high-profile combined training equestrian event). The other three were not doing much, except living in a spotlessly clean prison cell twenty-four hours per day. There were a host of rituals that had to be completed in these horse-prisons with no questions asked. Miss Manning did not tolerate any "slackness" as she called it. She marched around inspecting the stables, but we of course had many strategies of avoiding her. One I often used was to skip the girly talk and "coffee" in the tack room, to camp in the particularly deep straw in Kongoni's stable where I could not be seen and read a book. It was clear at that time, that my experiences of life were so different from the other girls' chatter and boy-talk, that I had more in common with my horse, however much I might have wanted to be part of the chattering, flirting group. There was an acre of yard to be swept every day after all the muck and wet straw had been wheel-barrowed from the stables and placed on a particular vertical-sided muck heap (there were inter-stable competitions on the shape of the muck heap!). After that, it was plaiting the straw in front of the stable door. This was done by grabbing lengths of straw and twisting it until it was possible to plait. The idea was to ensure no straw unwisely ventured into the yard. During the day we had to rush around with buckets to catch any muck that the thoughtless horse might be depositing in this spotless stable. There was tack (equipment) to be cleaned every day, and a particular way in which it all had to be done: taken to pieces, sponged, rubbed, dried, saddle soaped, polished, put together again, etc.

There were horses to "strap" (groom) and they must glisten, this took around three-quarters of an hour per horse per day; with six to do, it took up most of the day. Again, there was a special ritual to be observed with special brushes and sponges for special bits of the horse, done in a special order. Many of the horses hated this as they were pricked by the brushes we had to use, so they had to be tied up or they would bite and kick. I tried the brushes on me and found how they pricked and hurt ... ever since then, I make everyone try the brush on themselves before lathering into the horse with some particularly prickly brush to avoid the "silly" horse kicking or moving away as he tries to avoid it. I remember having to unlearn most of what I had learnt when under Miss Mannings' acute stable-management eye. It was tough, but an excellent training and we also had fun. One of the best memories are the enormous breakfasts around an AGA in a warm kitchen after the three hours' early morning work, stuffing one's mouth with sausages and marmalade together, fried bread and baked beans, eggs and toast, being with a group of girls who had similar interests, chatting about the horses, joking and teasing.

It was here in around 1954, that I first saw horses crib bite (grabbing something with their teeth, sucking in air and swallowing it). This is now known as a stereotype, and it was this among other curious behaviours of the different horses in these cells that fired my interest in the horses' point of view: were their polished, manicured, overprotected, boring lives really worth living? Until my questions were dismissed with a sniff and a sigh by all, I would ask why they had to live in solitary confinement where there can never be horse-chat? Why can they never make any choices? Why can they never move around at will? Why can they not eat as much hay as they like? Why do we have to have so many pointless rituals, but we do not when looking after dogs or cattle under our care? The response was the usual mantra: "because this is the way it is done". It took a few years of puzzling, but now I have an explanation for these rituals ... they are there to keep the squaddies busy; the cavalry kept a large number of horses in stables. The last thing that officers wanted was for the squaddies to have time to drink and whore; they must be kept busy and off the streets at all cost. This is easier with the armed forces' mantra: "ours is not to question why, just do and die", than it is for a bunch of adolescent giggling girls, but the same thing applied and was put into effect by the Miss Mannings of this world. What better way to achieve these ends than to establish rituals and pointless tasks that had to be completed daily: polishing buckles, boots, and harnesses, strapping horses, picking out their feet, removing their muck in buckets, sweeping yards, tending geraniums in pots and plaiting straw in front of stable doors, these tasks at least kept us busy! Horse owners who keep horses for pleasure, whether grand dukes or the plebs, have learnt the "proper way" from the military, and it continues today, often backed by fake "science" taught to anyone including vets, instructors, guides, teachers, students, competitors, equine scientists and anyone else around.

Some of these practices are not particularly harmful for the horse, but some of them really are and it took me another twenty years or so to find out which and why. Today, horse welfare has become a buzz-word and everyone believes they look after their horses in the best possible way without questioning these traditional practices, or even by inventing new ones. One example was around twenty years ago, when I was invited to visit a university which was starting one of the first courses in equine science to include behaviour and welfare. They were well funded by the RSPCA and other welfare organisations, and had decided to build some "state of the art stables" for their horses. Apart from other slightly questionable design features from the horse welfare point of view, the top door of these stables had been fitted with what is called an "anti-weaving bar". These are triangular bars welded so that the horse can put his head out, but cannot move it from side to side, the idea being to prevent the horse from "weaving". Many stables horses do this, in fact rather more than twenty-five per cent in surveys we have conducted and sometimes for hours at a time.[1] Weaving is a stereotype and a sign of distress, that is, it shows that all is not right for that horse in that environment. It is the same as thumb sucking, hair twiddling, rocking back and forth in human children.[2] In this new university teaching faculty, instead of considering the cause of this stereotype, and how to prevent it from forming, in order to improve the welfare of the horse, the faculty assumed that the horses would be distressed so "anti-weaving bars" must therefore be fitted as standard! This does not seem to be a very satisfactory way of addressing Horse Welfare and teaching it to others in a "centre of excellence"! Actually, I think in retrospect, the faculty were so ignorant about the practical keeping of horses, that they relied on following current traditional practices advised by their hired staff. These people had been taught by the Miss Mannings of this world, whose authority cannot be questioned. A sad case of affairs for improving the welfare of stabled horses indeed.

But there were highs at the stable where I was a working pupil. The zenith was a master class where dressage riders brought their horses for Robert Hall to comment on their riding and the way their horses were going and then ride them. First, the owner rides the horse to show off what they can do. Most of the horses did not perform very well, but then, the fierce, arrogant, unsmiling, black-clad Mr Hall would get on the horse. Within five minutes, the pair of them would take over the stage and demonstrate real cooperation with a constant conversation continuing between them. The horses from looking like country hacks, suddenly looked like graceful stylish diva ballet dancers, and Mr Hall remained quite still, apparently doing nothing. It took one's breath away. This indeed was the art of equitation, the result of a total understanding of each other's languages and a mutual wish to work together. I vowed I would do everything to learn this and be able to ride like

---

[1] MKW, *Equine welfare*, 1997, London
[2] MKW, *Behaviour problems of farm animals and horses*, 1977 [see Bibliography for full details of MKW publications]

that, one day. Some fifty years later, I discovered that this type of cooperation is called "collective intentionality" and is generally believed to be something that only humans can do. The classic example is playing in an orchestra where different individuals play different notes and instruments, but all aim for the best orchestral performance. There can of course be times when the individual does not particularly want to do whatever it is to achieve the goal of the collective intention, for example practise their part, nevertheless, the goal of playing in the orchestra may be sufficient to allow for the unpleasant parts. This is certainly true for many animals working with humans, they put up with doing things they do not particularly want to in order to please the human handler and continue to have a good relationship. There are also examples of collective intentionality in animal societies, for example lionesses hunting where each lioness does something different to achieve the aim of catching the prey. Collective intentionality is the aim when teaching and working some animals.[1] It may not often be achieved, but it can sometimes be demonstrated by working sheepdogs, elephants being ridden to move timber, draught horses, retrieving dogs, circus performances and even tourists swimming and "playing" with dolphins. But, perhaps the most sophisticated demonstration of this symbiotic "oneness" of humans with another species is when performing complex movements in the art of equitation. Perhaps it is the constant tactile contact between the two that enhances this intimacy and togetherness, like carrying a baby or a prolonged embrace of a loved one.

Unfortunately, riding is by no means always cooperative and harmonious, any more than sheepdog training or swimming with dolphins always is, but IT CAN BE. When one has perhaps only moments of total togetherness and awareness, it cannot be bettered. This is one of the reasons (albeit often not articulated) why so many people continue to try to improve their riding. There are a growing number of scientific journals on "Animal Human Bond-

---

[1] "Collective intentionality in humans and animals. Hoofed mammals & elephants consciousness, ontology and collective intentionality" at *Collective Intentionality VI Conference*, 2008, Berkeley, USA

ing and Interactions", but rarely do the authors recognise or discuss collective intentionality, characterised by cooperation and real unrestricted harmony; a mutual delight of animal and human working or performing together.

That evening was worth all the months of exhaustion, frustration, being sworn and glowered at by others, and occasionally the arrogant ferocious Mr Hall. That vision sold me for life on trying to practise and develop the "art of interspecies collective intentionality". But at the same time, there were and still are physical restraints being used on the horse (bridles and bits in the mouth, tight nosebands, spurs and whips to push him forwards). Surely, collective intentionality should encompass real "harmony" between horse and rider, which must mean that there are no physical restraints placed on the animal, because harmony cannot be demonstrated when the horse is in any way physically restrained. The "art" comes in being able to teach and do the movement when, where and how required, but without restraints of either horse or rider. Only this will show complete mutual cooperation and collective intentionality. One day we may get there, but so far real harmony between horse and rider is never displayed in competitions because of all the restraints that are permitted or which are in some cases obligatory.

After a few months, Kongoni and I returned to Granchester. The colonial administrative period was over, although my mother's heart always remained in Kenya, my ambitious father wanted to be in the centre of developments in his field, which had become wildlife conservation where Europe was one of the main players.

My mother found the quaint Granchester cottage and hundred square metre garden insufficiently stimulating for the long term. My father had his ambitions fulfilled by being offered the post of the first scientific director of the Nature Conservancy in London. Thus, a house, garden and farm within commuting distance of London would suit them both, so a farm search was launched. One May weekend, my father and I drove in his ancient Rolls Royce (which made me very sick, and had a cocktail cabinet) to Ashdown forest in Sussex, now a nature reserve, but then an open piece of common ground surrounded by birch and oak woodland that the gypsies frequented. We were on our way to see an old highwayman's hide-out in Sheffield forest. This entailed driving up a very steep wooded lane from which a stone house was glimpsed through the trees. It had been built as a fort against intruding customs officers by part-time highwaymen. The horses and cows could be herded into an open courtyard inside the house, so they were unseen by any raiding officers of the law. From the easily accessed roof, there was a superb view of the South Downs a few miles away so that the dust from the carriages to plunder could be seen, near enough to give the highwaymen time to jump onto their horses and hold up the coach as it came off the Downs into what was a large unoccupied forest. There was also said to be a secret passage under the woods for the smuggling of wool during the sixteenth and seventeenth centuries (we found it eventually, and put a duck down it, she came out around 400 metres away in a sunken lane in the forest). The farm and house were ancient, romantic, remote, and tumbling down and not far from stations to London. The farmer who owned it had been overcome with the work and lack of finance, the fences were in disrepair, the old Sussex barns collapsing, and the house cold, draughty, damp, and dark with cold running water and dangling electric cords. It looked just the sort of place that my mother would reconstruct and enjoy. So it was bought "lock, stock and barrel" and we installed ourselves.

A year or so later, I had left school and decided to spend a year on the farm helping my mother and enjoying the company of her dairy cows. She and my father had acquired a large golden-coloured poodle puppy: Canute, who I was to train with my father's help to be a retrieving gun dog. I had little finance, but hoped to find a horse to buy (Kongoni had been sold some years before due to my absences at school and the lack of interest from other family members).

My mother's dairy herd gradually took shape. Some of the cows were selected from the herd of mixed-breed cows which came with the farm, but she decided to specialise in Guernseys, gentle brown and white cows slightly larger than Jerseys, and with the same very creamy milk. To start with there were around twenty cows, but her herd gradually grew to sixty milkers and twenty or so heifers and dry cows.

We had an old-fashioned cow shed which the cows entered in their own order, and were tied in pairs so they could receive their rations in a trough in front of them. Their back ends with the teats were easy to reach from the passage behind them. Milking started at 4 a.m. so the milk could be milked, cooled and taken in churns to the end of the drive before 8 a.m. when the milk lorry puffed up the hill to collect it. The milk-woman ran up and down the byre, washing the teats, fixing the milking machine onto each cow in turn, re-fitting it when they kicked if off, stripping out the last of the milk, and then humping the milking bucket (a heavy stainless steel contraption) to the dairy where it was then poured into a cooler. This was a cold radiator, the milk running gently over pipes filled with cold water, and ending up in ten gallon churns. The churns had to be rolled out onto the box at the back of the tractor. When the milking was finished, the calves had to be fed milk replacement in buckets, the cows let out, the tractor driven to the end of the drive and the churns heaved onto a high platform from which the milk lorry would collect them, if one was on time!

Initially it was panic, we had to get up earlier and earlier, but gradually, like most dairy people, we absorbed the ambiance from the non-flustered cows and had it all done in time without irritating them. We learnt a lot about cows in these sessions, their attitudes to the world and how important it was not to convey a sense of urgency to them if you wanted to get the job done in time. This ensured that we had to develop quiet, efficient ways of milking, and a sense of pride in our cows. It was a pleasant two hours in the early morning and evening learning about the individual cows and their worldly ways, singing old Scottish ballads that my mum had taught me, synchronised with the hump and puff of the milking machine, the cows munching, stomach rumblings, sighing, cudding and gurking. I very soon learnt that cows are specialists at ensuring relaxation and enjoyment of what you are doing whether the sun was rising, the snow falling, or humans' hiccoughing. The warm, patient, gentleness of the cow byre is something that will always remain with you, and may be responsible for changing the way you live and what you do; it was the case with me.

But I missed horses. It was not long before my father insisted that I take Canute to a shooting day. This is a highly organised meeting of those who want to let off their guns and kill birds, but who are either too idle, scared of getting shot or mainly just incapable of finding the birds they have killed hidden in the undergrowth, so they have dogs to find them for them. They usually hire a bunch of local lads and lasses who are out of work, or press gang the family to be "beaters". The job of the beater is to walk in a line through the forests or across the fields accompanied by dogs and big sticks, shouting, making a noise and hitting things with the sticks. This is supposed to scare the birds (usually pheasants, woodcock, snipe, pigeons or ducks) who are in the wood to fly away in the direction of the "guns"; these are the people who have the guns who are lined up in the birds' flight path, they hope. The idea is that "the guns" only shoot animals on the wing, it is not "sporting" to shoot them on the ground. The more birds there are, and the higher they fly, the more fun it is said to be, and "good shoots" are places that have these qualities. "Good shots" are people (usually men but not exclusively) who often hit and kill the birds and they may then be invited to shoot all over the country. If, however, you are a bad shot you will pay a great deal of money to shoot, but, the plus side is that you buy an entry into the "shooting set". This is moneyed (or would-be moneyed) landed gents, "aristocrats" or politicians such as people from the Foreign Office, journalists, people "with influence" and/or money; a good move for contacts for the ambitious.

The dogs' job is to find and pick up the dead or wounded bird. S/he must not grab the bird tight and leave teeth marks, but must be able to find the bird in thick cover and catch it if it is still running, without inflicting further damage. S/he then carries it back to the handler who dispatches it, if wounded. It is the dogs' job to spot where the birds land in the thicket, and remember their location. They may be told to retrieve that particular bird as much as half an hour later. The best dogs can remember where four or five different pheasants fell, even when sent back to get them from 500 yards half an hour later. The dogs are expected to swim if necessary, burrow into thorny thickets, be good at following scents ("have a good nose") and be very persistent, never give up. They must always come back to the handler when called, even if a rabbit jumps up right in front of them. They must anticipate when they will be sent to search and what to search for and they must never "run in", that is rush off and try and find a bird without permission, or they may be shot (by mistake of course). They have to be very enthusiastic, athletic, and have a whole brain full of skills. It takes time and a good teacher for them to acquire all this, but some do and are then taken for granted, although they often exhibit more skills than the human "guns"!

As with any sport, there is a whole litany of rituals, which dog handlers, "guns" and "beaters" all have to learn. Enjoying killing things has never been something I could do, but, teaching the dogs and discovering how skilled and able they can become was humbling and widened my admiration for their mental abilities, loyalty and willingness to do it all for a pat or a smile. Today, we have probably only scratched the surface of what we could teach dogs to do, but first we have to believe that they will be able to do the mental tasks we give them. The next stage is to teach them step by step, as we do children. The vast majority of humans who train dogs have strongly held views on how they must be "the leader of the pack" for example, and also the limits of the dogs' mental abilities ... but perhaps things are gradually changing. There is now a sheepdog who is able to identify over 300 symbols and retrieve the object that they represent[1] and gradually the idea that the owner must "dominate" the dog is being replaced by "you must respect and encourage your dog".

---

[1] John W Pilley & Alliston K Reid, "Border collie comprehends object names as verbal referents" in *Behavioural Processes*, 2011, 86:2 DOI: 10.1016/j.beproc.2010.11.007

The "shoots" were primarily human social occasions. Conversation blossoms over the port, roast pig and whiskey at the "shooting lunches" which take place after everyone has got wet and tired as a result of staggering through holly and blackthorn bushes in the woods, or sinking into the Wealden clay. On one of these occasions, one "gun" happened to mention he had a horse, which he and his family were terrified of. They had obtained the horse cheap from a dealer as horse owning goes with shooting; it represents status, being in the know and having influence, whether you want to own one or not. The dealer had sworn, as all horse dealers do ("horse dealing" is not a phrase without foundation) that this mare was the perfect horse for them. In their ignorance, they found themselves with a very beautiful thoroughbred ex-racehorse of 16.2 hands high (height at the withers: where the neck meets the body, about two metres). She had been sold to a series of different dealers because she had reared over backwards with various jockeys and may have killed one. She was also a "crib biter" so was going cheap. No one knew much about the cause of stereotypes at that time, but it was recognised as a "vice". Other vices include things like biting, kicking, bucking or rearing, if the horse has one of these, the price is lower. She cost £40 and was the most beautiful bay mare anyone could acquire, with a small white blaze down her face, two white socks on her hind legs, and a beautifully proportioned athletic body and head. I rushed to my Post Office savings and drew out all my £40 savings; just enough to buy her. That was in 1958, and Stirrup Cup her registered name: Kathiawar my name for her, became the next greatest friend and mentor of my life, and the foundation of our Druimghigha Stud.

It turned out that she could jump almost anything and was very fast and very reactive to any change. It was Kathiawar who confirmed me in my route to continue to better understand the mentality of horses. If humans knew enough about horses' minds, maybe they could change how they kept and taught them and avoid them having a distressing life and psychological problems. Kathiawar would crib bite when she was frustrated, anxious, worried or nervous but when tired, relaxed, doing something she enjoyed or with enough to think about in her work and social life, she did not. I never cured her of it completely but did reduce it. I now know that as she was eight years old, it had become an established habit and "generalised" to be performed in almost every situation except when she was asleep. She taught me how to ride:- (1) remain relaxed, (2) balanced, and (3) understand and feel the rhythms of her movements. These are, I believe the fundamental rules for riding well, and beginning to cooperate and communicate with the horse, the "art of riding".

I took her around the local shows, jumping competitively. This gave us both something to prepare for which we both enjoyed. In the process, we met some of her past owners and dealers and it was possible to piece together some of her life story. She was well-known as a "dangerous" horse. She had been a special foal from an expensive stallion and a mother who had won plenty of money flat racing. But she had responded badly to the normal racehorse husbandry and training which consists of being separated from her mother at seven months old, sold to a training stable where she had been "broken" in the conventional way by having lads thrown on top of her, while being restricted and restrained by a whole variety of pieces of leather. Her response was terror, to get away and stop it all. Instead of the trainers allowing her time to get used to being ridden and understand the whole business, she had quickly learnt that it was all terrifying; enough to send her into a panic. She began to rear up, so she was beaten and further terrified. The next time they tried to mount her, she reared over backwards and almost killed both herself and the lad. This happened again and again as she was sent from one trainer to the another to become more and more terrified and invent new ways of being left alone. Eventually, she became too much trouble and too dangerous, moving from dealer to dealer, gradually becoming cheaper. Her story was like a modern version of Ginger in *Black Beauty*[1] but starting from a racing stable and ending up with would-be moneyed gents who were complete

---

[1] Anna Seewell, *Black Beauty*, 1872

novices with horses and had been persuaded by a dealer into buying her. When I bought her from them, no one was able to do much with her. Luckily, I did not know this, so I had no expectations and we became friends straight away. Her crib biting, the result originally of anxiety and distress, had became a habit that served to switch her attention onto self-stimulation and away from the usual terrifying things that constantly confronted her.

Riding her was an opportunity to develop the Robert Hall riding skills I had begun to acquire, and after some initial difficulties, she responded rapidly to being given time to get used to new things and having a two-legged friend to whom she could make suggestions, who would try not to annoy or frighten her, but also had suggestions. Her attachment to me was probably, in part, because she had no other horses around, and I had no other friend either, but it worked; we both had a friend with whom we could swop physical, emotional and cognitive experiences. Soon we were galloping manically around the countryside, jumping any fence that we came to. This turned out to be very useful as when trespassing, if you get caught, you get lectured and into trouble. If caught again, it would mean my parents as well as me being hauled up in front of magistrates. The answer: don't get caught. We learnt the highways and byways, after all, we only wanted to gallop about over fields and through woods, not to steal or damage. Now as a land owner I know how much trouble trespassers can be, particularly on horse back, but at the time, a fleeting gallop across a field to jump a barbed wire fence into a wood and disappear when shouted at from the farm seemed like good fun, with a touch of risk of course.

We went to shows in a very narrow one-horse trailer I had borrowed from an aunt. I made the mistake of not teaching Kathiawar about trailers before going to an event and she quickly decided, she would go in at home, but when in a strange place and we were both tired and hungry, she would not. I would often have to spend three hours trying to encourage her to go into the trailer after an event, and never did one of the "horsey set" ever offer to help this eighteen-year-old struggling girl. Perhaps in part because we did not have and could not afford the smart expensive clothes and equipment ... and we had often beaten them in the jumping arena. We went hunting where we were constantly shouted at for not wearing the right clothes, going the wrong way, jumping the fences that the hunt servants could not jump and so on ... My impressions of the landed gentry, their hunting and shooting and the horsey set, became rather tarnished after a year. As a result, I had nothing to do with them for about twenty years even though I always had horses. I shed no tears when they started being attacked for hunting foxes some thirty years later although the debate had very little to do with animal welfare, but a lot to do with the people who hunted and their manners. To give them their due, they have improved since hunting was banned, and may even throw a smile your way every now and then, if you are very lucky.

I wanted to ride Kathiawar (who had won all the gymkhana races we had competed in) in a point to point. This is a race across country not on a racetrack, jumping hedges and ditches as they appear. It was originally a competition for those who went hunting to see whose horse was best and fastest. My mother thought my taking part was a bad idea as the other women (there was only one race women could ride in), were likely to try bad stuff like "riding off" or "cutting out" before the jump and other heinous racing crimes that I had no idea about, so she said I could get badly injured, or even killed ... many died after all in the hunting field. There is no denying that they were a tough lot, riding manically out of control on enormous horses over fences with no idea what was on the other side, and continuing to do this two or three times a week until they were well over seventy-five, and originally, the women riding side saddle too! Their courage was admirable, their manners deplorable. Anyway, Kathiawar and I were banned from point to pointing, and in retrospect, my mum was probably right, things would have happened. The hunting set were probably much relieved when Kathiawar and I gave up hunting to run our own imaginary point to points with our own dogs over forests and moors that still existed then, escaping the wardens and farmers.

The year ended and university was imminent. I had chosen St Andrew's in Scotland as it was far away, on the coast, a small town and with a good reputation for biology. I would be away most of the time, but what to do with Kathiawar? I managed to persuade my mother to look after her as she lived with the cows. But, in order for her not to become too bored and waste her life, she visited a local pure-bred Arab stallion to foal the following year. The stallion was one from the Crabbet stud which was started by Lady Blunt who became interested in Arab horses when visiting Egypt with her aristocratic husband at the end of the nineteenth century. It is true that Arab horses are the most beautifully symmetrical, elegant horses in the world, and no one can resist admiring the sheer elegance and beauty of their faces and bodies, the grace of their movements and their apparent delicacy. Lady Blunt wandered around Cairo inspecting thin scraggy-looking Arab horses pulling carts, and with her Edwardian skirts flowing, rode around in the desert, taking risks and having adventures while living with and watching the Bedouins' Arab horses. She had an incredible eye for a horse, even though it might be terribly thin, abused, lame, or beaten up, and bought those she chose if she could. They were imported into England, and with feeding and selective breeding, they turned into the foundation stock of the British Arab, a hardy, tough, elegant small horse who few can resist. She founded the Arab Horse Society, initially mainly composed of upper crust Brits but they had good manners and welcomed anyone from any background as long as they loved Arab horses.

October finally arrived and I had to leave the farm, Mum, Canute and Kathiawar, and all the cows to catch the train north to Scotland, another new country.

During this chapter of my life, I had the age, the time and the motivation to live with and learn much about dogs, cattle and horses, how to begin to understand them and what they needed most in the world. I also learnt much from them about how to live my life, first to avoid unnecessary complications, always try to simplify in order to understand. I learnt what we could say to each other, how to say it, how to work together, have fun and amuse each other. Yes, humans could sometimes be amusing and fun, but they surround themselves with so many unnecessary do-s and don'ts, and symbols built on symbols, arrogantly patting themselves on the back for being brighter, cleverer, more intelligent than non-human mammals but thereby losing sight of where they were going and why. Humans were confusing, although every now and then, like with horses, cows and dogs, one did meet one who could make you laugh. Another trick that I learnt from my four-footed family was always to have a plan B or even sometimes a plan C. Putting all your wishes in one basket is risky, spread them about and some will come true some of the time.

## University & back to Africa

St Andrew's is a pretty old harbour town on the east coast of Scotland, cold in the winter and relatively dry in the summer. Presumably because of its generous sand dunes next to the sea, it was long ago chosen as a golfing centre. However, some centuries before its golf renown, it was the first tertiary educational centre in Scotland. This ecclesiastical centre trained monks and priests, becoming a university in 1413. The university dominates the town with offices, laboratories and student residences dotted throughout, and the majority of residents are employed by, or are dependent on, the university in some way.

As one might suspect with such a history, there were a number of rituals and initiations that had to be suffered by new students. Not all of these, at that time, were benign; the men students taking particular pleasure in rough and sometimes excessive teasing of new male students, and the senior girls, as is their wont, tongue lashing the new girls. Although many of the 16/18 year old students, who had often left home for the first time, suffered, it was nothing new for me and others from English boarding schools. We had to wear great red flannel coats with velvet trimmings as we were butted by the below-zero coastal winds between dormitory and lecture halls, cafés and the harbour. The "gowns" were welcome warmth, and had the added advantage of not needing to dress carefully underneath, something I was not too interested in. My degree work covered the natural sciences including geology and chemistry as well as botany and zoology. It was an unspecialised degree which is out of fashion now where it has become important to learn more about less. I am eternally grateful for this broad natural history background that St Andrew's gave me. It has allowed me to understand many different aspects of the natural world better, and particularly to bring knowledge from one area to another and to mix science with folk knowledge that comes from what the natural historians know. There were written exams every term designed to test the students' overall knowledge as well as whether they could write English, no guessing the right answers on forms. We had to write six essays on the trot, identify 200 plants or cut up a worm and show that we knew all its internal organs and how they worked. The teaching was by lectures and we had to make our own notes. The only way to find out what was badly explained was by talking to other students. There were three or four lectures every working day which everyone was advised to attend, but almost no tutorials. No hand feeding: either the student got down to learning or failed, and at the end of the degree we certainly knew our stuff. It was often a hard grind, at least when trying to learn scatty notes a few weeks before the exams. In a sense like practising the piano to learn to play, there were no short cuts, there could be no reliance on the Internet or blagging which can dominate tertiary education today.

My written English was appalling, and, like almost every student in the first year, I made every effort to blag my way through the first exams; studying was not my top priority in the student social whirl, however, this was not entirely successful. I failed physical chemistry which meant a trip back in September to re-sit. This gave me the opportunity of camping by the sea in the sand dunes, and swimming in the magical floodlit sea, lit by millions of tiny illuminating crustacea. It was illegal to camp in the dunes, but after one or two encounters, "the law" found it difficult to find the tent when it moved from where it has been spotted and they gave up looking. A couple of weeks with just the sea, the sky and crabs at low tide suited me well and encouraged the necessary exam swotting. Luckily, I passed and in the second year could continue with subjects I had chosen, such as geology, zoology and botany. One part of geology is mineralogy. This involves handling and carefully looking at pieces of rock which all look white, brown or black. But when we cut thin slices to make microscopic slides, the colours blew one away, rainbow on rainbow on rainbow. Then there was crystallography, the study of crystals of which there are hundreds of types. For this, it was essential to think and visualise in three dimensions. The fourth dimension: time, was added by geological history. Botany was not restricted to colourful flowers, we studied in depth every aspect of ferns, liverworts and mosses, their ecology, their cell structure, their reproduction, and of course who they are related to: the study of systematics, Linneaus's forte. Linneaus was a Swedish naturalist who lived in the eighteenth century (1707–1778) and studied every plant or animal he could find or was sent to him. He drew up a list of how they were all related to each other based on their detailed anatomy which was a really staggering achievement. Even today when we can assess the DNA of each species, the majority of his classification of species still stands.

The extraordinary varied beat of life that is the living world and its effects and affects on the whole picture, its differences, diversity, complexity and adaptability and however closely you look, the complexity remains or even grows with the microscopic details. It is truly mind blowing. I was privileged to do one of the last undergraduate degrees that covered both the whole picture of ecology but also the detail anatomy, physiology and systematics. St Andrew's had been at the beginning of the evolutionary story, and we had some world class lecturers. I remember a Professor Barnett, a fattish man who pottered between the benches in the labs, discussing the intricacies of flowers. He had a great admiration for my father's botanical friend Bad Uncle John who had discovered how many of the tropical trees were related to each other and had come up with a discourse on their evolution which fascinated us. But, he had only managed this by training monkeys in Singapore to climb the trees and throw down their blooms to him from the top of the equatorial forest. For organic chemistry we had a superb ancient lecturer, Professor Reid, who managed to make it all extremely simple, even the structure of DNA was possible to unravel, even though it had not long before been discovered. Of course, we had our fair share of very bad, boring lecturers too. Today, ecology degrees concentrate mostly on economics and anthropology rather than actually studying our living homeland. Perhaps it is this education which is responsible for the ever-growing ecological crises; where are those who are blown away by what they know and have discovered about the living world, and why are they not teaching the young? I did find some of these rare species of humans when I was asked to teach a field course in ecology at Cambridge in 2003, enthusiastic teachers who would work all of twenty-four hours to help their bright, hard-working students and had enthused them, hopefully, to follow in their footsteps. There are a few left, so like the threatened rhinoceros, elephant or toad, all is not yet entirely lost.

There were practical skills to learn too, a steady dissecting hand to cut up bits and pieces into tiny slices or, for example, carefully dissect out the intestines of a worm or to manipulate tiny bits of stem on slides under a microscope. Scrambling through sea ponds at low tide to collect specimens, or wandering through the rain identifying and counting trees in a wood. Field trips to the beaches only a stone's throw away to study their ecology were frequent. We would potter around the high and low tide marks, digging about in the sand and the rock pools, trying to understand the effects of the tide, the wind, the salt in the water, the location on the beach, the other species present and all the possible variables on what lived where and why. We were taken for a week to study an ash woodland, and another week to the high moorland in the Cairngorms, we staggered through mountains and moorland, streams and ponds and lakes, basked briefly in quick spurts of sun, but more usually bashed by the wind and rain, camped in cold leaky tents, rode on sick-making buses and walked and walked, collected, studied, chatted, ate and drank. It was great being with a group of people my own age who were seriously interested in the natural world and learning about it, and with tutors who could answer most of your questions and would help and encourage. We discussed and argued about what we had seen and identified and enthused each other even if the weather was often not helpful.

After acquiring some detailed knowledge and learning how to find out more (even though some of it may be quickly forgotten), how can anyone do such an all-encompassing study without ending up with an enormous admiration for the living world and where it came from, and its amazing ability to adapt?

In the early 1950s, animal behaviour was not something universities provided courses in, it was something those who lived and worked with animals knew about and practised. However, there was one bright lecturer at St Andrew's who specialised in the new study of neurophysiology and behaviour in sea anemones: Dr Horridge was studying why, how, when and where they waved their tentacles about, or contracted themselves into balls. He was my final year tutor and set us essays where we could explore ideas of the origin of life and the primeval soup it came from. These questions were stimulating and very relevant to my understanding of the differences between a living being, and a non-living one. We also discussed feelings, bodies and minds, even though, at that time "scientists" more or less believed that "minds" were exclusively human.

Apart from the rituals, human social life was sometimes fun. I made some friends in our female dormitory and had adventures swimming in the freezing sea, or climbing up to creep through small windows when back late at night from some harmless but illicit function. But the traditional dances and Ceilidhs (Scottish country dances) were agony. The boys and girls stood on different sides of the room and girls had to wait to be chosen by a boy before they could dance. I had put on plenty of weight from the UK food and insufficient exercise, and looked a fright, so, was never chosen, but learnt to hide behind curtains and tea urns. I also found the boys young and ugly and if some poor ugly spotty youth picked me as the last available, I was probably rude as well; it was sometimes difficult for me to think these boys were really just male mammals. I quickly gave up these social activities, but it brought to my attention my gross overweight, so I starved for six weeks, which is far the best way to lose weight, whether a dog, horse or human. I lost two stone (about ten kilogrammes) and by my third year had had two offers of marriage; one from a rather intense medical student, one from a young shy son of a Scottish laird who was to inherit an enormous Scottish estate. I went to stay with the Scottish laird and his family once to be "vetted", my father was known to them; but was I a suitable match? I remember an enormous house like Downton Abbey, full of very correct but terrifying people, miles of faultless lawns, a couple of rather remote dogs ... best of all were the very delicious biscuits by the bed. But, these kind of guys and marriage were not on my agenda. In those days it was even risqué to kiss, so there were no physicals either.

In my third year, Anna, a quiet medical student, and her boyfriend Peter, a very long, thin Dutch geological student, and I found a small cottage on the edge of the cliffs with no electricity, gas lights and a small garden looking down into the sea. It was the property of an old deaf fisherman's widow. She lived in one room and was glad of the £3 10 shillings rent per week we paid. We usually had an evening meal together which one of us cooked with a budget of ten shillings a week, but other than that we followed our own interests. Peter loved making boats, but he seemed to be somewhat scared of going to sea in them, and we never went sailing with him. I last heard of Anna and Peter, some thirty years down the line. They were living on the coast of Northern Ireland during "the troubles" and Anna was doctoring the IRA and any others who were wounded. Peter was still making (but not sailing) boats. I often wonder what they are up to, or if still alive: smiling, very shy, quiet Anna, who had to re-sit most of her medical exams, but I am sure made an excellent general practitioner. Long, thin, lanky, gentle Peter with his beard and floppy clothes.

How on earth did they fare in a war zone? In a sense, this couple, when I knew them, were similar to my four-footed family, where the squabbling of different groups of humans which ends in wars and killing each other, could not be less aligned to their world view. An alternative which they and my four-footed family taught me was whatever the provocation, try and get on and even invent ways of cooperating by ignoring inflammatory difficulties between people, and copy nice behaviour in order to have a life with the rest of the living world.[1]

Long vacations were spent volunteering with the newly invented international student camps in France and Portugal. The idea here was to try to get the young of previously warring nations to work together and respect each other a little, as it was not long after the Second World War. Particularly memorable was forest work in Alsace near the German-French border. The care of the forests had been neglected during the war years when they had become military hideouts for both sides. The forests badly needed thinning so that the ash, oak, beech and birch could grow strong. As usual, the few girl volunteers were meant to stay in the kitchen. I managed to wiggle my way out of this, by being a messy house cleaner and an atrocious cook, unforgivable inabilities for females in France. So I joined the 4 a.m. starts, walk of five kilometres to our working areas, then cutting and slashing trees and bushes until 3 p.m. before walking back to camp to sleep and chatter. We all lived in single sex dormitories, but my education concerning human males had begun and I soon realised that male humans were at least mammals, although they are different in both body and mind from me, but like with my horse or dog or duiker, I could begin to unravel who they were. Unlike non-humans however, men of my age, in their early twenties, seemed to be almost exclusively interested in sex!

It was spooky in this neglected large forest to come across ten-year-old remains of camps of British or American soldiers with the remnants of their canteens, tins of baked beans, bits of

---

[1] MKW, "Communication in a small herd of semi-domestic elephants," 2019

uniforms and boots and then further on in a valley, German boots, old food tins and bits of uniforms decorated with swastikas. We never came across any skulls although we did find the odd human bone, the remains of what a wild boar had devoured, I imagine. We were the sons and the odd daughter of the British, German and French soldiers who had killed each other in this forest, here we were working together to clean it up. Perhaps there was hope at least in Europe, for the future.

Even then the majority of students came from cities and when I attended a work camp in Portugal, everyone seemed to be frightened of the forests and would see monsters behind every tree. They would grab me as they rushed off terrified. I ended up being dragged along on my knees by them and was badly gashed. After this I thought I could hop along and do the vendange (grape picking) in the south of France and earn some money for a while. It was a vineyard in Chateauneuf du Pape country. The other pickers were all Spanish country women who had left their families and homes for a brief respite to earn money, and to get away from their domineering Spanish men. They spoke only Spanish, I spoke none; it was hand signal time. They were a lively, chatty lot. We had to work very hard, up at 3 a.m. to start picking just as it got light. Breakfast at 8 a.m. after four or five hours' work. Breakfast included wine, which was not a good idea as it was followed by another four hours of picking in the sun, before more wine and lunch, an hour's siesta when the temperature was in the late thirty degrees centigrade, then more picking until 9.30 p.m. and bed. We earned around twenty Euros each per day and signed contracts to work seven days a week. Despite the long hours and heat, these peasant Spanish women would chatter and laugh all day long as they advanced in a line picking the grapes and placing them in boxes without touching the bloom on each grape. Bunches of grapes are round and tapering, but somehow, without touching them, we had to rapidly place them in boxes so they were flat on top. How they managed to do this so speedily, while chattering and dancing along, I will never know. It took all my energy and attention, just to try to get a flat top on the box. They reminded me sometimes of parrots in the Congo, attacking fruiting trees in their hundreds, chattering, calling, dancing around as they gorged themselves.

Our living arrangements would not come up to today's European Health and Safety standards by a long shot. We all slept in a barn on straw, and because they were frightened of being raped, we had to have all the openings covered and fastened for the night. It was completely dark even at dawn. The toilet was a bucket in the middle that was constantly adding noises and odour to the night. Personal hygiene was not high on anyone's list of priorities, there was one tap for forty of us. The odour of the sleeping arrangements was overpowering sometimes, so at the risk of being raped (although I never understood why this was so great in the middle of around 200 hectares of vines), I slept in the vines, when the others would let me, which was not often. They were quite sure that I would come to harm. These Spanish women were the migrant workers of Europe, earning a crust for a six-week period in the late summer. For them it was their summer holiday, and they danced, sang, dressed up in all sorts of finery, and talked and talked. I learnt a little flamenco, and even less Spanish, but had the benefit of amusing them with my efforts. This perhaps was a little like a sympathetic human going to live temporarily with a pack of friendly dogs living under an unknown human's rules.

On another vacation, my sister had been asked to identify boxes of animal bones that had been dug up at an archaeological site somewhere in South England. The diggers wished to know what species the humans had been eating, how many, what age and how they had been cooked. It was a tall order identifying species from bits of bones, and Grizelda was more interested in viruses and microbiology which was just emerging at Cambridge. The bone package was passed on to me. I stared at this vast pile, believing the questions impossible to answer. But, with constant reference to a human skeleton, the bones gradually became identifiable. Pig bones were shorter, fatter and more rounded, cattle bones larger, heavier, stronger, dog bones smaller, lighter and angular. Even if most of the bones were just broken bits, they could still be identified. After about two months, I had a rough estimate of the species and numbers and it was a fair guess they had all been cooked over an open fire. The message here was how similar human skeletons are to other mammals, and if their skeletons are so similar why not the rest? Pay was not on the agenda, but I was mentioned in the eventual published paper in some obscure archaeological journal of which I was very proud. It was an interesting job, and came with a place to do it in, a sleeping bag to kip in, some friends who might feed me sometimes, and a few saved shillings to pay for the odd meal. What more could you want?

After graduation Africa pulled me back. My mum's farm was prospering, and she had a quiet, pleasant hard-working farm worker of whom both the humans and the cows became fond, so she could cope. Peter lived in the house and was devoted to the cows with whom he worked for a good thirty-five years. My mum insisted she would not keep two horses while I skipped off to Africa. Either, I must stay in the UK, or sell one of them. I did the latter. Beautiful Kathiawar went to a young woman in the New Forest. Her daughter Syringa or Padna Parameter (Perfect Wisdom: I was having a Buddist interlude) would be allowed to stay with the cows, at least for a year or so. To avoid her becoming bored and lonely without other equine company, she visited an Arab stallion and eleven months later would have a foal which would keep her occupied for a while.

There was no money for the voyage and budget travel for the young had not been invented. Hitch-hiking had though. I had frequently hitch-hiked to university in Scotland and through France and Portugal so thought I was a dab hand at it. Why not hitch to Kampala in Uganda? The overland trip would be through France, Italy, across the Med to Alexandria and then up the Nile to the Sudan, south through the deserts and the huge wet Sudd marsh in southern Sudan. Finally, to Uganda where I had been accepted to start research, studying the behaviour of an animal of my choice, and been given a scholarship by the Goldsmiths Company in London. I managed to persuade Angie, a close friend from our gang at primary school in Kenya and a qualified nurse, to come with me, despite the outraged disapproval of her parents. Luckily, they were still in Tanzania, so communication was long range by letter as phoning was far too expensive. We had already left when their outraged and disapproving letter arrived, forbidding her to go.

It was an epic journey at that time. Young middle-class naïve women really did NOT travel in unknown places by asking for, and accepting lifts, without men. We took about three months. On balance, most people were very hospitable and helpful, but Angie was a bit of a flirt, and got us into a couple of scrapes with men. Luckily, we were both fairly strong, well-built lasses, and learnt fast how to extract ourselves from such troubles. On one occasion we had been standing for hours in the sun in Yugoslavia. A large van stopped with two well-built young men in it. The rule was not to accept a lift if there was more than one man, but Angie insisted, and I was also hot and tired. One was a boxer and his friend, a wrestler. Of course, they had a convenient "aunt" in the nearest town who could put us up for the night. It resulted in us having to barricade ourselves in on the fifth floor of an ancient building in Split as they and other friends they had invited, got drunk and tried to break down the door. The window was the obvious escape route, but it was five floors up in an ancient building with a narrow, cobbled street at the base. Luckily, with our joint weights against the door and various pieces of furniture, the door held, and we survived unblemished to pick our way over their drunken sleeping bodies the next morning. Something that has always mystified me was how is it that women can be raped (have sex when they do not want it), but other mammals cannot, a cow, mare, elephant or bitch can just avoid it because they must stand still, but women and female chimpanzees or orangutans can just be grabbed and forced to comply, not a pleasant experience. We took a third-class ferry from Naples to Alexandria, brushing off Italian efforts at grabbing any piece of our anatomy, but when we reached Alexandria, Angie picked up what she believed to be a wildly attractive Alexandrian, and was arrested for kissing in the street! It took a couple of days and help from friends at the British Embassy to get her out. Ironically in the "free world" of today, she probably would still be there, unless her parents had come up with a large hand-out. I, anyway, was learning more about primate (particularly human) males.

The tourist industry outside Cairo was used to catering for upper crust visiting Brits, a left-over from their colonial administrative days, so there was the famous Gazira club to visit, and, more interestingly, Arab horses to meet and ride around the pyramids. We had a couple of days galloping around in the desert having befriended an Arab owner who had some difficulty with his horses. Then, we spent time struggling through the overcrowded tiny bazaar streets dodging through camel and donkey legs, attempting to prevent the goats stealing vegetables and fruit we had bought for the journey south. At that time not all the women were in purdah due to the efforts of educationalists from the British administration. As cities go, it was exciting with a mass of hand-made carpets, knives, bangles or anything else you needed to buy. There were many animals wandering around, and slightly terrifying shady businesses being conducted by Arabs in flowing gowns with shifty eyes and knives stuck into their belts to add spice.

We continued to hitch south on jeeps, boats and trains, and visited the temple of Abu Simble before it was flooded by the Aswan dam, wandered around Upper Luxor in the desert looking at ancient mummie sites and temples, learning about the human history of ancient Egypt and the invention of writing, something that only humans do and is considered by many the secret of their greater "intelligence". Then we spent three days on a hot and dusty train chuntering through the Sahara before being invited to a three-day wedding in Khartoum. After which there was a slow eight-day steamer through southern Sudan where we joined other third-class passengers, their goats and hens, to eat chappaties and chilli soup. The ship stopped every now and then to load and unload people, animals and goods. There was a tribe called the Dinkas who never wore clothes at many of the stops. The captain was anxious we did not take any pictures of such "primitive people"! Crocodiles, hippopotamuses and elephants would lumber slowly out of the way of this slow-moving overcrowded steamer. But, we saw nothing of the rumbling war between north and south Sudan that was taking place.

On arrival in Uganda, Angie travelled by train onto her parents in Tanzania. I started my post-graduate research at the University of Kampala. The campus was quite beautiful, with fruit bats nesting in the tall gum trees just outside my window, hedges of hibiscus and bougainvillea in profusion, and other familiar tropical plants which I had not realised how much I missed. I was given a room in the only female residence; there was one girl to every 500 boys at that time. The undergraduates in Mary Stuart hall were all very bright girls from tribal communities who had managed to pass university entrance even though girls usually were not even sent to primary school. For them, it was a cultural shock to live in a stone-built three-floor building with water, toilets and showers, tables, chairs, knives and forks with which they were expected to eat. There were a few white girls, daughters of colonial servants or Kenya settlers, or some doing research or studying from Europe. Being used to household equipment, we could help the tribal girls to become accustomed to such curiosities, and we spent a lot of time laughing and singing together. There was also a large contingent of Indian girls whose families ran most of the businesses in Kampala, one reason why it was such a thriving town. They were mostly vegetarian, hard working and with a whole bunch of contacts in the town; whatever you needed could be provided. The typical food served was *ugali* (maize meal cooked up to a glutinous mush), *matoke* (special sour green bananas mushed up), a very hot chilli soup and cabbage that had been cooked for at least a week. But the vegetarian Indian food was often delicious, vegetable curries, chapatties and rice, so most of us whites gradually became vegetarian.

Several of us decided to put on a play or review to which anyone from the hall could contribute. It was a fun time, learning about different tribal cultures and beliefs, and making friends. We eventually performed a review for the rest of the university, followed by beer and dancing; it was not professional but fun. At Makerere University, the first East African University, there were students who later became the first well-known East African writers and intellectuals, various British girls doing post-graduate work who became world renowned academics, and my friend Kathy who was the daughter of a Buckinghamshire jockey and married an Arab from Zanzibar. She came back pregnant every year and was studying English.

Right in the middle of the preparations for our review, the first ever group of Peace Corps volunteers arrived from the USA. These USA girls and boys had never been "off the tarmac," and most of them found Uganda unbelievably awful ... flies, wild animals, dirt, and nothing at all American. Being parochial (as many who do not travel are), they could not even understand the Ugandan English. Consequently, they found it more difficult to adjust than even the tribal girls and were not really motivated to try. They disapproved greatly of any white face (unless it was American) and considered that we must all be colonial bullies and monsters. However, the Ugandans of all colours, saw the colour of their ample money and shamelessly took them for rides because, although they were supposed to be volunteers, they were paid more monthly than the annual income of many Ugandans. Initially, we ran around finding cars or buses to take them from here to there, and particularly with visits to the hospital; it turned out most of them were hypochondriacs! Luckily, about half of them were sent home after a few months, and the rest sort of managed, although they became a great drain on the resources of the rural poor whom they had been sent to "help" by constantly requiring what were, for the rural people, total luxuries that were alien and that they could not afford to supply, things like sterilised food out of tins or packets rather than local foods and medicines; visits to hospitals and motor cars and transport. This illustrated the first problem of what has now become "the AID industry" which continues today. Idealistic youngsters sent from rich countries to "do their bit" for other humans, only to end up hindering them by causing them expenses they cannot afford, as well as dragging the locals into consumer dependency. The visitors may benefit but the benefit to the local communities is very debatable. Today anyway, volunteers usually have to pay; let us hope the price rises too! Similar problems have emerged in the tourist industry worldwide. Although usually, most tourists are sufficiently isolated from the local populations except for meeting one or two guides and going on a cultural day to the local village. Westerners today cannot imagine what it might be like to live as non-consumers. Villagers at that time had to be self-sufficient to find their own food, water and make their own shelter out of what was available. The urban westerners try to "help them" by giving them goods or money so that they join the economic rat race. The four-footed non-human communities, particularly those living without any human interference, are always self sufficient, and as a result they have to acquire a great deal of ecological knowledge or they will not survive; they end up good ecologists whereas the AID workers, unlike the tribal populations that they are destroying, are usually unaware of the local ecology except if it is too hot, cold, windy, dangerous or rainy.

My supervisor at Makerere was a gentle, quiet limnologist (fresh water biologist) who was somewhat bemused by my insistence on studying the behaviour of a wild mammal. However, he sent me off to the national parks in the west of Uganda to decide what species I wished to study and how. This meant a vehicle. I could not afford a four-wheel drive which is de rigueur for researchers, so I opted for a VW Beetle that I bought off an Indian friend, brother of one of the student girls. Later, with the help of her friends who were Indian mechanics, I managed to locate an ancient US jeep which from its condition, must have been dropped from an aircraft during the war ... it was of the less-going sort, so eventually, I found another for spares.

I set off in my VW Beetle to Western Uganda with binoculars, note books, Brownie box camera, stores of emergency food, tyres and petrol, and arrived after a few days at Mweya, the newly constructed centre of what was then the Queen Elizabeth National

Park on the Congo border. The game warden, a Mr Popperton gave me a *banda* (small grass hut) to live in overlooking Lake Edward. It was called Swank Cottage which I thought was rather appropriate at the time since it had running water and a primus stove.

It was a hot, lush, and gloriously isolated place, and there was no way one could forget that you were surrounded by real Africa from the continual honking of the hippopotamus in the lake. But, at that time, 1959-61, there were no written accounts of how to conduct a behavioural study, so I had to invent methods. The first thing was to choose which species to work on. There was so much choice: hippopotamus, elephant, buffalo, lion, leopard, cheetah, ant eater, and about twenty species of antelope, goodness knows how many types of rodent and several species of monkey. First eliminated were the primates, after all, since I am a primate, why study one who may have a similar mind? Then the predators were eliminated. I was more interested in prey species who have subsequently received less attention and are generally considered less interesting. Of the large mammals, that cut it down to rhinoceros, elephant, buffalo, giraffe and hippopotamus. The hippopotamuses were about to be "culled" (shot selectively) because they said there were too many. Rhinoceros, elephant and buffalo were rather difficult to find as they had recently been shot and were frightened of humans, and giraffe were uncommon in that area. That left the antelopes. Impala, bushbuck, sitatunga, Uganda kob, Grants and Thomson's gazelles, duikers and dik-dik, occasional oryx, and waterbuck. I finally chose waterbuck, *Kobus defassa*. That part of Uganda was on the border between two species, *K.defassa* and *K.ellipsiprimnus* and it turned out that they interbred (so were not different species). Waterbuck are beautiful large antelope weighing about 300 kilogrammes. They have long brown hair and the males have lengthy scimitar-shaped horns, the females none. The first problem was how to identify individuals. This turned out not to be so difficult, because the two races had interbred which was reflected in the individual shape of the white patches on their rears: *K.defassa* has an entirely white behind, but *K.ellipsiprimnus* has just a white ring, as if he had sat on a newly painted lavatory seat. Depending on their mum and dad, they had different amounts of white in different patterns on their rumps and I was able to identify around forty individuals with no need for modern technology such as immobilisers, fitting collars or transmitters. These technologies are de rigueur today for any animal study, but I could never have afforded them anyway. The disadvantage of all the technology is that because it is available, people do not observe so carefully and consequently become dependent on it. We could take a leaf out of the book of the animals we are studying in terms of learning better how to observe, rather than forgetting the little we know.

I guess if I had chosen one of the high-profile species such as lion, elephant or chimp, my career would have been very different. The post-graduate students who went to East Africa about ten years later chose high-profile species and became world famous for their studies, the best known being studies of chimpanzees, elephants and lions. Perhaps in some ways they were better studies, but by then, there was an established method. I am glad that I chose the lowly waterbuck. They taught me the importance of living by water in shady tropical Africa, how to live well in small groups and to admire their knowledge of the terrain and abilities to swim, jump, watch, play and hide when necessary. I had £400 for the year, a notebook, several pencils, a broken-down jeep, a box camera and lots of enthusiasm. I had been raised in Africa and so Africa with all its quirks and difficulties was really my home and I had the ability to survive and live well and joyfully, knew what animals and plants to avoid, which to befriend, how to respect and live in the natural world reasonably safely, how to tolerate constant frustrations because things constantly go wrong, when to drink the water, and how to speak the language and live with tribal African people. I wanted very much to learn and do a good job.

At that time there was a group of whites who lived at Mweya, the main warden and his wife and family, a white mechanic who was meant to keep the road machinery going with his gang, and various visiting scientists either from the UK or USA from time to time. The first group of UK scientists arrived about four months after me. They started by trying to do a grandiose survey of the ecology of the whole park, but although trained ecologists (one of them was Richard Laws who subsequently went on to lead the British Antarctic survey), they were ignorant about Africa. I made myself as useful as possible, translating Kiswahili, learning to identify grasses and other plants with them and starting their herbarium, and did a lot of fetching and carrying; but they never considered me a part of "their team". This has remained a mystery, since then I have had many enthusiastic people wanting to help and learn with my different research projects, and although not all have been terribly helpful, they have always been recognised and accepted as part of "our team", like my quadruped friends and the Africans I knew as a child, they not only accept members of their own species who joined up with them (though this may take some time) but even those from different species and cultures ... but not so these UK scientists.

It seemed that human social contact within the park was not going to work, so I went outside it. Nearby there was an African village in which some Greeks had taken up residence in order to trade and smuggle things in and out of Uganda to the Congo (it was only about fifty kilometres to the Congo border). I became close friends with some of this community. I remember well Uncle Basil, an ancient Greek adventurer whom I helped to drive vans and lorries in and out of the Congo. It was not always a good idea to ask what was in them, even though at that time they would not be either drugs or armaments, as all that came later. Maybe there were occasional jewels, but it was mostly bananas, mangos and salads from up-country Uganda, delivered to Belgians who did not seem to be able to grow anything. Uncle Basil must have been about sixty-five, he gave me a lot of advice about how to live well and have adventures without too much risk. It was, I suppose, a relatively innocent period in African history when tribal do-s and don'ts were still respected, and even wars were not dominated by firearms and drugs. I managed to crash his van into a flooded bridge once, but we extracted it and never breathed a word to the Mweya community who thought my actions were beyond the pale ... even though I worked hard, helping them as well as following my own animals everyday. I found out later that Greeks at that time were socially unacceptable, like Blacks or Indians. Humans certainly were/are confusing!

A delightful couple of biologists came from Kenya to kill selected hippopotamuses and study their meat. It was considered that the hippopotamus population was threatening the survival of other species. Dr Mann and his wife were German Jews who had fled to Kenya during the war. He was an exceptional physiologist, and his wife an excellent cook. We tried sweet and sour hippopotamus, hippopotamus steak tartar, hippopotamus curries and *pot-au-feu*, and any other way of cooking the meat we could think of, and then asked the local tribesmen and women to try it. I greatly enjoyed the company of the Manns, and we spent many glorious evenings having traditional Kenya "sun downers" of pink gin while looking over the lake, them teaching me and telling me stories, and all of us counting our blessings.

The culling of the hippopotamus was ethically difficult. There certainly were so many that there was very little grass for all the other species within three kilometres of the lake as they came out to graze at night. But should they be killed in a National Park, and if so how? If they were to be killed, how could we get the locals to eat them and, in this way, increase their protein intake which was far too low? Hippopotamuses were one of the species that most tribes consider inedible, and consequently taboo. It is true, the meat was not very nice, and they were fierce and difficult to kill without firearms.

If the growth of one species' population threatens the survival of others, what should be done? Such debates continue today. Should the animals and plants just be left to get on with it, or should they be managed and if so to what aim? The difference now, some sixty years on, is that nowhere is it possible to leave the animals just to "get on with it" because all areas are enclosed in one way or another, there is practically no wild left. "No management", sometimes called "re-wilding" of enclosed areas will result in over population of some species at the cost of others, exactly what has happened in the human population. If the aim is to conserve all the species in that area, then how, what, where, why and when can and should this be done? For hippopotamuses this "culling" problem no longer exists, since almost all hippopotamuses have been killed and now they only exist in some isolated populations in Africa. But even in these isolated populations they may be too many for the small area to support, so there has to be some form of management, whatever that is.

Today, although people take strong positions on these debates, there are no universal simple answers. But, as early as the 1960s they were hitting the headlines, even though, at that time, no one believed that wildlife would ever be seriously threatened. As a result, all decisions were skewed to helping humans. Today the same skewed decisions are nearly always made, wild animals and their environments are almost always sacrificed for what are regarded as "human needs", many of them trivial rather than life threatening which they generally are for the other species. Since more and more food is needed to feed the exponentially growing human population worldwide, and people have almost all become used to great variety, producing food primarily for the family, the district or the country is now a thing of the past. Trade is the buzz-word and if you have plenty of money and feel you need to feed the hungry, it is "food aid". In this way, everyone becomes absorbed into the consumer economy, and trade will grow. The result is that self-sustaining populations, hunter gatherers and peasants, are a thing of the past, and that trivial needs of humans, such as more jobs and more money, replace them. Both small and large businesses run by both men and women pat themselves on the back because they have increased the market. It has taken only one generation to lose both the knowledge and the will to live self-sustainably in the natural world and respect other species. As a result, wildlife has almost vanished.[1]

But, applying a generalised stance such as to "preserve species

---

[1] MKW, "Problems of modern agriculture," 1980, pp208-215

diversity" is not always conducive to conserving the living world, any more than many people's need to "help poor humans" achieves their aim. It is often believed that despite some six billion years of evolution, the living world cannot look after itself, it must have constant human involvement and be kept "tidy".

If, in each situation or place, the aim concerning how to integrate human and non-human needs is properly debated, and all conflicting interests represented, there could be hope. This will not happen however, until people realise that the wild world is important, not just for visiting, admiring, creating jobs and money, but for our own survival. There are ways of marrying wildlife and human needs and it is usually possible to find solutions, wherever there is a will. Enormous mistakes continue to be made in the name of wildlife conservation throughout the world, because these questions have not been debated carefully. The usual mantra employed today is to conserve a species because it will make money. In Africa, this usually means at best "eco-tourism", rather than "learn to live well with other species". But, eco-tourism itself is causing very serious environmental problems, some of which may not be reversible.

Tribal people had no money, no need for it and few of the things that the aid workers and consuming westerners believe are "essential", like running water, electricity and jobs. But, using their own initiative, they had shelter, enough to eat and drink, plenty of free time and, depending on their culture, more liberties than most city dwellers have become used to.

Improved medical care and justice for all are probably the only things that would have improved their lives. Much of the colonial service was devoted to these improvements, even though it is usually believed that British colonialism was only about making money for Britain. There were thousands of colonial servants who worked their lives away helping in the British colonies with medical and veterinary care, agricultural and forestry improvements, and in particular, justice for all. The clock cannot and should not, in many cases, be turned back to tribalism and its values, but we should pause to think what we could learn from each tribal society for the future, before they disappear, just like we could begin to learn other ways of living and viewing the world from the perspective of other species, before they vanish too.

After about six months, when I was beginning to have some idea how to record the behaviour of the waterbuck, and getting to know my population individually, it was Christmas and I was invited to Soroti, to the east and north of Uganda, where my sister and her family were living. Her husband was a community doctor. Although recently qualified, he was running the local hospital, learning to do surgery and generally holding the medical health of the area in his hands.

I drove to Kampala and then north in my dusty VW Beetle, but the rains had started and the rivers were flooded. We arrived at a flooded river some 300 kilometres from Kampala, a day and a half's drive away, to find queues of lorries, Land Rovers and cars waiting. Many others had gathered to see the fun, and were standing chatting and shaking their heads, and in the African way, waiting for the river to go down, which probably would not happen for a week or two. It was two days to Christmas, and I was going to cross the river come what may. After stopping and assessing what was happening, I drove as fast as I could into the river. Luckily the VW Beetle was light, small and more or less floated. Everyone was so amazed that they laughed and laughed, and ran up to push the stalled semi-floating car through, Christmas presents and all. The VW and I merrily, if rather wetly, continued on our way and were the only vehicle that got through until New Year.

After New Year when I was getting ready to drive back to Mweya, I received a letter from the "chief warden", Mr Popperton, telling me I was not to come back. I had no idea that there had been so much antagonism against me, or why. I thought I had done what was asked, helped them all when I could and had generally been well behaved. It was long after that I found out that Popperton's wife thought I was after Popperton and had thrown a wobbly, so he and Laws decided I should go. Popperton was a red-faced, red-haired sort of man from the UK, whom I had no attraction to whatsoever, but particularly idle women usually felt threatened by other young women then, particularly in Africa ... "Are you married, or do you live in Kenya?" was the usual question asked. It turned out also that Laws and his British assistants were finding their "African feet" and did not need me to translate or run errands anymore.

But my study was beginning to take shape and I was really enjoying it. Twelve-hour days sitting in the bush watching and following waterbuck, trying to understand who they were, how and what they communicated and how they organised their society, the odd adventure with hippopotamus, buffalo or elephant really was how I wanted to live. So, I talked to my supervisor and asked him to interfere on my behalf, which he did with the result that I could go back, but not to Mweya, to Ishasha, the south end of the park which was much more isolated, and where there were no wardens and no Greeks, just Uganda National Parks' scouts. Actually, I was delighted, it would be much more fun in the middle of real wild Africa. Before I went, a USA girl about my age called Joannie Waite who also rode horses and had lived an unconventional life, so was capable of adapting to most situations, announced she would like to come with me, so we set off this time in the ancient jeep.

It was an adventurous journey with several break-downs and much waiting around on dusty roads, but after about a week we turned up at Ishasha, pitched our tent right on the lake shore, made a table and a couple of chairs from bits and pieces and set about finding out where the waterbuck were. I gathered further experience of how to study waterbuck helped by Joannie, a great companion and friend.

One particular incident lives in my memory. We had been out watching waterbuck all morning and with the growing heat at midday, decided to return to camp for a siesta and go out again later in the day when the animals would be more active. Driving happily home towards the camp (there were no roads), I managed to put the jeep into a donga (deep ditch). We got it out eventually, but then found that the steering was broken. We were about twenty kilometres from the camp, and the long walk was not a happy prospect. I thought back to Aunt Helen, her purple nose with a drop on the end and her barked instructions on how to wheel a trolley: "Keep the moveable wheels behind Martha". Suppose one was to sit on the bonnet with the steering wheel between one's legs, press the accelerator underneath with a little help from one's friend, perhaps we could go backwards, more or less in the direction required without much steering. We managed, and several hours later came in sight of the camp. Horror of horrors, there was a contingent of eminent visitors from the International Union for the Conservation of Nature, high-up Uganda politicians and the head of Uganda National Parks all sitting down to a large camp luncheon served by scouts and servants at our table. The snag was that the table, which was furnished with a white gently flapping tablecloth, was right in our path. With no steering and legs too short to touch the brake in time, we smashed the table and lunch to smithereens; eminent scientists and politician had to jump clear ... What on earth were two young white women doing driving a Jeep backwards around in the National Park? It took a lot of explaining, but we parted amicably, the newly appointed Bugandan head of National Parks thought it was amusing and was pleased to hear about our research, and we spent the rest of the afternoon enjoyably chatting with them all.

Six months later, Joannie had left to take up art (eventually she became a famous artist in Kenya), and I realised that I had a lot of data but no clue what to do with it. I travelled back to Kampala where my supervisor had invited me to meet and perhaps get some help from one of the first ethologists (those who study animal behaviour) who was visiting Africa: Professor Thorpe from Cambridge. After much burning of the midnight oil and reading anything I could get hold of concerning studying animal behaviour in the wild, in trepidation, I went to see him. It was a great disappointment. Although he knew that I was a young student trying hard to learn how to study an antelope in the wild, he found it difficult to even talk to me. I showed him my notes and asked him questions, he told me not to waste his time. In retrospect, after many years as an academic, I think his behaviour was unforgivable ... how could one try one's best to discourage someone from doing research in an area you were interested in if they were enthusiastic, energetic and determined and not an idiot? Perhaps he thought I was an idiot so a week of depression followed, but Beadle, my supervisor came to my help again. He knew little about behaviour, but he found out that a Swiss called Rudi Shenckel, who had done the first ever study on wolves in the wild, was coming to the University of Nairobi to teach for a while. I wrote to him, and he invited me to meet him in Nairobi. He acted out the behaviour of any animal or human he was talking about and talked in Swiss-German-English very fast. He was small, energetic and had a grandiose sense of humour. After our first meeting, I realised that he could, and would like to, help me tidy up my act, and set about helping me do some serious research; I transferred my studies to the University of Nairobi, back home to Kenya.

Rudi managed to get me permission to camp at the Athi River end of the Nairobi National Park where there was only one scout but plenty of waterbuck next to the river. By this time I had developed my African minimum-equipment, long term camping skills and together with Mbogwa, whom I employed to look after camp and do odd jobs when I was away all day looking at waterbuck, we built an enviable camp, with a grass roof over the tent, a lovely veranda facing the Athi river, and even a bath with a glorious view, which was filled up with hot water before I jumped in, to lie and contemplate the great African continent. It was my home for six months, while I got to know the waterbuck individually, and finally published my first papers on how they behaved socially, carefully guided by Rudi whom I would go to see in Nairobi whenever I could tear myself away.

I met a cockney couple on one of my quick trips to Nairobi who were living in Karen, a suburb of Nairobi. He was a mechanic in London, but they had been after adventure, so they had bought an old ambulance to drive across the Sahara. It had been a two-year journey and they had ended up in Karen, a beautiful suburb. They were renting a large rambling settler bungalow with an enormous pepper tree in the garden. Because they had had so much hospitality on their journey, they wanted to return it to other travellers, so I would stay with them when in Nairobi. They sometimes gave parties and at one of these, I met Dennis, a rather overweight South African journalist who was an amusing conversationalist. Eventually we became "an item" and we even got engaged, although I was still living at Athi river and continuing my research and not going to change this for a while. He had been married twice before and was in Kenya as a freelance journalist, he had ex-wives and kids in South Africa but had had to flee to avoid being locked up without trial for reporting the bad prison conditions in apartheid South Africa. He gave me a ruby ring, which I lost when watching a particularly interesting waterbuck, but an Indian friend in Nairobi came to my aid, and helped me get a replica made with a piece of red glass. I don't think Dennis ever knew ... but who knows, perhaps his ruby was a piece of red glass made into a ruby ring by a resident Hindu too, a bit like the story of our marriage as it turned out.

The lessons learnt from this chapter were that in Africa (at that time), there was plenty of wildlife living their lives, some even unaffected by humans. Being with them day after day, it became clear that they lived much more complex lives than is usually imagined. Each has his/her own mind and experiences, always investigating and learning, but cautiously, in order to stay alive. Clearly, we could find out more about each species by carefully watching, recording and thinking, if we bothered. Many animals at least in the newly created National Parks had little opportunity to learn about humans, since humans were rare at that time in the parks and they had previously learnt to avoid them. But, slowly, the waterbuck near the camp became accustomed to us and stayed around to watch, rather than rushing off into the bush ... this made my studies infinitely easier because all the time could be spent with them, rather than looking for them. Waterbuck males took up territories near the river, but these territories often had mobile boundaries, depending on the food available and the presence of other males or females. The females wandered through these territories (and sometimes further away from the river if there were no predators) at will, alone or in groups of from two to four with their young and adolescent offspring. When they were feeling randy and in season, the male would rush towards them as they moved through his territory, and begin to court the female with little chases and circles. The female would decide if she wanted sex with this or that one, and might wander through the various territories. This would cause mayhem between the males who would then fight on the territory borders. Eventually she would decide which one she liked best, and allow him to court her and have sex, while her companions grazed and played around. Some males had more visitors than others, maybe the females chose a male because he had the most sought-after territory, maybe it was something about his appearance, or maybe it was just because there were other females with him and they liked female company.

Waterbuck never went very far from water, preferring the cover of the riverine areas and I guess the cooler temperatures, and they would be at home wandering through water up to their waists. They were very elegant, long brown-coated and moved gracefully, walking quietly or when startled they leapt away, bouncing over vegetation in their way. They grazed some of the 100 species of grass I had identified but they also browsed a little in the trees around. Despite being preyed on by any of the large predators, hyaena, lion and leopard that were in the area, they were surprisingly un-phased for an antelope. They would stand and stare if anything unfamiliar, like my jeep, came along, and after a while when used to it, some of them would allow me to get within ten yards.

But it was evident even in the early sixties that the vast areas of Africa, with their hundreds of species living undisturbed lives, were doomed by human population rises and consumerism. This worried me greatly … the main interest anyone seemed to have in the wild world, other than the odd passionate researcher, was how these areas could help humans have more money and be "developed". Even then it looked probable that some of the large animals, particularly the predators, would be wiped out from most areas, and that before we knew almost anything about them. I had also learnt that the easiest way to find out the mental attributes and the possible world view of a species, is to live with them and have common experiences with them. The only way of finding this out was to live closely with them and have a range of experiences with them. For example, one of the young female waterbuck who lived near our camp could not resist peeping at me while I was in my bath, so I tried to attract her nearer with collections of fruit and vegetables she might like. There were also a couple of territorial males who allowed the jeep within a metre or so of them as they grazed and observed me, so I could follow what they did very carefully, and they could watch me too. In this way we began a recognition of each other's emotions, and even little exchanges.

This chapter of my life was a great introduction to helping me achieve my aim of better understanding non-human mammals. It had been a steep learning curve but had set the tone for the rest of my life, as new and different questions arose from what had been learnt.

## The Katangese war, wedding & other African adventures

Towards the end of my research grant, Rudi Shenkel went back to Switzerland and my field work was finished. To work out the results and write up, I moved into my parents' abandoned house at Liloni, twenty miles from Nairobi, where much of my childhood had been spent. The garden had more or less disappeared under grass and bushes, the swimming pool/water tank was empty and leaking, the tennis court used for goat grazing, and all the wattle home-made stables collapsed, eaten by termites. But the house was solidly there, standing atop of the Rift Valley gazing towards Mount Kilimanjaro and its snows.

In those days there was of course no Internet or Microsoft Office to help, it was slow clicking typewriter writing. But, since there was no scientific information at all on waterbuck, referring to others' studies which required hours in libraries, was minimal. The males' territories could be mapped, and details of the known promiscuous females who wandered between each worked out. Of prime interest to me was their communication, how they used their postures, tails, ears and head movements to convey what they were feeling, and what the recipients did in return. After months of rewriting, my first paper was sent off to one of the only scientific journals publishing ethology, and much to my amazement accepted for publication. This was a confidence booster, although I don't believe anyone ever read it, until the next postgraduate to work on waterbuck began his studies.[1]

It was useful to have learnt observational recording techniques, and how to do some number crunching, but more important for understanding the world of the waterbuck, were the common experiences we had together. Nevertheless, publishing the paper added some scientific knowledge and respectability which is also valuable.

One important problem was to find out in more detail what waterbuck preferred to eat because in this way it would be possible to assess whether there was enough food for the different populations in different areas.

Research in this area was developing fast and a research ecologist, Dr Stewart, had developed a way of identifying the plants an animal had eaten by recognising the different species in the faeces. He was based at the Natural History Museum in Nairobi. Louis Leakey was director (the anthropologist who, with his wife, had discovered *Homo erectus*' bones in the Oldewai gorge). It was around 1962 and Jane Goodall had started work with him, before she went off to study chimpanzees, but I never met her. Before Dr Stewart's work, the only way of finding out what plants the differ-

---

[1] MKW, "Spatial distribution & sexual behaviour of the waterbuck," 1965, pp199-204 & CA Spinage, *A territorial antelope. The Uganda Waterbuck*, 1985, Academic Press, London

ent herbivores were eating, was either by being very close to the animal to identify the grass at species level from each mouthful, a time consuming and often inaccurate assessment, or to shoot the animal and empty the stomach. This was the usual way for ecologists, but I could not bring myself to kill or arrange for an animal to be killed for what appeared to be a trivial reason, although many ecologists told me that it was imperative. However, Dr Stewart's method only required the collection of faeces ... and then their careful analysis in a lab. The trick was to identify over 100 types of grasses and about fifty bushes from tiny indigestible pieces washed, sifted and placed on a microscope slide. With a magnification of about 400x, each species could be recognised because it had a slightly different cuticle, that is skin cells. By learning the minute different characteristics of the cuticle of around 150 species, a rough estimate of what the animal had been eating could be made. Dr Stewart lent me his reference slides of different grasses so I could compare the samples. I laboured away, gradually becoming familiar with more species. The snag was that digestion, stage of growth, time of year and age of faeces also contribute to whether the estimate of their preferred grasses was accurate. But, eventually there was enough data for me to write another short paper on the food preferences of the waterbuck which was published in a new journal, called the *East African Journal of Wildlife*.[1]

The discovery of how each species dovetailed its eating preferences with others so that all the various species could live together on the grand African savannahs was an example of the phenomenal symbiosis of the natural ecological world. Round about that time two Americans[2] had concluded that because of the range of different species with different eating habits on the African savannahs, these dry savannahs could maintain indefinitely one and a half times the weight of introduced cattle and or sheep

---

[1] MKW, "A preliminary investigation into the feeding habits of the waterbuck ..." 1966, pp153-7

[2] RF Dassman & AS Mossman, "Commercial use of game animals on a Rhodesian ranch" in *Wildlife*, 1961, 3:3, pp7-14

and goats, even when they had improved the grasses, poured on fertilizer, fenced it all and supplied water. So much for agriculture "improving efficiency and production"; maybe we could learn something from nature about producing food for humans! They also discovered that the weight of the bugs and beasties working underground turned out to be twice as high as all those of the elephant, rhinoceros, hippopotamus and antelope and all the rest that the savannah carried above ground. Perhaps it was symbiosis (living together for mutual advantage) that kept the living world going,[1] rather than competition between them.

At that time, the Belgian Congo had become independent. Murders, rapes, "ethnic cleansing" between tribes and any remaining missionaries or Belgian settlers were frequent. There were no police and anarchy ruled. This breakdown and chaos was, it turned out, mainly due to the Belgians having broken down the original tribal structures without replacing them with any Western administration and beliefs. They had done little about schooling, there were only about ten graduates in the whole country, and the number of secondary schools could be counted on two hands.

Nevertheless, British and French investors had serious financial interests in the copper mines in what was then Northern Rhodesia (Zambia) and an adjoining part of the Congo: Katanga. Copper mining had attracted investment of much European money and many Belgians as well as Katangese to work in the mines. With the help of the mining companies, Katanga was marginally better organised and certainly richer in foreign capital than the rest of the Congo, so with the help of the Europeans, Katanga had decided to secede from the Congo and declare itself an independent state. The head statesman was a charismatic, large, very black Katangese called Tshombe who charmed everyone. But, succession was a bad idea since Katanga had most of the wealth, so the UN had sent Swedish, Indian, and UK troops to try to reintegrate Katanga into the Congo. But, the situation was further complicated, as African wars usually are, by the French Foreign Legion and many

---

[1] L Margoulis & D Sagan, *Microcosmos: four billion years of evolution from our microbial ancestors*, 1987, Harper Collins, London

young South African whites (as well as a sprinkling of adventurers and wanted criminals) who had hired themselves out as mercenaries to the Katangese, aided and abetted by the mining companies who had a strong vested interest in seeing that Katanga continued to function and remained apart from the chaotic Congo.

No freelance journalist could resist the possibility of scoops and adventures in Katanga covering the war, and a bevy of "stringers" (journalists who stay near the action and report for a range of papers and radio stations worldwide) made their way there to report the ensuing war. Stringers are usually conversant with the local conditions and often appointed as they save the media moguls money who otherwise would have to send their own reporters. My fiancé was encouraged by his mentor George Clay, to break out on his own and become a "stringer" in Katanga. After a few weeks there, it was clear that the "war" was going to hiccup along in its chaotic way for months, so he invited me to join him. I made some enquires on the state of play of wildlife in Katanga and found that there was a large National Park just north of Elizabethville (Fungurume), the Katangese capital. The park was in chaos, the wardens had not been paid for years and it had been invaded by people from the Congo who were slaughtering and eating the animals, and it was rumoured, each other: cannibalism. It sounded like a worthwhile adventure: if I joined my fiancé, it might be possible to visit the National Park and report back to IUCN (International Union for Conservation of Nature), since no one really knew what was happening in the National Park. As usual there was no money, so I hitch-hiked from Nairobi, through Tanzania to what was then Northern Rhodesia (Zambia) and finally to Elizabethville through the Copper Belt. It was a long hot trip. The memorable part was a lift with what must have been one of the first large trucks taking materials south. The driver stopped for me because he needed someone to keep him awake on the long, hot, boring drive through Tanganyika as it was then. This had to be completed in a couple of days and nights, he was alone and the roads were full of holes, they were corrugated (like a corrugated roof that you drive over, which is normal with mud roads) and they were either very dusty or muddy. The driver was Italian. Since all Italians it seems like opera, I sang opera so badly as we rolled along that his teeth were kept on edge and he could not fall asleep. On the third day, descending the southern escarpment to Northern Rhodesia, he finally admitted that he was defeated so at 2 a.m. I struggled into a sleeping bag and kitted down under the lorry in the hope of discouraging wild dogs, jackals or hyaenas from eating me. It worked and the following day in Ndola, I waved goodbye to the sleepy Italian, and obtained a lift with a journalist going to Elizabethville 300 miles away in Katanga.

Eventually, dusty and tired, we arrived at the hotel where my fiancé was staying to find that there was no water or electricity, and the only food in the restaurant was frogs legs flown in on a Belgian plane from Brussels. Journalists are not accustomed to making the most of things, they usually stay in the smartest hotel in town so after a couple of weeks, I had to get away from their endless moaning and found an extraordinarily cheap bungalow for rent on the outskirts of the town next to a golf course. Only after we had moved in did we discover why it was so cheap ... the front line between the Katangese and UN troops was at the bottom of the garden. The UN troops camped there turned out to be the British contribution to the UN: a Gurkha squadron. They had settled into termite mounds, making little gardens and individual rooms decorated with statues of the Buddha, and little kitchens and bedrooms, all hacked out in the inside of these abandoned enormous termite mounds which the ants had taken probably tens of years to build. The Gurkhas (a regiment of the British army who come from Nepal) are renowned for their military prowess. They are small, savage and silent and always carry kukris. These are large bent knives that they use for everything, particularly, in this case, creeping up and spiking the drunken Katangese troops rolling around in their tents only 100 metres away, I was glad that we had the Gurkhas nearest us. I had little to do while waiting for further military developments so befriended the Gurkhas, admired their little gardens and took them soups and vegetable stews, sometimes swopping these for dried up UN food rations. I weeded our garden, mucked out its little pool and learnt to touch type with the help of a diagram in *Teach Yourself Typing*.

My French was poor, but it had an African twang and it was a lot better than the efforts of most of the English-speaking journalists, so they asked me to translate at press conferences. These were held at Government House where Tshombe resided. This had been a splendid colonial palace and garden made for a Belgian governor, but it had become like a comic opera set. There were guards at the doors who wore costumes that must have come from Bertram Mills Circus. As we entered, they lifted their arms out sideways holding swords, and attempted to click their heels together in their old tennis shoes without laces. They wore tight black breaches and scarlet jackets which had soup stains and mouse holes in them. Then we entered a large living room with dead rubber plants in pots and Christmas decorations. There were enormous sofas with startling white covers. Tshombe, who usually appeared at the press conferences, was good at press public relations. One essential here is always be late so that the journalists, well primed with the alcohol provided, are on your side. In this case it was very good wine that must have been imported from Belgium years earlier and whiskey and gin which were unavailable in the town. By the time he arrived, everyone was quite happy, including me. On one occasion, as I began my translation sitting next to him on the white sofa, I knocked over my glass full of red wine ... red stains all over the pristine white covers, perhaps it was symbolic of what was to happen, but Tshombe chose not to notice and I continued my stumbling translations, and was even invited the next time!

On one occasion we were invited to dinner with the Indian Army of the UN contingent. This was a real luxury as we had delicious curries such as only the Indians cook. The evening began with the officers betting high stakes that none of us could eat a whole chilli. After watching a particularly elegant Sikh salivating as he chewed his way through with eyes watering, I decided to swallow mine whole. It was quickly over, and they had to concede that I had won ... I suffered a few hours later, but it was worth it. We then were taken on a guided tour of the regiments' kitchens. There was a whole range of stoves with many different dishes being prepared by the "coolies" whose job it was. Then we dived into the deep freeze. There were dozens of imported frozen sheep carcasses, stamped on their rump the date they were frozen: 1939, a good twenty-four years previous. Nevertheless, the curries were delicious.

The smart, smooth Swedes were one of the other military contingents of the UN. They were spooky, they seemed to be involved in every racket in the book, had their hands in every type of technology and smuggled anything that could be sold at a profit to anyone, while appearing to be serious, upright, good citizens from a very civilised society ... it was some forty years later that I learnt how Sweden had cooperated with the Nazis, when I was given a house to stay in when running a workshop on horse behaviour, welfare and teaching. The cottage had swastikas all over the walls. On questioning the owners they proudly admitted that it was used by Göring when he came to visit his Swedish mistress ... perhaps the trilogy of the *The Girl with the Dragon Tattoo* is not imagined, and after all the author did die quickly after it was completed. This whole war smelt of a Faustian farce, but eventually it proved also to be bloody.

About six weeks later during which everyone was expecting hourly the war between the Katangese and the UN troops to start, it did. The UN troops, due to bad press, had to be shot at first by the Katangese before it could start. Since the Katangese were not the most brilliant shots, and were usually drunk (they had a special pocket sewn into their military trousers to put a beer bottle in), it took some time to establish that they had intentionally and with orders, shot at the United Nations' troops. When this happened, it was in the night at our bungalow; the Katangese began a barrage of shooting towards the Gurkhas' termite mounds. Suddenly, bullets were whistling through the bedroom. We crawled under the bed, wincing as each blast hit the walls, and remained trapped there for a couple of hours. Then the shooting ceased and there were motor car noises. Tshombe had been brought up to the front line by various ambassadors and was standing in front of the Gurkhas facing his own troops who were shooting towards him. When Tshombe was shot at by his own troops, he admitted on tape that it was his troops who had started shooting and he was quickly taken away, sweating liberally; and the war began.

In the morning, we found the "battle" had moved to the adjoining golf course where glimpses of Katangese soldiers could be seen running away, and the Gurkhas creeping after them. In the absence of any professional photographers, it seemed a good idea to try and take some photos to sell to various newspapers, so I crept about trying to get "action shots", without much luck. The UN troops around Elizabethville (Fungurume) took over the town after about three days and most of the Katangese troops ran off into the bush, but the white mercenaries were holding onto small enclaves. The main press contingent moved along the road to another town. A rather sophisticated French journalist and I jogged along together discussing books and plays while every now and then jumping into the ditch to avoid being shot by roving Katangese or mercenaries. The astonishing thing was that at the end of the day of leaping about and crawling in ditches, the Frenchman looked as if he had just stepped out of a taxi in Paris, while the rest of the press looked as if they had spent at least three weeks in the trenches. The Frenchman agreed to buy my photos for Agence France Press, but I could not take the picture which was the "scoop of the day"; I realised then that being a journalist was really not, and never could be, my scene; someone had to do it, but it would not be me.

We were advancing into a small town where many Belgian miners and shopkeepers lived, when we came to a roadblock manned by UN Indian troops. As we arrived, a VW Beetle car drove fast towards the barrier and instead of stopping, drove straight through. Two of the Indians on guard opened fire and 100 yards further on the car stopped. On going up to the car, we found two dead or dying white women in the back, and the driver was pouring blood from a head wound. He rushed up to me, presumably as I was the only woman present, and went down on his knees pleading *"je suis innocent, je suis innocent"*. It turned out that he had panicked instead of stopping at the barrier. The guards thought he must be a mercenary and fired, killing the women in the back. Of course, the press had a scoop, UN involvement in Katanga was not popular in Europe and here unarmed women were being shot by them. I could not take the picture, it seemed too callous to point my camera and take a picture of a suffering man who had just had his wife and friend killed. Somebody had to do it to tell the world, but it would not be me. Another journalist got the "scoop"; a picture of the man on his knees next to me, and earned several £1,000s. But, I did have my arm twisted to drive to Ndola with the "copy" (at that time copy could be sent by telex from some places, but that did not include Elizabethville). The VW and I set off on the 300-kilometre drive only to be stopped just outside the town by the Katangese troops. Stupidly, I managed to pull out the wrong press pass: a UN one, not the Katangese one, I really thought my days were ended then. But by showing them a Polaroid camera someone had lent me, taking their photos, printing them a copy and giving them to the Katangese soldiers, they stopped riffling through the car and prodding me with their bayonets. Rather than me being shot or stabbed, they all fell about laughing and invited me in for a large glass of whiskey, obviously relieved off the last traveller. This type of encounter was repeated a couple more times before I eventually managed to drunkenly hit the tarmac in Ndola, drive the VW to the post office and the telex machine.

Press work did not somehow have the glamour it was trumped up to have. It involved an enormous amount of waiting, usually drinking with other journalists, and living almost full time with them for weeks. They can be good conversationalists and they do a job that must be done, but I found them either very cynical (Don Wise of the *Daily Mirror* was famous for saying "my readers who move their lips as they read"), or they seemed to be blind to what was happening around them. They would interpret the incident through the eyes of their paper, not their own. At that time, the UN troops were always at fault, whatever actually happened. My press days ended there, even though I had managed to earn enough from my photos to travel by public transport back to Kenya. It was a homocentric interlude that lasted me a long time. I was glad to get back to waterbuck and my parents' small farm.

Other adventures emerged. Kenyan independence had been set for a year ahead and many white settlers were scared and leaving, often in a hurry. They were turning their animals, particularly their horses, loose and then driving away, usually to South Africa, although some with more money made it to Australia. Joannie Waite my artist friend turned up again, and we decided we could capture some of the horses left behind, drive them to our farm near Nairobi, check that they could be ridden and then sell them to new aid workers and other "helpers" and volunteers who were arriving in droves in Kenya. They wanted to "integrate", or "experience" the life that was lived by the middle-class whites at least, and this meant playing polo. However, there were few polo ponies for sale, but of course they could buy them from two nice trustworthy white girls! Once the word got around that we were looking for horses, many farmers contacted us and told us we could have theirs if we came and collected them.

We rode into the Rift Valley, to Naivasha and Nakuru, to Naro Moru and Nanyuki on a couple of horses we had been given, and collected a herd which, with many delays and adventures, we eventually managed to get back to the farm at the top of the Rift Valley. We camped each night and attempted to corral the horses, but they usually escaped and galloped off into Africa, which meant spending most of the next day finding them, separating them from a herd of zebra, or collecting them after being stampeded by a tower of giraffe. At that time, there were still large groups of wild animals all over Kenya and about one quarter of today's human population, which was restricted to small hamlets. Neither of us were deterred by the difficulties or exasperated by our erratic horses: falling off, being run away with, our tack breaking (we had borrowed some rather rummy saddles and ropes). We revelled in the risks, adventures and the hardships, running out of water or food, snakes in sleeping bags and what not.

After about three weeks, we arrived at Liloni and the small herd of about eight horses was run into one of our grassy fields that my mum had cleared. The next few weeks we spent teaching them to be ridden, and to play something like polo. Joannie took up with a lonely neighbour, but came over to help daily, and eventually we had a few ready to ride down to the polo club in Nairobi where I knew the president (a very influential Indian who ran a fun multi-racial "reading group" in Nairobi). We rapidly sold a couple of the horses to the "pinkies" as we called them (white people who came to Africa and became very red in the sun, easily picked out from residents because of their pinkness). We collected what to us were good prices, but to them very little for a horse. Although neither the horses nor the new owners had ever played polo, now at least they could try. After all Horse Dealing is Horse Dealing, and none of our horses had any major problems that we concealed, at least not that I know about. I don't remember that we heard of any horse or rider being badly hurt, but we have no idea how many stayed the course.

We did sell a rather nice grey called Picasso to the girlfriend of a new serious Scottish vet who was doing his "good works" as an AID worker and had just arrived. Picasso was blind in one eye, but fair do-s, if the vet did not spot it well, honestly! The woman fell off at a jump so when I asked her if she had turned Picasso's head so he could see the jump, she looked baffled, but actually Picasso had cleared the jump with room to spare, even though it was on his blind side, so she was to blame. He also had a large growth on the side of his head ... they must have seen it, but perhaps the vet dismissed it as benign and still paid the full price.

With my share of the money earned, I bought a cow to milk, and a few fat tailed sheep, and we kept several of the mares for breeding. I found a pure-bred Arab stallion whose father had been imported from England, one of the Crabbet line. He was the other side of the Athi plains, about 160 kilometres away, three years old, untrained, a bay and beautiful, his name was Xenophon. I bought him and had an exciting three-day ride back. By the time we arrived, he knew a lot about being ridden, and I knew a lot about him. We had ups and downs, but eventually learnt to enjoy each other's company. He was the first stallion I owned, and there is something special about having a close relationship with a stallion that is quite different from mares. It seems often as if stallions need a close association with another feeling, living being because, generally, they do not obtain this companionship from mares who are too involved with each other and their offspring to care much for the stallion, other than when they are feeling randy. So stallions often have to stay outside the group, look at the others from a distance, perhaps longing for company. Stallions therefore often need companionship, and a human or indeed a cat, goat or dog who likes him, and will spend time with him, is a friend indeed. I did not know this at the time, so I was somewhat overawed by Xenophon's friendliness and compliance with me and owe him much gratitude for his teaching and the fun we had.

To try and make the books balance, I found a job teaching biology to bright sixth formers at the Kenya High School, the girls' school, which was where my sisters had been some ten or more years earlier. It had been a school for the daughters of colonial servants and white settlers. But, since independence was looming, girls of all colours were now accepted. Our dynamic A level biology class consisted of three white girls, daughters of settlers, three bright Indian girls, daughters of shopkeepers and garage owners in Nairobi, and four Kikuyu girls, daughters of vicars or the emerging middle-class Kikuyus who were bright and ambitious. Girls with both their sex and race against them, but determined, as always in Africa, to enjoy their lives. It was a fun class. We went on field trips, visited the waterbuck and the new orphanage at the Nairobi National Park (which has now become so famous), studied the ecology in forests and plains and covered the syllabus as laid down by London for A levels at that time. All of us were bemused when it turned out that the students had to learn about oak and ash trees, primroses and daffodils, a whole variety of plants that did not grow anywhere in Kenya and there was no mention of acacias, gum or wattle trees, bananas or any of the animals and plants all of us were familiar with. We staggered on none the less. They learnt systemics, anatomy and basic physiology and eventually, all passed their exams, some to go on to universities.

My fiancé continued to be a journalist and travelled about a lot, but when he was there, we discussed how and where we were going to live when we were married. One idea was to run a sort of back to nature *banda* hotel in Zanzibar. I contacted Kathie from the University of Kampala who had finished her English degree at Makerere and was now back with her four children and Arab husband in Zanzibar. I had booked tickets on the train to the coast, when I heard over the radio that there was a revolution in Zanzibar. This was a socialist demonstration against the Arabs who ran Zanzibar which previously was a British Protectorate. The Arabs had unquestioning racist policies and ran a somewhat brutal regime even though it was overseen, at least, by the British. They had also tried, unsuccessfully, to put the women in purdah. The revolution organised mainly from Tanganyika, had already killed many of the rich Arabs, and many of the women and children had been sent off on dhows to Yemen and other Arab states. Goodness knows if any of them ever got there. Kathie and her young family of four children under eight, and Arab husband vanished. I tried to find out what had happened to them, but there was general chaos on the Kenya coast with refugees by the thousands flocking in and no trace. We cancelled our trip and the idea of running a guest house was put on hold.

Kenyan independence loomed, and a general feeling of joy and optimism pervaded the country. It had been arranged that the International Union for the Conservation of Nature (IUCN) would hold a big international conference during the independence celebrations in Kenya, to emphasise and highlight the importance of conserving the wealth of Kenya's nature. My parents were coming, so we decided to get married at the same time. They arrived a couple of weeks before the wedding, my father and I attended meetings in Nairobi, while my mum arranged for the garden to be brought to heel, the house to be mended and painted and everything to be tip top for a wedding reception. We were married by the District Commissioner in his baggy white shorts, moustache and long socks and then had a blessing in Nairobi Cathedral where Mum had arranged some of her musical buddies to play and sing. The blessing was impressive, sauntering up the aisle of Nairobi Cathedral on the arm of my father, but the Archbishop or whatever he was, had a stutter and the marriage vows went adrift: he managed to marry Dennis to Charles! Then we all adjourned to Liloni for the reception with a *mishwi* (whole sheep cooked over the fire). Most of the guests, as I remember, were delegates to the IUCN conference so discussions on wildlife conservation took precedence, which was as it should be. Nevertheless, about fifty local Kikuyu women arrived in full tribal Kikuyu costumes. They had remembered my mum from some ten years back and had decided to come and dance for us. My mum danced away in her tangerine dress and enormous floppy hat with the *bibis* (Kikuyu women) with their dread locks, and skin wrap arounds, coloured beads and bangles, and George Clay, the best man, filmed it all. We rode away on horses, Xenophon and Picasso, to catch the overnight train to the coast for a traditional Kenya coastal honeymoon. Dennis, with South Africa out of bounds for him, had decided to give up journalism for the time being, and take up second-hand car dealer for one of my mum's cellist pals in Nairobi.

George, my husband's long-term South African journalist mentor, continued to rush from war to war and three months later was killed by a stray bullet while reporting from the Stanleyville (Kisingani) insurrection. It was a sad time and I was pregnant. We decided to wait until the baby was born and then leave Kenya at least temporarily, to study for degrees in the UK. Dennis wanted to finish a degree which he had started in Johannesburg, and I badly wanted to continue with my research, study for a PhD, and catch up with my four-footed and two-footed family. It was a wrench leaving Xenophon, our ridgeback dog, and other friends, but change was needed, we boarded a plane with a new baby and far too much luggage and arrived back in the UK.

This was a predominantly human chapter of life. But the natural world continued to be unrivalled in its interest to me. Adventuring with horses and the odd human friend in the natural world and the experiences acquired was a constant desire throughout my life which included study of different species and getting to know them, travelling with some: horses, dogs and a human friend or two. Xenophon, the first stallion I had much to do with, had taught me many new things about equines and their wish to cooperate. But, the swings and roundabouts of outrageous fortune control one's life; from time to time, some things have to be placed on hold.

## Animal communication, animal welfare & eco-farming

It was back to my family's farm in Sussex, the Guernsey cows, my mum, her farm helper, my itinerant father, my filly Syringa and Canute the poodle. A few weeks re-acclimatising and discussing where we should go to study, my husband for a BA, me for a PhD were followed by endless visits and interviews at universities. Eventually we were both accepted at the University of Sussex, one of the new universities which intended to re-examine aspects of academia. The first change was that everyone was on first name terms, a previously rare occurrence, secondly there were innumerable discussions and informal meetings between people from different disciplines, and a great variety of new courses; but at that time the idea was also to keep the standards of scholarship high and unimpeachable. It was conveniently near my parents in Sussex, and with limited finance, we eventually found a tumbled-down farm cottage, barn and a couple of acres sold at an auction within £50 of our maximum price and with no chance of a mortgage as neither of us were earning; it was a country cottage and a place to start to build a multi-species community.

For a birthday present, while I had been in Africa, my mum had sent Syringa to a local horse trainer to "break", which means to train to ride or drive. The continual use of this term is odd in a country which is renowned for its animal care. The other curious use of language in English-speaking countries is that they rarely give horses a gender, dogs yes, but horses among the "horsey set" are always "it" ... they might as well be talking about a table. In the case of Syringa, the trainer luckily did not "break" her, rather, she had been quietly ridden, although her knowledge of the world was still sketchy; she had a lot more to learn. It was lovely to have her and appreciate her beauty; a dark grey beautifully proportioned mare with a large eye and a slightly Araby head. She was bright and happy to learn, and it was great fun to help her develop and use her abounding skills; and therapeutic to bury my head in her ample shoulder when distressed or worried. We rode around the forests, met the last of the itinerant charcoal burners and gypsies camping by their burning smoky heaps on Ashdown Forest. Minority groups were being "cleaned up" so that Ashdown Forest could be turned into a Nature Reserve where each species survived by "grace and favour" of the surrounding human bourgeoisie, tripping delicately through the mud in their green wellington boots. It was to become a valuable possession, assessed by environmental assessors as worth tens of millions of pounds!

I was interviewed in London to do a PhD by a grey-haired academic who subsequently became famous in the study of sociobiology (an examination of the evolution of animal societies). His name was John Maynard Smith, originally an engineer, who had, somewhat mysteriously, been chosen to be head of the new department of biological science at the University of Sussex. He was a typical academic, absent minded, immersed in his work, charming, delighted to discuss any issue that was of mutual interest and capable of dancing with the best of them when it was the time. Richard Andrews, who I had written to in the US, was coming back to the UK to head the sub-department of animal behaviour and he was to be my supervisor. His recent research on primate communication, including ideas on the origin and evolution of the smile particularly interested me. How, why and what large mammals communicate was the subject closest to my heart, and at last I was accepted to study this for a doctorate.

Studying domestic animals was looked down on by "proper

biologists". It was generally believed (still is by some) that they have been changed genetically so much, that they hardly resemble their wild counterparts in behaviour and therefore are of no interest. In East Sussex, there were few large wild mammals left: aurochs (the ancestor of domestic cattle), European bison, wolves, bears, wild boar, wild sheep and goats, and all the deer except for a few roe deer, had been exterminated by the sixteenth century. Preswalski horses had vanished and been replaced by feral ponies such as the Exmoor and Dartmoor. Foxes, badgers, grey squirrels, and many rodents were still hanging on but were difficult to find and even more difficult to study. Therefore, I decided I would try and develop a hypothesis on the evolution of the hows, whats, whys and whens of domesticated animal communication and then test these on a host of species of wild or captive non-domestic mammals. This study was to include their vocalisations, but also changes in their postures, ear movements and tail movements. At the cocktail party launch of the new department of biological sciences at Sussex, I remember being asked by an imposing female professor of psychology what I was doing and replying that I was "counting tail movements" of cattle, sheep, dogs and horses. Needless to say, she fell about laughing and in those early days, I was far from being able to defend the study!

There were cattle on my mum's farm, and she was encouraging. I had Syringa with us, and other horses were available in nearby stables. Dogs and cats were around, sheep, goats and pigs could be found on farms. This was a start. Other species could be observed at several zoos. With a small child to care for, it all had to be carefully slotted together.

At that time there were no organised child minders; it was socially unacceptable for mothers with young children to work and even my slightly avant-garde family thought it was shocking that I would not be staying at home with the child. But for me it was not an option if I was going to stay sane. It did seem unfair, as in the early sixties, husbands and fathers were still entitled to do what they liked when they liked, while women were expected to look after and bring up children. If, in addition, women wanted to do something else, even if it did earn money to support the family, well, that was up to them, but it must never interfere with their "natural duties". Multi-tasking became a necessity, since within a couple of years I had a new baby, but with practice one gets better at it. Another doctoral student in our department who became one of my lifelong friends, and is now an eminent academic, Lesley Rogers, discovered during her DPhil that female chicks were better at multi-tasking than male chicks, which was very relevant to my lifestyle at the time! It is obvious that multi-tasking would be selected in females because mothers have to multi-task to suckle babies, look after them, see they are warm and comfortable, nurse them when sick, find food for all, protect themselves from predators, as well as go about their own normal lives. No surprise therefore that females are usually better at it, although, of course males can learn ... and some females never do!

In line with the new idea of working mothers, my new baby was allowed to be in his pram in the garden outside bio-chemistry on the ground floor. If he squealed too much, someone would ring me up to collect and comfort him, and I was also allowed to bring in my dog who would lie quietly under my desk. Life was active, stimulating and generally happy. Weekend respites at home with the two boys, Syringa, a pony I had bought for my eldest son, and our dog Sneefus, a Labrador all sorts we had adopted from the pound.

The time was right for change, society was in a state of flux, challenged by the ideas of flower power and anti-war, and the University of Sussex biology department was an intellectual hotbed. Many who later became some of the most influential twentieth century biologists, psychologists, thinkers and experimenters had been chosen to be part of the new biology faculty or were graduate students. There were ecologists, botanists, experimental psychologists, zoologists, microbiologists, neurophysiologists and ethologists and we all met for coffee mid-morning. In line with the new informality, anyone could discuss anything with any of the others. Social events such as Christmas parties involved important professors doing the conga and getting drunk with unknown graduate students as hierarchies were dissolving. The greatest joy for me, was that there were excellent specialists in many fields and it was easy to move between them. This allowed the development of a nascent multi-disciplinary approach to the study of animal worlds. It may also have been instrumental in the phenomenal intellectual success of the department in its first twenty years.

Towards the end of the four years of study and research, on one of the rare family outings, we went to Wales to visit one of my DPhil colleagues with whom I had shared a lab for four years. Mike was a delightful, dour humorist, and one of the rudest people I have ever met; not once in four years sharing a lab did he ever say "Good Morning" or greet me in any fashion. But he was fun and witty. After completing his DPhil in which he discovered a great deal about guinea pig behaviour that was unknown, he decided that academia was not for him, so went off to Wales to live on a small farm and suffer from "cold, malnutrition and cripple up his body with hard labour", as he said. I'm not so sure if all three came true as I think his women looked after him too well, but, like all of us who work hard on farms, his body is probably at least partly crippled by now, after fifty years.

His wife, Lucy, was one of those beautiful women whom everyone turned to stare at as she glided around with her auburn hair. She had some history of psychological problems and failing her degree, but was clearly very bright and had been taken on as a technician in our department where she was helping to find out more about vision, I think, in various invertebrates. She was interested in horses too and helped me take photos to illustrate postures for my doctorate. A few years later she had left Mike, and had written one of the first books outlining the normal behaviour of horses. As a result, she became one of the founders of the movement to improve horse training. When we visited Mike and Lucy in Snowdonia in the sixties however, the couple had rented an old, draughty, damp cottage and a patch of soaking paddock. Lucy was related to the remains of the Bloomsbury Set, intellectual movers and shakers of the 1920s-40s who had taken up residence nearby. One of their ancient members was architect Williams-Ellis who had built Portmarion, an Italianate village on the wild Welsh coast. His wife was a keen innovative gardener and an organiser.

Later I was invited by his wife to give a paper at a conference on the future of agriculture and food. This was the first of its kind in the early 1970s and involved agriculturalists, nutritionists, cooks, environmentalists and many others. I remember particularly the culinary developments: someone had invented a very nutritious (but horrible) concoction from algae, reputed to be a complete diet; somebody else produced delicious home outside-reared Organic hams; another some peculiar sprouting broccoli served in a sweet and sour sauce, and there were hundreds of types of grain, biscuits, and cakes, some nice, some nasty. After speeches and speeches, we all attended a slap-up conference dinner consisting of a range of weird and wonderful culinary inventions: Food for the Future. It was around thirty years later that anyone followed this up when the disasters caused by modern agriculture began to be so severe that they could not be ignored.[1] Today in France such an event has become an annual occurrence ... but little has changed in how the majority farm, although consumer pressure is changing what they buy and eat, mostly to ensure that humans are not poisoned by agricultural chemicals, never mind the effect on the rest of the animals and plants as a result of "normal" farming techniques.[2]

It was October and the time for gathering the Welsh ponies from off the hills to sell as that year's crop of foals. There were hundreds of nine-month old foals being auctioned for around £5 each. At that time, some were still being exported live to France to be fattened up and killed for meat, and a few others were bought to become pit ponies in the coal mines. A dry and homely Welsh pub drew our party out of the rain, except for me. But when a couple of hours later they emerged with beer-rich smiles, I had bought six

---

[1] MKW, "Problems of modern agriculture," 1980, pp208-215
[2] https://www.nytimes.com/2018/10/31/world/australia/australia-wilderness-environment-gone.html

small wild equine balls of hair with very nifty hooves. Lucy said she would take five of them to train as children's ponies and sell at a profit, if possible, over the next year. But, one pure bred Welsh Mountain filly was reserved for me. Her name was Aderin, and she eventually became the pony mentor of my children and many others, was mother to several famous competing ponies (Aisha Evans, fourth in 160 kilometre (100 mile) race and best condition prize) and lived until she was forty years old.

Weaving practical experiences with science continued to form my doctoral thesis which explored the minds and emotions of non-human mammals by studying the origin and evolution of their communication. All mammal species have feelings which they make clear to anyone who can read them, but how this is done varies. I had the advantage that I had spent most of my life with non-human, non-primate mammals who were my friends and therefore had become aware of what they might be feeling and could predict what they might do next. Nevertheless, nearly everyone is able to tell when an individual of another species is feeling frightened, hungry, fearful or aggressive. Even people who have never seen a pig before can tell that much about a pig.[1] This is because we can read that the pig is behaving similarly to the way we might behave when feeling the same way. In effect, the mental state of the pig can be read by other pigs and by another species of mammal by watching them. The meaning of many of their gestures and movements are clear and "explicit", even though they may not be exactly the same as in a human. For example, when a horse puts his ears flat back, our immediate response is to move away; we feel threatened because we intuit that he is feeling aggressive even though we do not have ears that move. Ask any child who has hardly ever seen a horse what that horse is feeling when he sees this, and you will get the same answer: "he is angry". There are also some similar postures and movements that have come to mean different things depending on the species. For example, why does a cat wag his tail when he is

---

[1] F Wemelsfelder, *Animal boredom towards an empirical approach to animal subjectivity*, 1984, Proefschrift, Leiden

cross, a dog when he is happy and a horse when he is anxious or annoyed? What is the original cause of wagging a tail and can we explain how tail wagging has come to mean different things in different species? I think we can, one of the subjects covered by my thesis and resulting publications.[1] Another example, when a bull puts his head down and points his horns at you, you know he means to butt or charge, and generally you run ... but how do you know this, and why does he do this instead of running at you straight away?[2]

Humans of course have human speech which is different from the speech or vocalisations of other species. In fact, human language is often still believed to be the key to our "cognitive" or mental advances.[3] Human language can of course say all sorts of things by using particular words without reference to any context, and often without reference to any emotion felt. It is highly symbolic, and symbols are built on symbols until we evolved advanced mathematics. In many ways human language resembles a computer programme, hardly surprising since they are both written by humans. The same word means the same thing in any situation. The result is that human language is understood on the telephone, read from the computer or in a book. People believe that non-human vocalisations are just a simple form of human language. For example it was originally believed that a cow gives one type of "moo" to call her calf, and another type when she is hungry.[4] But is this actually the case? If not, what is it that dogs, horses, elephants, lions or any other mammal say to each other by making vocal noises? Why and how have these noises evolved, and what do they mean? Eventually after recording and analysing hundreds

---

[1] MKW, *Some displays of canids, felids and ungulates with particular reference to their causation.*,1969; MKW, "The tail movements of ungulates, canids and felids ..." 1976

[2] MKW, "The visual displays of eland," 1978; MKW, "The social organisation of a small herd of captive eland, oryx and roan antelope," 1978; MKW, "Individual differences in a small herd ..." 1978

[3] e.g. CDL Wynne, *Do animals think?*, 2004, Princeton University

[4] Fraser Darling, *A herd of red deer*, 1954, Cambridge UP

of different calls of different species it was evident that most of the species of hoofed mammals, canids and felids, make the same call in a whole variety of situations. Unlike human language, it is not possible when you hear the neigh of a horse, the grunt of a pig, the trumpet of an elephant, or the "moo" of a cow to know whether he is lonely, hungry, sexy, welcoming, lost, or perhaps just frustrated and annoyed. The calls recorded in many different situations (I measured ten different parameters of each call from sonograms), were all the same from each individual animal. Several years of measuring, listening, recording, observing, discussing and thinking about this finally brought me to the conclusion that what that call tells you, is not whether the caller is saying this or that or feeling a particular emotion, rather it tells you how aroused or excited he is, but it does not tell you why.

The particular emotion he is feeling when he gives that call only becomes evident when one looks at the context in which it is given. For example, you hear Daisy the cow giving a long loud call. Is she separated from her calf and distressed? Has she seen the tractor coming towards her with her late breakfast and is delighted? Or does she see the bull over the fence and is feeling frustrated and randy? If you are another cow (or any other species) what you do is look around at what is happening in order to know why Daisy is giving this call. In other words, the particular meaning of the call can only be read from the context, that is what is going on around. I called this "context dependent" communication as opposed to the language of humans which is "context independent": "a dog" is a dog in any context, but the same moo, neigh, nicker, trumpet, bark means that the animal is excited, but why and consequently what he is feeling, depends on what is happening around and how this is read by those around.[1] We can use expression in our language which to some extent is similar, so you can say "dog" in a long gloomy way, indicating perhaps that you don't like dogs, or joyfully, indicating that you do, but dog still means dog without any expression. In both human and non-human mammals visual communication, things like facial expressions, ges-

---

[1] MKW, "The vocalisation of ungulates," 1972, pp171-222

tures and other movements are often "context dependent", but some have become "ritualised" to mean particular things which is why a tail wag by a dog means friendliness, and in a cat "keep away, I might scratch".[1]

Context independent talk is highly symbolic and therefore can result, among other things, in mathematics, building space craft and conveying information independent of any emotion. But does this, as has often been argued, really mean that we are superior in all mental attributes? At that time, in the 60s, this question was not debated. Today, it is beginning to be but often with a bias to assuming that humans are superior in everything mental. In the 60s these discoveries for me were exciting and interesting, and began to set me thinking on the similarities and differences between my quadruped friends' view of the world and mine. One of them being, how aware they are of everything that is happening around them, and how unaware humans often are because they are either talking or "thinking" in context independent ways.

Then again there is the matter of emotions and feelings which were and still are a tricky one for science and scientists to get their heads around. For example, when writing up my thesis, when I mentioned that a horse was frightened, my supervisor put a red line through it. It was not permitted in ethology or neuroscience, to attribute emotions to non-humans. In psychology or ethology at that time, if an animal had an emotion, it was best ignored or given it a pseudonym, today this is slowly changing, well, at least for primates and dogs![2] This is not because we have been interpreting and using the emotions of our animals to live with them and teach them for centuries and consequently know that they have these emotions, it is because one or two scientists who have now become "big sellers" have invented experimental tests and published the results which tell us "the truth" (as if we did not know!). However, they have not and I doubt if they ever will be

---

[1] MKW, "The tail movements of ungulates ..." 1976, pp69-115

[2] e.g. Jeffrey Moussaieff Masson & Susan McCarthy, *When elephants weep*, 2019, Delacote; Frans de Waal, *Mama's last hug: Animal emotions and what they tell us about ourselves*, 2019, Granta

able to unravel exactly what rats' fear feels like to me, and after all, my fear may be very different from your fear. This is called the Private World Argument, so I doubt that "science" or scientists have actually found out anything on this subject that we mammals who live in the mammalian world did not know since Xenophon (350BC)!

My results clearly showed that the central most common message communicated between non-human mammals is to convey their general emotional state: their state of excitement or arousal, and the type of emotion experienced has to be interpreted from the context. The majority of communication between and with ungulates (hoofed mammals), canids and felids, is all about emotion, they are not tag-ons, they are a central part of our being.

There are, however, a few vocalisations or noises such the bark of a pig or snort of a horse that are "context independent". In this case they are the result of being startled and indicate "beware there is something odd out there". Vervet monkeys and chickens have subsequently been shown to have different calls when they spot different types of predator: a leopard, or a lion if a vervet, or in the case of chickens: an air or a ground predator.[1] But the specific meanings of the majority of their vocal calls are still context dependent. The same call is used in different situations and tells about the state of arousal of the individual. When the context is understood, then it is possible to tell if the communicator is happy, unhappy, friendly or what emotions he is feeling.

Animals also make noises as they wander around in the world, breaking sticks, sighing, tummy rumbling, and so on. These noises usually mean that they are relaxed, going about their normal business, but when they become aware of a predator, such as a human hunter creeping up, they must be aware of the noises they make, and can stop making them, if they choose. From the sudden silence, others may get a message "beware, there is danger abroad". Studying these noises is another fascinating study that points to the fact that the animal must be self-aware, is aware that

---

[1] e.g. RM Seyfarth, DL Cheney, "Meaning and mind in monkeys," in *Scientific American*, 1992, 267, pp122-129

he is making the noise and can stop it ... more food for thought.

It has been suggested by those who study people who cannot talk, that this emphasis on context independent language dominating our world view can be a handicap to considering other world views. It often controls our human thinking and it is difficult to "think outside the box". We jump to the conclusion that "normal" humans are just better at communicating than someone who is deaf, dumb or blind and omit to look carefully at the other different communicative skills they may have learnt.

It was many years before I started to consider these questions in any detail, but it was the results of studies, experiences with animals and my particular emotional life at the time which was responsible for the non-conventional academic career that followed. From age six, communication between and within species has remained a central interest to me, and greatly improves the possibility of understanding another species' world views.

Post-graduates, particularly those without funding such as I was, were often given priority to help demonstrate in practical classes and conduct tutorials for undergraduates, for which they were paid. I did as much of this as I could, but there was not enough to keep the kids and I financially afloat, so I opted for teaching adult education courses. These included almost anything within my field that I could dream up. The minimum number of paid up students or participants had to be more than ten, and they voted with their feet. So, if your course was not interesting enough, they gave up coming and the course collapsed. As a result, teaching adult education classes taught me to teach both humans and non-humans. People who know very much more about the subject than you do may turn up, and you need to be aware of this, respect their knowledge, and allow them to show it. Unlike university or school teaching, the class will not continue if you do not keep the interest of the participants. Neither the adult education students nor the animals are doing the course to advance their careers, earn more money, or pass exams, they are just doing it because the teacher has interested them in it. So it must be fun and interesting, and the teacher must keep it so. It turned out that teaching adult education classes was very useful in learning how to improve non-human mammal teaching.

After a few years and making some mistakes, I had learnt enough to be able to keep the students who signed up for the classes, and with classes almost every evening, could just about make the books balance. I taught how to use botanical floras to housewives who had no idea of botany, but they eventually found it more fun than just using flower pictures. We learnt about the natural history of Sussex, its nature reserves and coast line and identified things on the beaches. We organic-gardened and visited gardens all over Sussex to discuss the history of gardens. We discussed world futures, energy use and abuse, environmental ethics, the origins and evolution of agriculture, animal welfare and factory farming, horse behaviour and how to teach them better and improve their welfare.

Almost any subject I wanted to learn more about seemed to be acceptable, but boy, did I have to learn. I remember one time talking about energy and trying to outline how nuclear fission worked to find that I had a nuclear physicist in the group who knew very much more about this than I. All I had to do was turn the class teaching over to him for the rest of that evening, and learn more about it as the discussion abounded. I did weekend courses, we went on visits and trips, and did practical things. It was exhausting but fun and challenging, and it allowed me to finish my PhD with the money earned. But it cut back on evenings with my sons, which was probably a mistake and they certainly resented it later. Most importantly, it taught me a lot about teaching both humans and non-humans; "voting with your feet" is rare in academic institutions and when animals are taught, they are almost always physically prevented from going away. If I could interest them enough to stay without being restrained, it must show that they enjoyed it.

In the 1950s after the war, the Ministry of Agriculture encouraged and financed almost any agricultural development that would produce more food for humans whatever the costs: financial, environmental or ethical. After all there had not been enough food in Britain during the war and this, governments believed, must not happen again. This approach was still being taught in agricultural colleges in the 60s, yet there was no longer any shortage of food in the UK. A few people, however, were beginning to consider energy, environmental and ethical concerns that had raised their ugly heads as a result of recent agricultural developments. A minority were coming together in France in a society called *Nature et Progres*, and in the UK with the Soil Association. Organic Agriculture, which was first put in place to try and address some of these problems, was considered an unprofessional silliness and dubbed "muck and magic" farming. Yet, one of the very first exponents: Howard, a member of the Indian Colonial Service, had shown the benefits in terms of its net production for feeding the poor in India in the 1930s. Lady Eve Balfour, followed in his footsteps and developed and ran her own experimental farm in the UK writing up the arguments and results for all to see.[1] She was a lady of strong views and influence and had collected around her a small group of thinking people who were to change the current beliefs in agriculture. These people, as today, came from many backgrounds, some believing that modern chemical farming was going to poison them, some from concern about the life of animals, some from an interest in wildlife and ecology.

The environmental problems of modern agriculture were summarised first by Rachel Carson in *Silent Spring*.[2] She had evidence to show how animals and plants were dying out as a result of the application of uncontrolled pesticides and herbicides. Then, Ruth Harrison published *Animal Machines*[3] a couple of years later which showed what was happening to animals in intensive modern agriculture. A government economist, EF Shumaker, poured fuel on the smouldering fire of alternative futures in his book *Small is Beautiful*,[4] showing that even economically many of the

---

[1] EB Balfour, *The living soil & the Haughley experiment*, 1976, Universe, New York

[2] R Carson, *Silent spring*, 1962, Houghton Mifflin, Boston

[3] Ruth Harrison, *Animal machines*, 1964, Lyle Stuart, London

[4] EF Shumaker, *Small is beautiful*, 1974. Abacus, London

current agricultural developments made no sense. Then in the early 70s, The Club of Rome was established to look carefully at the problems of the world and predict world futures. They published *Limits to Growth*,[1] pointing out how and why there are limits to "development" and human exploitation. These books were best sellers. Several others were also beginning to examine the state of human culture and the environment and suggesting alternatives. Thus, at least the people who read books and newspapers were becoming aware of environmental problems, the effects of the human population explosion and how unsustainable current human "development" and in particular modern agriculture was for the future. This was in the 1970s ... and both adult education and university students were asking for courses to try and find out more, so I volunteered to teach university courses that were being asked for by the students, as well as adult education.

Sadly, during the next forty years these problems were pushed to the side-lines in the all-encompassing pursuit of money and the worship of humanity and its technology which would somehow solve the riddles and also allow galloping consumerism, industrialisation, and exponential human population growth (which threatens life itself) to continue unabated. The answer to all problems was more money = more industries = more jobs = more "development". Environmental problems were turned into economic costs and benefits, nuclear power expanded because of problems created by humans such as acid rain and scarcity of fossil fuels. Poverty was redefined as having less money than your neighbour and not being able to buy a car or a new TV. Finally, in 2013 Jonathon Porritt, a respected environmental leader of Green Peace for many years, said in an interview that, "technology will solve all our problems". Even the Greens had bought it! The Anthropocene had won ... but the living world or God or whatever, will eventually tell us otherwise.

Today, it is difficult, if not impossible, to remain optimistic about

---

[1] DH Meadows, DL Meadows, J Rander & WW Behrens, *Limits to growth*, 1972, Universe Books, New York

the future of our great and beautiful natural world which has continued to be more battered, ignored, and destroyed, whatever the political or economic system. We are living beings who are dependent on the living world. Most humans are aware of the environmental problems predicted as early as the 1920s by ecologists, and the cause of these being exponentially growing human population and consumerism. The results are already evident in political instability, wars, conflict for resources, energy, water, food, space and the wiping out of other species. It is still believed that this economic, consumer boom which everyone wishes to join, just needs the odd tweak here and there for it all to come right, and development and growth continue *ad infinitum*. On top of this, the universal religion of anthropocentrism will somehow solve all the problems. We continue to live and to die, but ... we die rich ... in possessions ... Even if our experiences are mostly "virtual". Perhaps we must turn the clock back and begin again to learn from other species around us.

The only problem which the governments did anything about in the 1970s was the welfare of animals on farms. As a result of visiting and working on many farms to do my DPhil, I had firsthand experience of the way in which animals were treated as machines. So, together with my sympathetic supervisor, after completing a doctorate, I applied for a post-doctoral research fellowship to study in depth the behavioural problems of intensively kept animals, and review what might be known about their causes. The Agricultural Research Council gave me two years' funding for this study, and some three years on, the resulting book was published.[1] This, with a couple of others' studies, launched animal welfare science for which, due to popular appeal, government research money was available and by the 1980s, academic departments established. Departments of animal welfare were started in veterinary schools and animal behaviour and welfare groups were established in universities in UK, USA and Europe. Professors were appointed, an academic growth industry was afoot. The training of veterinarians is supposed to reflect their

---

[1] MKW, *Behavioural problems of farm animals*, 1977

Hippocratic oath which is to "relieve animal suffering". But, until the 1980s (and often still today) veterinary science was almost exclusively concerned with physical suffering. If the pigs or hens were well fed, in good physical health, grew fast and made money, then nothing was obviously wrong ... but, if animals had minds and feelings, then surely they could suffer mentally too. This being the case, even when physically healthy, their lives can consist of continual suffering.

There were one or two ethologists and veterinarians also putting two and two together, a South African agricultural researcher in Edinburgh, David Woodgush, a vet in Liverpool, Roger Ewbank, and another vet, Andrew Fraser in Edinburgh. They got together and started the Society of Veterinary Ethology to encourage further research into the behavioural problems of farm animals. This society has now grown to thousands with members all over the world. Animal welfare science is indeed an academic industry with large amounts of money spent on hiring and firing, paying for research and conferences in many countries. Today, in order for companies to gain bonus points with consumers, animal welfare science is predominantly financed by interested industries or by animal welfare organisations, both of which have vested interests in different conclusions.[1]

In a few countries, there has been a ban on some animal-keeping practices which are particularly terrible, such as keeping hens or veal calves in small cages for their lives. Although these bans have been the result of consumer pressure, the icing on the cake "evidence" comes from scientific investigation. When kept in restricted, crowded, difficult conditions, however well-fed and physically healthy, they may suffer. But greed, money-making and feeding humans takes priority over animal suffering. So ironically, despite the enormous amount of research on animal welfare in the last fifty years, there has been remarkably little change in the lives that farm animals live whether intensively raised or not, although

---

[1] Chapter 9, Symbiotic animal management in MKW, *Animals in circuses & zoos*, 1990; MKW, Hayley Dixon, "The pretence of Professor Stephen Harris as an 'expert witness'," *The Telegraph* 03/05/2018

there has been a considerable advance in the technology accompanying how they are looked after, and the drugs they are given.

In 1969, I was awarded a doctorate and was missing many aspects of East Africa, so a trip to see what developments there had been in studying animal behaviour in the field was my self-appointed reward. I had saved enough money for my sons, aged five and three, and my own fares, and we caught a plane to Nairobi to visit old friends, and make new ones. We travelled to the Serengeti and found it had developed into a large research station with around twenty graduate students and researchers working on a variety of animals. We visited Amboseli and contacted Cynthia Moss who was just starting her elephant studies, we visited a young fellow who was a post-graduate researching the behaviour of elephants. He was green behind the ears in terms of surviving in Africa at that time, but he had plenty of funding and a large camp with helpers and servants in a beautiful spot next to a bubbling river. I went out with him in his new Land Rover to observe the elephants and see what he was doing. But, much to my horror, he spent the afternoon trying his hardest to make a couple of young males charge the Land Rover. He was obviously getting his kicks from driving fast through the bush being chased by an elephant. This did not strike me as the best approach if one was trying to find out about their social life and behaviour. It was clear, however, how relatively easy elephants were to study, easy to find, easy to identify and because they were a high profile species, relatively easy to find funding to study. This young man eventually became one of the chief pundits on elephant behaviour. He has done some good by raising awareness of them for their conservation, although looking through the dark glass of 2019, it does not appear that elephants will last much longer, certainly not in the wild. We all bathed in the bubbling brook while we were at this camp, only to be informed afterwards that three people had already contracted Bilharzia from it ... no one had thought to tell me, and my kids were small and unlikely to have any immunity. So, on returning to the UK, the first port of call was to the Tropical Diseases Hospital and luckily we were all free.

During this two-month trip, we had some full-on African times. When in the Serengeti, for example, one evening I left the children in camp with Jane (a young Kikuyu helper) to go to eat and talk with the researchers in their camp. When we were nicely sozzled at the end of the meal, Jane came rushing in in a panic: the safari ants had found the tents and had set up a strong safari line into Pip, my youngest son's bed, to bite him. Safari ants have been known to kill babies and small children, so I rushed out to try to redirect their long "safaris" with paraffin, at last it worked and they changed their route. Another time, the VW Beetle I had hired, having carried us through many miles of bumping no-road safaris, rivers, floods and thorny savannah, finally broke down on the way back from Manyara in an isolated dusty place. I crawled under the car and realised that the sump plug had come off and all the oil from the engine had drained away. What to do? A couple of safari lorries came along after an hour or two, so I asked them for a lift to the 200 kilometre-away garage, but they told me there would be no one coming back that way for three days, so that was not an option, I could not leave the children and Jane for that time. I burrowed under the car, to have another look and think what I could do, when I heard a very Oxford English accent addressing me. I backed out to find a Masai warrior standing on one leg and leaning on his spear in the normal way, dread locks and ear lobes dangling. He introduced himself as an Oxford graduate which from his accent I could well believe. He was one of the first Masai to go to secondary school I suspect, but after Oxford where he distinguished himself, he did not like the life-style or the values, so, I felt wisely, had decided to return to his tribal Masai life which was happier, freer and a lot more fun. I told him my plight, a broken-down VW Beetle, two young children and Jane a young Kikuyu girl. He smiled and produced a bar of Sunlight Soap, a blue soap sold by the foot which was sought after for doing the laundry in the rivers. He told me to melt it down with a little water and stuff the mixture up the sump. I did this and after adding oil which the Safari trucks had given me, we happily continued on our way. It was a month and some 800 kilometres later when I returned the car to the Hindi who had hired it to me before I realised that I had never repaired it properly ... he laughed and said it would probably hold until the car finally came to grief anyway. I wonder what became of that Masai warrior and whether he managed to persuade others to stay with their culture and not abandon it for money and goods.

In Samburu, north Kenya, there is a nature reserve that has a marvellous clear pool in the middle of what is very dry savannah land. The pool is called Buffalo Pool. We camped nearby and after a morning of travelling after giraffe and glimpsing some black rhinoceroses, we were so hot that we jumped in and cooled off among the hippopotamuses. That swim in a glorious cool pool with hippopotamus grunts and a view of the snow on Mount Kenya remains with me today as a memory of real Africa. At that time the National Parks were not stiff with visitors and rules. Twenty years later when I visited the same pool, it was lined with tourists and guides confined to their Land Rovers ... the very thought of getting out of the car, never mind bathing in the pool, shocked them to the core ... but I doubt whether the hippopotamuses had changed their attitude, although they were now used to being photographed and stared at by humans.

A day or two before we were due to go back to the UK, I received a letter to tell me that I had won the post-doctoral fellowship applied for, so I had a job to return to. Maybe it was the hippopotamuses who had wished it, who knows, but the tribal Masai kept on reverting to my mind, what about just staying in the bush for ever? The children were against it ... so it was back to Blighty.

We returned to Mongers Farm, our cottage and barn in East Sussex. Its reconstruction was complete and we, or I, needed more land for my growing quadruped family's needs. Several months later after a great deal of searching we bought a large Victorian mansion in desperate need of renovation but with a glorious old flint-built Victorian farmyard and seventeen acres in an idyllic setting: nestled in the South Downs on the river Cuckmere. We moved in to camp in a couple of sprawling cold, draughty rooms, and installed the horses in the Victorian farm stables with hayloft above. Sneefus, greatly satisfied with the library of smells both inside and out, took up residence next to the enormous fire, the only heating we had. The children were welcomed at the village primary school within walking distance across the river and fields, so all was set for the next ten years of hard labour in the Sussex Downs, or at least that is what I had in mind, but the sticky root of Africa also drew me back.

This period of my life was a steep learning curve. It taught me a great deal about different mammals and how to find out more, how to do research and the scientific method, whether this is by experimentation or observation. It also, of necessity, taught me how to bring up and support a young family as well as pursue my driving interests of learning more about the minds of other species. Although the raising of my sons was probably faulty (as my sons keep telling me), I learnt a lot about mammals from them. I also learnt "scholarship", how to find out things and weigh up their importance both mathematically and philosophically. It taught me more about the hierarchical world of humans, the curious ignorance of many academics about the practical reality of the animals they study in their labs and books, and the disproportionate power of human beliefs. It taught me to study inter-disciplinarily around the subjects that interested me, and to observe the ins and outs of both human and other mammals' lives. In fact, it really started the ball rolling on how to study non-humans' different mental attributes, and of course it was fun with a good few emotional hiccups that had to be swallowed or brushed aside.

## South Africa, people, politics, eland & blesbok

My husband had a serious itch to return to South Africa, and because he now had a UK passport as a result of marrying me, he could do this without risk of being shut up in jail without trial. He had graduated earlier and gone to work in Fleet Street, but he obtained a post on a South African paper in Pretoria so I eventually found a research post in a South African university which would allow me to continue studying large mammals when I had finished my fellowship. This was in an Afrikaans university, the University of Pretoria where a mammal research institute was being established, and they badly needed an ethologist, who were few on the ground at that time. I accepted the post.

The boys and I left Milton Court, the Victorian mansion, six months after moving in. We travelled to South Africa on a Union Castle boat with Sneefus and a second-hand Land Rover which this time I had allowed myself for wildlife research, a great luxury. We made friends on the voyage with cockney immigrants from East London; white immigrant families who were being financed to go to Apartheid South Africa to live a suburban whites' life with sun, servants, swimming pools, braaivleis [barbeque] and plenty of money; such stuff as they had only dreamed of in workingman's London. The three-week journey on the boat passed pleasantly with plenty of events, and time to read and sit in the sun now and then. This time we did not go through the Suez Canal or stop at Port Said to be entertained by the Gilli Gilli men and their chicks, but there were plenty of fancy dress parties, deck coits, large meals, and, of course, Sneefus to be taken on wanders around the deck early in the morning or late at night when few other people were about. As usual, he made friends with everyone and was soon allowed where no dogs were meant to be, including swimming in the pool, at least first thing in the morning, a secret between the deck steward and Sneefus.

My first introduction to South Africa was the view of Table Mountain, and the strange and exciting Mediterranean vegetation on islets of land awaiting "development". We disembarked, and the first human social encounter was something of a shock ... I began to wonder about the wisdom of our immigrant friends' decision to abandon the UK for the "good life" in South Africa. The whites were "white" enough, but their culture was weird to say the least. This first visit was to the house of a rich white suburban South African woman. The house was in a magnificent site, built on a steep slope overlooking the sea with a swimming pool, garden, views, servants and an enormous spotlessly clean house full of expensive furniture and art. We had the obligatory braavleis (a barbecue where the men cook, and usually burn, enormous chunks of meat on an elaborate grill over a fire). Everyone stands and sits about the swimming pool drinking large gins and talking loudly about their problems with the servants, while the children swim and squabble.

This white woman shocked me to the core. She announced, and everyone agreed, that "the South African doctors are the best in the world". Did people actually say that sort of thing? How on earth did she know? Apparently, this sort of statement was normal in South Africa. It was just after one of their doctors had transplanted a pig heart into a human. The human had lived for a few days, then died, as predicted, because, at that time, the medics were incapable of countering the host body's rejection of strange bits from others. We had frequently discussed the pros and cons of transplant surgery at Sussex in the tearoom, and I was aware of the serious ethical concerns that were arising as a result of these operations. Was everything else in South Africa going to be "the best in the world"? It turned out that the answer was YES, as maintained by the whites, whether you supported the Nationalist Apartheid Government or opposed it as the Progressive Party did. There was no room for discussion.

My husband had hired a smart house with swimming pool and servants in a "whites only" suburb of Pretoria where the neighbours were Afrikaans and I had difficulty understanding their English but it was assumed that I would fit into this suburban life. My husband flitted off in the morning well-polished and ironed by the servants to his office at the Pretoria *Times* where he and others like him would report what was happening in South Africa. Colleagues, their wives and children were welcome for braaivleises and cocktail parties at the weekends, as long as the conversation stuck to fashion, servants and illnesses, and of course they wore the right clothes. Any animals around were, like the servants, only allowed to politely say good day, and perform for their masters before being sent back (probably feeling much relieved) to their kennels.

This could not last long. I found a *pandokkie* (a small primitive hut/house) to rent with fifteen acres of grass in the high Transvaal between Johannesburg and Pretoria, no light or running water, but Jonas and his wife, Rebecca, turned up looking for work a few days after we moved in. Jonas was an older man, around fifty with white hair and a slight stoop. He was originally Tanzanian, but as a young boy had walked to Johannesburg through Zambia, Zimbabwe, Botswana and so to South Africa in order to work in the gold mines. Many young African adventurers had done this between the wars as the streets in South Africa were believed to be lined with gold and apartheid had not yet reared its ugly head. Jonas spoke about fifteen languages, including Swahili, English and Afrikaans. He and his gentle wife, moved in to organise the house, garden and animals. With his help, it was not long before we had a couple of horses and a house cow, had planted a dry garden, and dug out a mud pool so that we could cool off in the summer. Rebecca did the washing, ironing, cleaning and even the cooking with a little help, and was kind and gentle with the children (who were not always kind and gentle back!). Jonas was a real philosopher and it would be difficult to find experiences of war, friendship, work or travel that he had not lived through. He was shy of whites of course, but with prompting, he talked and I listened or asked questions, a real latter day Socratic philosopher ... I wish I had spent more time with him now.

The University of Pretoria, where I was working, was also weird. I was the only woman in the faculty, and the only person who had studied animal behaviour, so I was commandeered to help a group of male post-graduates with their research and theses. Many of these men subsequently became important in the universities, national parks and wildlife conservation in South Africa, but at the time, they were very naïve concerning ethology and unaware that the behaviour and minds of the animals they studied might be of importance for their conservation. This surprised me as I had previously read *The soul of the ape* (written 1919, published 1969) and *The soul of the white ant* (1937) by Eugene Marais, books by a South African that were way ahead of their time. The thoughts of this far-sighted biologist were a long way away from anything I encountered concerning animal minds in South Africa in the later half of the twentieth century!

There are rituals and traditions in any mammal society, but those of the Biology department at the University of Pretoria were creepy. First, it was compulsory for all faculty to meet in the tearoom to drink Rooibos tea and talk to other faculty members at 10.30 sharp every morning. The tea was good and the chats were useful, but the rituals were extraordinary. When the head of department arrived, we all had to jump to our feet, click our heels and more or less give the Nazi salute. It was an Afrikaans University where everyone who was not a Nationalist (that is a right wing apartheid supporter) was likely to be shut up in jail or deported. After the informality of the University of Sussex, this was a shock. The first time it happened, I rolled about laughing ... for which I was given a very direct caution. Nevertheless, science of the natural world has a unifying effect, and I found several interesting Afrikaans colleagues, and we carefully avoided political discussions. Soon I had a British male student who was sponsored to do a PhD with me. His thesis was to be on the ecology and social behaviour of the eland, the largest, most stately and grandest of the African antelope. There was a population to study in a nature reserve around 300 kilometres away where he took up residence. I would visit him in his camp and sometimes take a couple of undergraduate or graduate students along to teach them how to do ethology fieldwork.

The eland is one of the most beautiful of the African antelopes which had not been studied at all and I had for years wanted to, and there was also a small population in a nature reserve near the university. They lived with some blesbok, a smaller antelope vaguely related to wildebeest and hartebeest. They also had not been studied so I embarked on studying their behaviour and communication near the university while collecting and delivering children to schools. Eland live in small herds of between three and ten individuals, mixed males and females, and they wander around with no fixed address. But the blesbok males take up relatively small territories next to each other and await the coming of the females, a bit like the waterbuck; only they have smaller territories. One of the questions was: were there some males that the females preferred over others, and if there were, were these the biggest, those with the largest horns or in the nicest territory? Why did the females choose one over another? Immobilising animals to mark or transport them by shooting them with a sedative was just beginning at that time in South Africa, and the immobilising team immobilised a couple of males on whom we put a harness fitted with a coloured chalk. The colour rubs off on the female when she is mounted and indicates which males had been allowed to mate with her.

Before I had unravelled why some males were chosen by the females, politics intervened.

There was also a zoo in Pretoria right near the school where I had to pick up the boys each day, where I found a group of eland, roan antelope and oryx, all large horned antelope kept in a paddock together, and so also studied how they communicated with gestures, postures and movements, both with their own kind and between species, a continuation of my previous research on communication which outlined "the language" used between these species.[1]

The research, our mini-farm and the kids were fun, but I wanted more contact with colleagues. One Afrikaans lecturer at the university was really keen on rodents in the desert and he became a firm friend. We also joined trips with the local ornithological society admiring the astonishing beauty of the birds. But horse talk was lacking. One day I saw a notice advertising a Lippizana horse display. The Lippizanas are horses that were bred by the Spanish Riding School which was transferred to Vienna in the sixteenth century by the Habsburg Monarchy (1572). They are small, stocky white chaps with large heads but they come to life

---

[1] MKW, "The visual displays of eland," 1978, pp179-222, MKW, "The social organisation of a small herd of captive eland …" 1978, pp32-40, MKW, "Individual differences in a small herd of captive eland …" 1978, pp44-55

and beauty when moving, particularly when ridden well. Robert Hall (where I had been a working pupil) had been one of their first English pupils and although I had never visited them, I rushed to see the performance and stayed to talk to the small, energetic owner of the stables: George Iwanowski. He was Polish and had been head of a cavalry division when the Germans invaded Poland in the Second World War. Before the Second World War, it was customary for the cavalries of many of the central European nations to send their very best riding officers to Vienna to study with the Spanish Riding School for a while. When they returned to their various countries to train their own cavalry, they would be able to teach the squaddies to ride better, and consequently survive longer in battle. George was one of the talented Polish cavalry officers who had been in Vienna before the war and had returned to his unit when the war started. He had tales to tell of how, as the German tanks advanced through the Polish cavalry ranks, he galloped about with his men shooting the Germans through the turrets in their tanks. Of course, he was eventually caught and put in an officers' camp by the Nazis where they planned to shoot him. He managed to escape in a laundry basket, but instead of escaping to another country, hid in another laundry basket to re-enter another prison to meet up with his head officer, a captain or general who was also a prisoner of war, in order to get his orders! Luckily, the general (who was shot a few days later together with all the rest of the officers) advised George to go to the UK and join the Free Poles. He clambered back into his basket, and somehow managed to get out of Poland. He then visited Vienna on his way west and procured some Lippizana horses which he smuggled to the UK to save them from German destruction. After further adventures, at the end of the war, the South Africans volunteered to take him and his horses, and they were all shipped to Cape Town. Over the next thirty years, he, his wife and show jumping daughter ran a superior riding school giving displays and training riders to ride these white stallions. Interestingly, his riders were all women as he said they could tune in easier to the horses than men. This delighted me as in Vienna, it is only men who can be trained for their displays. On a recent visit to Vienna, I found that you could become an *uber reider* today if you were female (and there was one female riding), but you had to be a certain height and weight, and curiously, to have a neck of a certain length! Ah well, even today I would not make it.

George was a delightful person, if a bit partial to young women. I had no trouble convincing him that I could teach him the science of horse behaviour and communication, if he would teach me to ride better. For the next six months, this exchange continued and I rode the high school Lippizanas and the young stallions under his sharp eye, but alas, because other things intervened, I never joined his team of riders who eventually became well known all over South Africa and other parts of the world for their displays. George was one of the best teachers and riders I have met, and we remained friends until his death in 2011 in Poland, where some of his family's land had finally been renegotiated back to him after the Soviet land confiscation. Finding George and the Lippizanas added much to my skills with riding horses and their mental abilities, watching other experienced horse-tuned professionals and practising more, added a positive attribute to the South African experience.

Wandering around one day on our gelding Sunny on the high veldt with thunderstorms and forked lightening scattered around us, there was a sudden hiss and a large spitting cobra raised his head and spat just by Sunny's front legs. Luckily neither of us got venom in the eye (which quickly sends you blind). It was my first encounter with a spitting cobra who were common in that area. Sneefus was with us, but darted off with his tail between his legs, luckily. Mostly snakes get out of your way if they hear you coming, but for some reason this cobra did not, and Sunny was unfazed. The other thing was that during the summer, there were frequent thunder and forked lightening storms over the highveld. You could see where the lightening had struck, often here and there and sometimes too close. Actually, when living among different species and storms like that, there is not really much point in worrying. If you can't avoid the snake or the forked lightening, it is the luck of the draw whether you survive or not. During the six months or so we continued with these expeditions, it turned out we were all lucky.

At that time there was no TV in South Africa, and the papers were government censored. The result was that these often bright, white English or Afrikaans speaking students had no idea what was really happening in the rest of Africa where countries were becoming independent. The only information they had came from rumour and government reports of nuns being raped and whites cut down with machetes. We often talked after lectures, and they were fascinated to find that I had direct experience of post-colonial Africa and that it was not all anarchy, murder and rape as they had come to believe.

Then politics intervened, at around 4 a.m. one day we were awakened by Rebecca with the early morning tea, to find armed South African Special Branch surrounding our *pandokkie*. They were rough, rude, red necks, and demanded to see passports, and any other identification. It turned out that although my husband now had a British passport, his visa had expired the day before and they had descended quick sharp to terrify him and his family because of his past record attacking the government in the press. Because he was now a British citizen having married me, they could not put him in jail without trial which clearly irked them and ensured their bad manners and their nasty and frightening behaviour. Luckily, he had renewed his visa the previous day and showed it to them. After strutting around, bullying Jonas and Rebecca relentlessly, and throwing all our furniture and things around (actually we did not have that much so it was not too serious), they jumped back into their 4 x 4s and roared off. It gave one a real taste of what a police state was like.

The next landmark was a dinner party we were invited to by the anti-government white group called the Progressive Party which consisted mostly of extremely rich people such as the Oppenheimers, and well-off writers and intellectuals. This was held in a splendid palace somewhere in Johannesburg. These people did some good and had some very dedicated activists (who were mostly in jail) but many who were not in jail, were so rich and had so many contacts, that they were more or less untouchable by the government. As a result, they talked late into the night about how awful the government was, although there was little agreement on what they should do about it. During the day, they made money, often by employing large labour forces on very low wages on things like sugar plantations or mines. At this dinner, we had the usual elaborate food and large number of courses. I was sitting next to a rich sugar plantation owner who was groaning on about his labour force. After forty-five minutes of listening to his ranting, while eating the dessert, I asked how he could remain part of the Progressive anti-government Party when he clearly was only interested in keeping some sort of apartheid in place so he did not have to pay black labour the same as white labour and could consequently increase his profits. At that time, white labourers were paid around three times more per hour to do the same job as the black labourers, a similar discrimination that had been in countries for centuries (and still is in some countries) between women and men. However, at this dinner party, on hearing my remark, probably stated with a certain amount of heat, there was a deathly hush. Then, after around twenty seconds and much cutlery clattering, the same conversations as before continued around the long table and the question was ignored ... However, about three people came up at the end when we were dressing to go away and congratulated me on speaking out, perhaps we should not tar them all with the same brush. In retrospect, there was probably much more to all this than I realised. I was politically naïve and inclined to put my foot in dangerous waters. Nevertheless, hypocrisy is hypocrisy. A few days later, I visited another member of this group who lived in a large house in an expensive suburb of Johannesburg and was mildly sympathetic to my question. We talked into the night, while she kept one of her servants back to cut her toenails after we had finished eating, she was rather fat and could not do it easily herself. He lived in Soweto more than an hour's bus journey away, but I don't think she even understood that she was preventing him from having even half a night's sleep as she ranted on about the terrible apartheid government.

Within a few weeks of this dinner party, I arrived in my laboratory to find a letter signed by the head of the institute to say I was sacked. In academia this sort of thing does not happen unless of course one has committed murder or something equally bad. If one has a teaching or research post, it is normal to trust whoever gave it to you, and it cannot be terminated until the contract date is reached. But, being naïve, I had not bothered to have a contract, the word of the director was sufficient.

My sacking was not, it seemed, because I had neglected my duties, rather the reverse since I had been teaching courses and helping post-graduates during much of my research time which I had not been employed to do; so why was I being sacked? The slimy English-speaking head of department slithered away from my questions, but, some years later, I found out that he had been pressurised by the government to get rid of me as a potential threat to the thinking of their nice young Afrikaans Nationalists. The slimy cove had made no effort to back me up or explain ... Although he said he was not a Nationalist supporter, it appeared that he would do anything, including wrecking people's careers, to butter the hand that fed him. If he is still alive, like many other whites in South Africa he will have switched horses by now! But horses never switch their loyalties to their riders, unless they are seriously beaten up, and even then not often.

Although the animals and, in the end, our living conditions were interesting, I disliked the South Africa of that time. There was a tangible atmosphere of mistrust and hatred between blacks and whites, and the dishonesty and hypocrisy in both deed and thought by even those who were supposed to be anti-government was distasteful. It was unpleasant, even the climate in Johannesburg was unpleasant, hot and wet in the summer and cold and dry in the winter. As a result, most weekends or whenever possible, the children and I escaped to Swaziland, Botswana or Lesotho, all independent states surrounded by South Africa where the atmosphere was different and familiarly African.

On one visit to Swaziland, I visited a Nuffield Project. The idea was to teach the children in schools out in the bush to garden better and grow their own food so that they would be able to feed themselves, and their families, when they grew up. They were short of teachers, so I applied as a volunteer. Since I had been sacked, there was nothing I could do but drop my research in South Africa, even though I was concerned for my PhD student and various graduate students whom I taught and supervised. Luckily, they ended up in the sympathetic and helpful hands of one of my Afrikaans colleagues who knew how to play the political game a great deal better than I.

The Nuffield Trust was delighted that I volunteered to help so they sent me and my two boys to a bush school some 100 kilometres from the nearest town in the mountains of Swaziland. It was a joint primary and secondary school, and we shared a hut with a young woman teacher and her little girl. The boys joined their appropriate classes and I taught anything they wanted me to teach, as well as starting a garden. The pupils at schools throughout most of Africa have to pay, so many parents take their children away for months at a time to help herd the goats or earn money before there is enough for the next year's school fees. The result was that there were children from eight to twenty in my oldest son's class, and from six to eighteen in the six-year-old's. All the children walked to and from school from their small farms, some had to cover as much as fifteen kilometres, and then back in the evening. They only had paraffin lamps to study by at night and were given a great deal of rote learning homework. But they recognised how lucky they were to be at school and were extraordinarily anxious to learn. There were no discipline problems at all, but, since all the teaching was rote learning, they had trouble learning to think about their work and how to do anything new like growing and eating new types of vegetables.

My eldest son, eight at the time, had been transported rapidly from a very posh whites-only English-speaking South African fee-paying school, to a somewhat disorganised, all-African poor bush school, living in a mud hut with a corrugated iron roof, sharing his meals with a little black girl and her mother, and attending a school in another large mud hut where the roof leaked and he had to share a desk meant for two, with four other boys and girls. There were almost no books or writing materials. But despite no running water, and all the other problems and hiccups, everyone was always washed and clean, took pride in their uniforms, books, if they had any, and their work. They listened avidly for hours to teachers, and probably also to Sam, who certainly has the gift of the gab and I am sure embroidered his life's story considerably. But he found the whole experience disturbing and was anxious to re-join a more conventional school. My six-year-old was, at that time, more happy-go-lucky and he rarely attended class, but wandered between them being petted and fussed over by the teachers, and not learning to read and write. One morning, Sam got himself into a rant because many of the girls and boys, some more than twice his age, could not restrain themselves from reaching over and touching his straight brown hair. It was so different from their fussy black curly mops. He was a somewhat irascible child, and it became too much for him and he lost it. Although all the teachers were sympathetic and we had a few laughs after about it, perhaps we needed to move on.

During our stay, I was asked by a rich landowner who owned an estate in Swaziland to do a consultancy on how he could turn his Swaziland estate into a nature reserve and what animals to re-introduce. This earned me enough money to establish a small library at the school, "Ekukawani Secondary School Library" which was appreciated, but probably no longer exists.

Before we left, I had also received a letter from a neighbour in Sussex. My mare, Syringa, had been cut to pieces in some barbed wire, although her yearling colt foal was unharmed, he was reported as being "unmanageable". The local zoo personnel had stepped in to tend her, but they were unable to devote the staff to the attention that she would need if she was to survive, and I was far away and could not immediately raise the air fare to return; so the vet's diagnosis was that the mare should be put down and her yearling Anglo-Arab son, Baksheesh, the third generation of our breeding, was placed in a livery yard where he was shut in a stable and considered dangerous ... anyone who had seen his mother being cut to pieces in barbed wire might indeed be traumatised.

Too many things were wrong, we would have to return to the UK. We said our goodbyes with regret, to all the staff and particularly to Maria with whom we had shared a cramped hut. I had enjoyed my time in Swaziland and would have liked to continue the project, but the children were full of glee to be going back to Sussex, Sneefus was coming but he would have to go into quarantine for six months on arrival. We loaded the Land Rover and set off back to Johannesburg airport.

A few days later, we arrived at the enormous, empty, cold Victorian house and untended garden, the children hopped off to the village school, Sneefus had been arrested and taken to the quarantine kennels, and the colt, Baksheesh was returned. He was a long-legged, beautiful roan going white, athletic, excited individual who seemed to have wings attached to each foot.

Living financially on the edge without a tenured university job had taught me that I needed to hedge my bets, there was always a plan B and C as well as plan A. Plan A had failed, so now it was Plan B to be put into action. This was to apply for another post-doctoral grant to work on animal welfare problems. The welfare of veal calves confined to crates was being discussed in Parliament; perhaps a research project measuring their welfare would be funded by someone. The most likely organisation was the RSPCA.

I suppose the major message I had learnt from this chapter of my life had been stay away from human politics, it clearly fosters hypocrisy, dishonesty, and disloyalty, or at least that was my experience. These are not characteristics of my family, either two- or four-footed. It was a tough lesson to learn. I had learnt quite a lot more about horses and their world view, and had begun to learn to ride so that the two of us could more often experience cooperative collective intentionality.[1] I had learnt a lot about eland, and other antelope and their different social contracts, but

---

[1] MKW, "Collective intentionality & social ontology of equines & elephants," at *Collective Intentionality VI Conference*, 2008. Berkeley

also the similarities in their communication and other mental attributes, and I had experienced another very different part of wild Africa.

## Milton Court Farm, horses, cattle & animal welfare

The boys and I arrived back in the UK just before the spring. We moved back into the enormous Victorian house, which was cold and in need of plumbing, electricity, furniture, and warmth. We spent most of the first days collecting firewood by scavenging around in our woods to make a fire in the enormous open fireplace, which consumed tons of wood, and all the heat went up the chimney. We camped around the fire with all the blankets and duvets we could find; at least the red flames and glowing embers were comforting, if not warming.

Baksheesh, the Anglo-Arab eighteen-month-old stallion, grandson of Kathiawar was delivered in an old trailer. He leapt out, nose flaring, eyes popping out, beaming, and immediately started dominating my life with his leaping, shouting, squealing, cavorting and the liberal use of all four legs in little-known horse athletics. He was certainly an "over the top" young Anglo-Arab stallion, and like his grandmother before him, as a result of being confined in a stable for several months, he had developed a stereotype. This one was called "weaving" and is similar to one that children show when confined in their highchairs too long: they swing backwards and forwards or, in his case sideways, and sideways with his head over the stable door. Stereotypes, "ticks" have long been acknowledged in horse owners and called "vices"; but within the last half century they have been identified as signs of distress: demonstrating that all is not right in that animal's world. In stabled horses, stereotypes are very common and predominantly indicative of boredom, overfeeding and lack of exercise. Curiously the weaving that Baksheesh was prone to, the result of being confined to a stable for long periods, is quite acceptable in modern horse keeping! It is recognised that it will be common, so that state of the art stable construction provides an "anti-weaving" bar ... in other words, it is assumed that a stabled horse will weave and therefore be distressed. Instead of changing the husbandry so that the horse is not distressed, the idea is to stop the horse weaving physically. Ironically, if similar behaviours are seen in cattle, pigs, goats or sheep (which are not infrequent in intensive systems), the owner can be reported for not fulfilling the European codes of practice on animal welfare and could be taken to court for "cruelty", but this is not so for horses.

Rarely do horse owners, even today, question WHY the horse might be distressed. They put it down to "race" or "genes" ... even though racism is not politically correct in human mammals. Racism is the belief that different races have different genes and behave in particular ways because of their genes. It used to be believed that if you have black skin you will be stupid or violent or run very fast. We now know that although there may be "instinctive tendencies" for some individuals to be better at some things than others, this is not tied to race, it is mostly the result of lifetime experiences combined with, and sometimes dominating, "innate tendencies". The illogicality that racism is unacceptable in humans but quite okay in non-humans has escaped many people, both scientists and lay people: why is human mammal behaviour not controlled by their genes to the same extent as non-human mammal behaviour? It is clearly daft to believe that for some reason weaving in horses is "in their genes" rather than the result of their lifetime experiences. Whatever breed they are, they can become distressed in inappropriate environments, and weaving is a sign of this whatever their genetic background (although some individuals may be more prone to develop it than others). Weaving is always a sign of distress, whatever the race. At that time the vets were doing extraordinary things to prevent stereotypes: they were cutting the muscles in horses' necks so they could not grab a post and suck in air ... I was at one stage taken around a clinical horse ward to see these post-operated horses!

Since Baksheesh was young, it was unlikely that his "weaving" had become a deeply engrained habit and consequently it could be eliminated. To do this, we had to make changes in his lifestyle, keeping him outside on grass where he could eat and move around at will, and having some other equines to live with. We could give him more exercise to help use up his excessive energy and develop his muscles and body. Perhaps, most importantly, like children at primary school, he needed to learn to do new and different things with both his body and his brain. He could learn to go around in circles at different gaits, walk, trot, canter, and learn the words for these different gaits; this is called lunging. He could also have more to do with humans to learn about them and begin to understand their language better. Whatever we could dream up to stimulate his brain and exercise his body would help. The result was I spent around thirty minutes each day teaching Baksheesh different things, and then around one hour using up his energy lunging or free schooling (this is without a line, asking him to go around me in circles). As soon as I thought he could carry me, I began slowly to get on his back. Since he was a great deal stronger, larger and heavier than me and very excitable, it was obvious that if he had a fright, all would go awry and I would probably be injured, so one needed to be quiet and cool, and confident to give him confidence. Thus, the quiet approach was as much to help me survive as it was for him to learn without fear. It is of course the approach of the traditional European school of equitation, which is very unlike American Western horse training.

In Western horse training, the cowboy must dominate the horse and he arrives at this by terrifying the mustang out of his wits by throwing things over him, tying him up tight, and making terrifying noises. Once the horse has "given up" or rather gone into a sort of frozen, suspended animation of fear ("learnt helplessness" it is called), it is assumed he will be so "broken" that he will be docile and do whatever is asked. About twenty years after my learning from Baksheesh what a foolish and dangerous approach this was, the first of the natural horsemen appeared in the west of the USA. He was a man called Monty Roberts who was charismatic and had very good public relations. He found that if you chased the horse around long enough in a round pen, he would want to come to you and then follow you around rather than have to go on galloping around and around. This was called "join up" and became a sort of symbol of how the "gentle" approach to horse training was possible even with Mustangs in the west of the USA. Monty's publicity was so good, that he managed to sell himself to the Queen and did demonstrations for her. This really surprised me as I would have thought coming from a long line of British country folk, she would already have understood that a quiet, clear approach to teaching a young horse is always going to work better. The very first horse whisperer, today he would be a "behavioural consultant", was Rarey,[1] who was brought in to be consulted by Queen Victoria. Today, most of the "natural horsemen and women" use a somewhat similar approach, although not always with the same success!

With Baksheesh, being quiet and cool was just common sense. I was working by myself and did not want to end up in hospital or have a horse who was so "broken" that he could not exercise choices, or be himself. In addition, I had never heard how cow boys and girls trained their horses ... but I had seen how badly they rode them in the Western films! So, Baksheesh and I learnt together something about communicating, while also trying to have fun together, and after about six months, he would carry me on my rounds of the livestock on the farm, slowly as we were both continuing to learn about each other and the world we lived in. There is nothing new in the techniques we used, it is just obvious common sense when dealing with another mammal with similar but also slightly different mental skills to mine.[2]

Then a chance encounter with a busy, small, grey-haired determined lady called Ruth, who walked into the yard one day, began the expansion of our "stud". Ruth asked me if we wanted to buy a big grey Irish cob mare very cheap because the veterinarian had declared she had leukaemia and would not last long. I went to see

---

[1] John Solomon Rarey (1827–1866), *The complete horse tamer*. 1870
[2] MKW, *Equine Education*. 2004 outlines this in more detail and why

Sheba, and she seemed to be just the ticket, an old, big, heavy grey mare who anyone could ride and enjoy her company, and very cheap. I bought her for around £50, less than the abattoir's price. Ruth then brought her other horse Paddy, an enormous piebald Irish draught, to live in one of our paddocks. In the beginning, I think she wanted to keep an eye on Sheba and see that nothing untoward happened to her, but we became firm friends. She had strong opinions but was kindness itself, although her loud, determined voice, which she did not hesitate to use at any time could frighten people and horses! She was devoted to animal welfare having accompanied her husband, a quiet gentlemanly headmaster, to Cairo. There she had worked tirelessly with the organisation SPANA to try and relieve some of the overworked draught horses and donkeys. She lived in Eastbourne with her retired husband and grey parrot, both of whom loved her, but the parrot hated her husband, and bit him whenever he could and Ruth was not looking.

Sheba became the first mare to breed with Baksheesh. The leukaemia was a myth, it turned out. The real reason that Ruth wanted a home for her was that she could not catch her, one indication from that loud rasping voice that Ruth was on her way was enough to send Sheba skidaddling to the other end of the field, and as Ruth approached her, do a turn about, gallop and be off again. We had no trouble catching her however, and she lived and bred foals with us for the next fifteen years, by which time she was twenty-eight. We now had one young Anglo-Arab stallion, one old Irish mare and a Welsh pony filly, the foundation of the Druimghigha Stud.

Backsheesh became the resident stallion living with a herd of mares and his weaving stopped, except when he was expecting food. Horses are, of course, social animals and although some of their social needs can be supplied by humans, they really need to be with each other at least some of the time, and preferably with unrestricted contact so they can touch, smell, cavort around, learn to like or hate each other and form relationships. But stallions, particularly young ones, are not to be trifled with. It was not all easy, there were many stallion behaviours that had to be coped with. One was, that he liked to grab and hold things with his front teeth, your arm, clothes or anything else he could grab. The conventional approach to this is to strap the young stallion up with ropes and leather and even bits that hurt when he opens his mouth. I rather felt that his behaviour was not aggressive, it was just over-enthusiastic play and his motivation would not be stopped by tying him up. A better approach would be to allow him to do a version of this where he could cause no harm or hurt. At that time, I had a thick ivory bracelet that someone had given me from Africa (before ivory was taboo and the market shown to cause poaching). I always wore it and it was not long before he learnt to grab this and wave it and my arm around which he was allowed to do, in preference to grabbing my arm or hand. Then I substituted this for a piece of wood which he would grab and throw about; in this way we solved the "grab the person when being led problem" without strapping him up in various ways. Young stallions also often throw their front legs about gleefully, while squealing loudly. This usually happened when he was saying hello to another horse. If you were in the way, it could be very painful, therefore, rule number one: "do not stand in front of a stallion when he is saying hello to another horse". But if you prevent him saying hello to another horse, then he will show his frustration by standing up on his hind legs and waving his front legs around in the air (rearing), or many other unacceptable athletics, so the best thing to do is always allow him to say hello to another, but make sure you are out of the way. If this leads to further excited leaping around, that can be stopped by leading him away. Gradually over the years he learnt that he could say hello to any other horse, but only politely. Most stallions never learn this because they are never taught, they are just stopped from doing anything, even saying hello to another.

    I had to learn that it was silly to restrict or terrify him into doing what I wanted, if I made this mistake, it ended in my tears, he was bigger and stronger. I also had to learn that it was the unfamiliar that usually gave rise to his fear, leaping around or refusing to do things, so the answer was to allow him to get used to something, slowly and without fear. He had to learn that some things were not and never would be tolerated by me, and one of those was he was never to hurt me. Over and above this, the trick I learnt was to invent ways to motivate him to want to do what I asked when I asked, by explaining why in some way or another. If for example, he did not want to cross a bridge or a stream, then first he must get used to being near it, and not frightened, then I would cross it again and again in the hope that he realised that it was okay, and would imitate me, and perhaps pick some delicious grass the other side so that he knew it might be worthwhile crossing over. It was also important, as with a child, not to ask him to do things at a time or place that would be unreasonable or foolish.

We both had to learn that when we did the things that the other wanted, it was fun for both. We had to know each other well enough to predict what the other might be feeling or thinking in order to know what might be the outcome, and guide it in the way we both wanted. One way of doing this was to recognise that neither of us wanted to be hurt or frightened. We both, after sorting out our own "social contract", wanted to please the other and enjoy each other's company, where we both understood the rules of our association, so that effectively, we both became moral agents; that is, both of us knew when we were breaking the rules and we could decide to obey them or not on each occasion, but the easiest way, and that which gave most pleasure, was to obey them and so we both benefited. If the rules were broken by either of us, then the other had the right to return the nastiness with nastiness. For example, if he stepped on my toe intentionally, not by accident, or was in an irritable mood and lifted a hind leg perhaps to let me know he had the intention to kick me, I had every right to tick him off verbally. If ever he did kick me intentionally (which he did not), and hurt me, I had the right to be very angry and hit him back. But if I lost my temper and hit him for any reason, then he had every right to hit me back, or in some way do something that I did not like. It was a cooperation: "do as you would be done by" and "be done by as you did"; Charles Kingsley's moral rules for children in *The Water Babies*. Neither he nor any other animal I have associated closely with has ever taken the sulks after being thoroughly and roughly corrected for breaking the social contract. Every animal I have known has been very upset and distrusted a person after they have broken the social contract by losing his/her cool. Perhaps we should learn from them how to behave post-conflict, and to understand who has justice on their side. Liberal humanists sometimes find this difficult!

When riding, I had to hone my skills to ensure perfect balance and relaxation before we could even begin to really communicate. Gradually, I got better, practising what I had learnt with Robert Hall and George Iwanowski, and eventually we enjoyed galloping around everywhere and anywhere as a twosome. We would gallop off into Sherwood Forest and along the South Downs. Ever since learning these guidelines with Baksheesh, I have used and developed them, and they have served us all well. I hope we have also taught people as well as horses that this approach works, it is a great help in living and working joyfully together.

Despite their stallion behaviours, males, either stallions or geldings, are often easier to have strong emotional ties with than mares. This may be because males, castrated or not, when living with mares are usually excluded from fraternisation with the mare group; they are not "one of us". The mares meet together under shady trees discussing foals, an equivalent of a knitting circle or an equine Women's Institute. The stallion or geldings usually wistfully watch the group but are not allowed to join them. But, just like human males, they are extremely interested in other males, especially stallions although the geldings may be rather frightened of them. They develop a sort of men's club and often are more interested in each other than in their own mares socially, again a lot like human males! Stallions when they are the only male with mares (the "harem" stallion, which is the way they generally will arrange themselves) have friends among the mares, but they may be somewhat lonely, not being able to associate closely with the group. Geldings also are usually not permitted by the mares to be too chummy, particularly if the mares have foals.

If a human can fill this gap in the males' lives, then they are often delighted to develop a profound intimate relationship. If you cultivate this, observe the mutually developed social contract and remain their loyal friend, there is nothing they will not do for you, so your mutual loyalty and admiration grows. Mares, like some women, are sometimes emotionally too involved with their offspring to really want to have such profound relationships. But if they do not have infants or other mares to fraternise with, they too are often pleased to develop close familial relations with a human or two.

The first priority is to put in place the rules, or "ethics", of your mutual contract: what can and cannot be done to each. This can vary, it is up to the individuals to decide. Once these limits are understood, whether by a horse, dog, elephant or any other mammal, each individual knows where he stands and knows his expected code of conduct in that society, whether it is multi-species or just with his own species. Even if the person is brusque and hard, the relationship will be much firmer than with someone, however kind and sweet, whose behaviour is not consistent and therefore does not understand that a social contract must be built. When a horse or other large mammal knows where he stands, he appears to give a sigh of relief, even if life is tough. This is why there can be so many routes to establishing a good relationship with another person. The only golden rule is to establish the basic social contract and then uphold it, whatever it is. Some people like their horses to rear around, remain jumpy and difficult, some like them not to be like that, it is up to the person to put in place a social contract and the horse will go along with it as long as he benefits in some way, physically or psychologically.

During the seventies, the fad was to raise your children in a "non-frustration" environment, in other words they could generally do whatever they liked so the parents were in a constant state of flux not knowing what they should do and how they should behave. The children grew up muddled and self-centred, and parents worried; it was no fun for anyone, much like some humans raising or living with their dogs, cats or horses. We learnt this as a result of giving lessons with our horses to our friends and their children. Once a year we ran a week's camping where everyone (including parents) was expected to ride, swim, run, eat, wash up, clean and look after a horse. If the camp was going to be fun for all, and we were going to do all these things, then it had to be run like a military camp: up at 7 a.m., clean your horse and self, be inspected, eat breakfast, do the washing up, rush out to get your horse ready to ride in five minutes and so on all day. Much to my amazement, all the children who came from very "alternative", "non-frustration" backgrounds loved it, and as a result, we were pressurised to run them year after year. The children, like the horses, responded well and with pleasure to having and obeying the rules of the society, even though they had little experience of having them previously. It is a general mammalian characteristic and pointed out to me how relevant much of the educational and developmental research on human infants and children was for the education of non-human mammals.

It was soon June, the loveliest month of the British year. We visited the village fair at Firle Estate, a country house. It was warm and sunny. The informal Capability Brown landscaped parkland and gardens were ablaze with early summer blossom, all the trees had new lush pleated multi-green leaves. The slightly misty South Downs towered on the horizon and the cattle were a-grazing and dozing in the fields ... all was well with the world. There is nowhere like the British countryside in June ... it sums up the living world's tranquillity and peace. It places one's own problems and hiccups in perspective ... life and beauty goes on, with or without "me". A sobering and salutary lesson that is learnt by all who are intimately connected with the earth, the living world and the sea. Perhaps because of a current lack of understanding of the primary importance of the living world rather than the individual in the grand order of things, there is an escalating growth in unhappiness and mental problems among the rich, consumer-orientated Western human societies. Long ago, I realised "I" was an often miss-shaped stitch in the living world's fabric and nothing more.

Although the farm and stud were developing slowly, it was

unlikely we would be able to live on its income. The publication of the book on the behavioural problems of farm animals and horses meant that I was approached by the RSPCA. The RSPCA had money to pay a post-doctoral fellow because their members had become aware that the conditions in which calves were kept were suspect. We negotiated, and eventually agreed that I would do a study on the behaviour of veal calves and see where and when they suffered, but also study the behaviour of calves living with their mothers outside in conditions as close to how they had evolved to live as possible. The scientific committee of the RSPCA agreed, and we set about buying half a dozen cows in order to study in depth their maternal behaviour and the behaviour of their calves. The next-door neighbour allowed me to study the behaviour of his herd of sixty black cows that ran with their calves on the Downs in order to draw up an "ethogram" for *Bos taurus*, domestic European cattle. An ethogram is a record of all their behaviours when they are left to their own devices in their own social groups. At that time, no other scientist had done this. A farmer in Kent allowed me and my assistant to study his veal calves who lived in crates.

The boys had left primary school and were at a school in the nearby town, so we had to have a collection and delivery rota. Then there were parents' days, cricket matches and choir concerts that must be attended as well as quality time to be spent with the children in the evenings when they got home. Time was short as we had around five horses, six cows and calves, hens, some very fierce geese, Sneefus and a Labrador puppy, Simba, and the vegetables and fruit to manage. I am sure I did not spend enough time with my children, but the advantages, from their point of view, were that they learnt to live with many other volunteers and students who came to help in one way or another and were all treated as members of the family. Like most mothers, I had learnt to multi-task and also had plenty of energy.

Two days a week we went to Kent to camp with the veal calves in their stinking sheds where they all lived in individual crates. They could not turn around, even scratch all parts of their bodies, or move a step. They were fed excessive amounts of milk substitute low in iron so their meat would be white and they would die of anaemia if kept beyond the twenty-four weeks of their intended life. They were given no fibre in their diet so could not even ruminate. The result was that they licked their own hair where they could, and ruminated on this. When they were killed, hair balls were found in their rumen and often ulcers in their stomachs. On release from their crates to be transported to the abattoir, they could not walk ... it was as if they had just been born and were having to learn to balance and which foot to raise next as they wobbled towards the transporter. To be thorough, we had to do twenty-hour-hour watches. It was cold, wet, muddy, stinking and depressing; generally horrible, watching these suffering youngsters all through the day and night. The owner was nice enough and lived in a lovely old house with a beautiful garden and pond. How could he justify his money-making when it was resulting in such suffering? But, as I have subsequently learnt, most people justify doing awful things with the mantra, "they have to make a living". It is then universally agreed that it is okay, and quickly becomes a "belief of convenience" ... "the animals do not really have such a bad life, they are well fed and their illnesses looked after. What more could they want, and after all it gives rise to jobs, so helps humans, so it must be good". This was the most unpleasant research project I have ever done, but probably one of the most important. If anything was going to change, someone had to show that these calves have a terrible suffering life and that it is quite unnecessary.

One eight-hour day a week was spent with the fifty black cows on the South Downs, whatever the weather. Sometimes it was blizzards and blustering winds, fine for walks in the fresh air, but not so fine for sitting recording cow behaviour for many hours. First, all fifty black cows had to be individually identified. The African experience of identifying individual antelope from natural markings had taught me to observe well; what with the differences in their udders and teats, ear splits and tail differences, it was not as difficult as it might have been to tell the individuals apart.

Today, this is easy as each cow has large, usually yellow, plastic tabs with numbers on them in each ear, but not then. The observations were of how each cow behaved towards her calf, which varies, just like it does in human mothers, some spend more time licking and touching their calves, others less, some called a lot to their calves, some did not, some calves wandered further from their mothers, some did not and so on. Also as they matured, like in humans and other non-humans, the relationship with mother changed. It was the age of the calf and the experience of the cow, how long they ate, where they went, when they slept, how they arranged their social life and so on, which determined their differences. The numbers were eventually crunched and what turned out to be five years' work was published in a monograph.[1] It was a very interesting study, and I learnt how to "think cow", as well as how to stay warm, alive and making notes in a blizzard for eight-hour periods on the South Downs, and a friend Susan de la Plain helped me edit it so it was presentable for publication.

Our own cows' calving and maternal behaviour was carefully studied, and I needed to be there at the birth. Nine times out of ten of course, having been up all night watching a cow, a quick dash for a cup of coffee resulted in the calf popping out, but in the end, I had a sample of around twenty births and details of how long mothers licked the calf, how long it took for the calf to stand, how often the calf fell over, how long to find the teats and suckle, to walk and how much they smelt or vocalised to each other. In fact, everything I could see that the calf and the mum did for the first three hours after birth, and how long they did it. A calf unlike a baby is precocial (born at an advanced stage of development), and within three hours has to come to grips with being out of the womb, stand up, walk, find the teats and suckle. The calf has innate tendencies to do these things, but still has to learn to do them and every calf, like every human, learns to do them in slightly different ways. Some fell over more than twenty times in their efforts to stand, others stood within the first hour and only fell over a couple of times, then some could balance well, others were

---

[1] MKW & Susan de la Plain, *Behaviour of beef suckler cattle*, 1983, p19

very wobbly. Mum usually licked them with her very rough tongue, which could cause them to topple over yet again, but in the process the calf was learning who mother was and she learning who her calf was. Then they had to learn how to organise walking with four feet. Most difficult of all, the calf had to find mum and find the teat. Once the calf found the teat and started to draw some colostrum out (the first very thick milk), efforts would be reinforced, and so the calf would quickly learn where the milk came from, the teat, and find it again. If mother kicked the calf when searching, the calf might give up searching near her and start suckling on a piece of string or a projecting piece of wood. It was a rapid learning curve, the calf having about three hours to complete the learning, or they would not be able to obtain the immunoglobulin (helps with immunity) which after three hours is too big a molecule to be absorbed by the stomach: the learning race was on.

All our calves were healthy and managed it, even though sometimes one was biting one's lip and wanting desperately to help. Thereafter, it was daily half-hour watches to record mum and calf. All the data then had to be analysed by hand. There is a lot of folk knowledge about the early behavioural development of the calf and maternal behaviour of the cow and a recognition of their individual differences. Recording it all in detail meant many sleepless nights, but could confirm the folk knowledge, and separate it from "folk belief": things that you are told and believe. A week after birth, the perfectly formed, sleek little calf would be leaping and jumping around, playing tag and top of the castle with their peers.

If, one day, the raising of calves in veal units was banned, which we hoped for, what would happen to the thousands of calves born in dairy herds where the cow must have a calf every year to produce milk? At that time there was a good market for them as they were bought to go into a veal unit. Calves were fetching as much as £150 and dairy farmers could not afford to lose that. Perhaps this market could continue if the calves could be fostered or adopted onto beef cows grazing out and about. There were many herds of beef cattle raising their own calves out at grass on the downs and moors, could they not raise two calves instead of one? The keeping of single suckler herds had to be subsidised as they did not pay for themselves, so would it be possible to get a suckler cow (a cow that normally suckles her own calf which is raised eventually for beef) to adopt a second calf so she would become more economic, and an extra calf from the dairy industry (which otherwise would be slaughtered or go into a veal unit) would be mothered and raised with the cow's natural calf? The reason why no one had done this before was because of the mother's behaviour: beef cows did not want to have a second calf to look after and suckle, they only want their own and they made this very clear by butting and kicking anyone else's calf who came too close. However, if we knew more about maternal behaviour and how cows recognised their calves, it might be possible to teach them to tolerate, foster and perhaps even adopt another young calf to raise as if they had had twins. We had a small herd of beef cows and we had learnt a lot about their maternal behaviour, so we could try.

First, we tested whether they recognised their calf by sight, smell, vocalisations, taste or touch and discovered, hardly surprisingly, that they used all their sensory modalities and recognised their own calf as theirs within the first three hours of birth. The calf, however, was often less discriminating, just as long as they could get a suckle. The calf would make an attempt to suckle from any cow, until he had learnt (by being booted by the wrong one) who was mother or who would let them suckle. The idea was to see if, shortly after birth, when she was learning to recognise her own calf, we could confuse the cow to believing that she had two; after all some cows do have twins and manage to mother both of them. We had to buy-in the calves for adoption so they were usually at least one week old, and consequently, behaved slightly differently from a new born calf, so the cow could easily tell by their behaviour which was hers and which was the bought-in calf. The introduced calves were usually larger and black and white; our cows and their calves were usually brown, so there were obvious other visual differences. There were also scent differences. To start with the new-born calf smells of the amniotic fluid of the mother. We found that if we blindfolded the cow for the first three hours after birth so she could not see the differences between the two, covered the introduced calf in her amniotic fluid to confuse the scent, washed the natural calf and then covered both of them with eau de cologne to ensure they smelt strongly the same, she was less likely to take against the introduced calf. But the behaviour of the older calf remained a problem, so we finally realised that if the calf had been fed recently, the calf would be much more interested in sleeping quietly for the first few hours with the new mother and brother or sister. This meant the calf did not rush up to try to suckle when introduced, so was less likely to be booted by her for not behaving like a new-born calf. Once both calves had suckled from the cow, things were easier although the cows varied in whether they continued to try to boot the adoptee away or not, and the calves varied if they found out how to suckle without being booted. For example, if the introducee learnt quickly to suckle between the back legs, or with her own calf but in such a position that when she turned her head, she smelt her own calf not the introducee, then she would stand much better. The adoptees also varied in how persistent they were, some gave up quickly and even when hungry did not try hard, while others would go on and on trying until they found a way.

The cows also varied, for example heifer cows (those having their first calf) were much easier to teach to let the introducee suckle because they had had no experience of motherhood before. Cows who had single-suckled several calves, were much more difficult to get to accept the presence of a second calf and let them suckle, as a rule, which showed us how important learning is in motherhood.

The aim was that after about a week, the cows with their two calves would join the other cows and calves outside on the Downs and the cow would look after both without our overseeing any suckling. At the end of nine months, we would then bring in two calves per cow. The calves had experienced a much better life than in the veal units, the cow had been a more or less willing partner, and we had more or less doubled our income. The first year of the research was difficult, but, with a lot of supervision during the first week, all the cows learnt they had to at least tolerate the introducee, and the introducees learnt how to behave with mother. After a few years though, it was as if they eventually learnt the "expected behaviour" in our herd, and most of them took to it easily, every year. But the interesting thing was that if the heifer calves who eventually became mothers, had been double-suckled themselves, they would adopt a second calf easily. In other words, it was their lifetime experiences that largely controlled this aspect of their maternal behaviour; it was part of what they had learnt "at their mother's knee" so to speak. After several generations, and about 200 introductions, we did not have to use blindfolds or eau de cologne any more for most cows, they took to it and it became part of what, eventually, became the "Double Suckling Culture" of our herd.

This research was written up in scientific papers but also in the popular farming press, *Farmer's Weekly*, and we had hopes that others would follow our lead and start double-suckling their beef cows. No such luck, it never took off, probably because it required a lot of time and skill during the first generation of introductions of the second calf. But, for us it was a breakthrough and allowed us to survive economically on our various farms for the next forty years, and it was this that has largely financed our subsequent research, including on elephants. Once we had the second generation of double-suckling cows, we never had a failure and had less and less to do at the introduction of the second calf.

I worked out all the statistics and wrote up papers and also a book on the results of the maternal studies, and of the calves in crates.[1] A year after the publication of the veal calf research, Compassion in World Farming decided to take the Catholic Church to court for cruelty because they kept their veal calves in crates. The calves cannot walk, gallop around, play, sleep, be

---

[1] MKW & S de la Plain, *Behaviour of beef suckler cattle*, p195; MKW, "The behaviour of confined calves raised for veal ..." 1983, pp198-213

mothered, suckle, eat, explore or socialise as normal youngsters. In addition, they are purposely raised anaemic and fed no fibre so cannot ruminate. The state veterinarian maintained that the church's veal calves were in "good physical health", and therefore not suffering. In other words, the state veterinarians believed that these calves did not have a mind which could make decisions and choices, or emotions where they could be happy or distressed. This was in the late 1970s. Top barristers were hired to prosecute, and experts called, but nothing dented the "beliefs of convenience of the chief veterinary officers"; even though all vets swear to "relieve suffering in animals", and the magistrates went along with the vets' decisions. Ten years later, consumer pressure was brought to bear and raising veal calves in crates was finally banned in the UK and Switzerland.

The cows we owned were not very special. They were a mix of beef cows we had picked up cheap because we had no grant to buy them, so we thought after a few years that, perhaps, if we reorganised a little, we could turn our small herd of beef cattle into a highly productive, economically viable, organic double-suckling herd. We decided to specialise in the biggest beef breed, South Devons, which would have sufficient milk to feed two calves. With care, I had managed to put aside enough money to buy two heifers and a young bull, so we visited pedigree breeders in Devon. South Devons are very big, gentle, brown, curly coated, originally dual-purpose cattle with plenty of milk. At that time there were still a few herds of South Devons being milked, too. Our three pedigree South Devons joined the clan on our small farm, and over the next twenty years, they bred enough pedigree heifers for us to have a full-pedigree herd of ten double-suckling South Devons, and because of their double-suckling, we managed to live, and develop three farms and several research programmes, on the money they brought in.

The organisation WWOOF, "willing workers on organic farms", had been started by a woman I met in Portugal on an international work camp. Her idea was that many young people would like to get out of town at the weekend, and find out where their food came from, help on organic farms, socialise and have "alternative" musical or meditation weekends. Our farm became one of the first hosts. We also had several young women who wanted to learn more about horses and riding who came for six-month periods as working pupils. There was further a demand from youngsters who wanted to become self-sufficient, retire from the cities and the rat race to the country to learn how to farm on a small holding. The flower power era had idealised an organic small holding life, usually in some lost valley in Wales, where John Seymour[1] the head of the small holding movement, was farming. As it was, because of their lack of practical skills, most of the small holdings and communes were failing, largely because they could not grow enough food to live on. The Department of Agriculture had labelled organic and/or ecological farming as "muck and magic" and it was having a bad press. Therefore, it was necessary for someone to develop and run theoretical and practical courses for people who wanted to run and live on organic small holdings successfully, as no agricultural colleges or universities were doing this at the time. So we started a course which evolved and developed into a diploma over the next forty years. Now agricultural colleges and universities have taken over, but we still have many wanting to do our course which particularly teaches people how to do it all. They could keep going financially, and still sometimes have time to stand and stare.

Finding out more about ecological agriculture, animal husbandry and ethics in organic systems was never far from my mind, so in 1976 I applied for and won a Churchill Memorial Travel Fellowship to visit organic farms throughout Europe. This was generously sponsored and although organic agriculture was just beginning in Europe it was interesting and instructive. Most of the farms I visited were bio dynamic, followers of Steiner. My sons, a student and I travelled around in an ancient motor car, visiting and chatting to farmers in France, Holland, Denmark and Germany. I learnt a lot and also found that animal welfare was not a priority on organic farms. Finally, after writing reports, there was a presenta-

---

[1] John Seymour, *The complete book of self-sufficiency*, 1997, Corgi

tion of medals to all the Churchill Memorial Fellows of the year. The Queen Mother presented them, smiling sweetly at each of us and passing a remark, she was a real professional and made everyone feel honoured. My mum came to wish me well which I was proud of; she also had just won an award for the best dairy farmer in East Sussex. The Churchill Memorial Fellowship confirmed me in my interest in changing the face of agriculture, away from money making and increasing production at any environmental, welfare or economic cost, towards trying to live and farm in some sort of symbiosis with the living world, while being efficient and being able to feed many people.[1]

It also put me in touch with the Soil Association and like-minded people who were challenging the current developments in agriculture. But all the answers to where, when, and particularly how agriculture should develop, had not been very carefully thought through. There was a lot of thinking to do to develop the tenets or principles of this new agriculture and, also, it would need experiments to test whether these were practical.[2] Trying things out at Milton Court farm by myself was hard, and I also had two small children. Various volunteers and students turned up, but I needed someone who would be good at doing things that I was bad at, would stay for a while and not require a large salary as there was no money. So, I placed an advert in the *Farmer's Weekly*, and a very long, thin, bespectacled, bearded fellow turned up in a bright African shirt. He danced around on his very long skinny legs and did not at all look the part of a good solid farmer, more like a well-meaning but drifting hippie. At the time Dutch elm disease was killing many of our elm trees, and we had by law to cut the dead and diseased ones down and burn them. This meant a chain saw. I had come by a second-hand one but had not had the time to learn how to use it. The long thin fellow picked up the saw the next morning, fixed it to go, and danced off to cut up the trees ... I was impressed, he cut them down, stacked and sorted them ... even

---

[1] MKW, *The behavioural problems of farm animals*, 1977
[2] MKW & CC Rendle, "Ecological agriculture ..." 1986, pp101-133; MKW, "Cattle husbandry in organic agriculture," Colloquiu,. Kassel, 1986

better, it turned out that he had spent his youth mucking about fixing broken-down cars and had even studied mechanics, or at least aeronautical engineering. He was interested in, and able with, machines, where I lacked both skills. He had also been teaching maths at a bush school in Ghana as a VSO so had some African background. On returning to the UK, he decided that farming rather than teaching, was more up his street so had answered the advert on the off chance. I offered him the job for at least a few months, as a volunteer, to see if he liked it and if it would work out for both of us. Well, it did, and forty-five years down the line, we are still developing and running farms together, we have a son, a couple of grandchildren, and we even got married three years ago! Best of all, I think I have convinced him to enjoy and admire the company of our four-legged family as much as I do, which has allowed all of us to expand our knowledge and understanding together.

During this period, it had been confirmed for me that both horses and cattle were great teachers of how to live. Cattle taught you how to adapt to even extreme environments, to be optimistic and make the most of the moment, even though it might not be the most pleasant. In other words, how to accept the world without being gnawed at by ambition, social status, more money, more goods and so on and how to try to live symbiotically with the natural world. When you can more or less do this, it is astonishing how much that world brings to your understanding and mind-boggling awareness of all its interconnections. Horses teach you how to destroy or build another's confidence, to observe and to register what you have seen rather than just chattering ... among many other things.

Above all, the interdependence of each on every other was increasingly apparent. As the Chinese sage said: "the beating of the wings of a butterfly in Peking will affect the temperature of the water in your bath in Sussex" ... or something to that effect. I had learnt that even if many humans were hypocritical and irresponsible, among them there always were some who were not. Perhaps there was not as much difference between humans and non-human mammals as people make out; they all love and hate, learn, experience things, pretend and think, and have their own individual personalities. Some are to be admired, some loved, and for all of us, there are some one does not like much. Perhaps also a careful look at different species' knowledge of the world and the way they live in it, would help all of us live better.

## Ecological agriculture, horse, cattle & human welfare & behaviour.
## Can a multi-species community work?

For the next six years, Chris, as the long thin dancing fellow was called, and I developed the definition and principles of ecological agriculture, and then attempted to put them into practice on our seventeen acres in the Cuckmere valley.[1] We had many WWOOF-ers and students who came and went, helped or hindered, amused or bored us. My boys grew up, went to secondary school, and I had another son with Chris. "Nitty Norah the Hair Explorer", as the local midwife was called (she also conducted anti-nit ceremonies at the local primary school), bustled in to arrange it all. This meant clambering up a wooden ladder to a room we had made in a roof space, and demanding hot water and plenty of towels. She took it all in her stride, as delivering gypsies in caravans on the moor, and farmers in remote parts of the Sussex clay-stuck Weald, had been her lifetime work. After the second bottle of gas had been heaved up the ladder, the baby still had not appeared so she finally rang the ambulance and Jake, our son, was born in it as it wound its way towards Eastbourne hospital. He was a strapping boy who had no option but to fit into our lifestyle and was a cheerful soul all his infancy.

By that time, we had bred some beautiful South Devon cows in our double-suckler herd, which was expanding to ten cows, their

---

[1] MKW, *Ecological agriculture. Food first farming*, 1992, London

calves and a bull. We had also about twenty home-bred Suffolk sheep most of whom had twins each year; we had built barns, planted more trees, and created a small nature reserve.

Increasing the domestic species' diversity involved having pigs, geese, turkeys and ducks. When we became attached to the pigs emotionally, they were difficult to kill or eat. I had taught piglets to be led and live with us while I was studying their grunts and squeals, and they had proved quick learners and delightful persons to have around, but they grew very large and ate a great deal of the same sort of food as we did; so they cost a lot to keep, but, other than being killed or joining the circus, they cannot do much to pay for themselves; there is no wool or milk to speak of and they are not much fun to ride and no good for transport. But, they do produce a lot of piglets and plenty of meat.

While worrying about whether it was ethically acceptable to keep animals whom we became fond of whose only use was as meat producers (which meant sacrificing their lives), we finally decided on a compromise. We would keep the sows for their lifetimes so they would become friends and family. Their babies, however, would go for pork or bacon when they were the appropriate weights. We would not become familiar with them or them with us. Even if their lives were shorter than their mothers', our obligation was to give them lives of quality. They could spend the winter, spring and summer digging with their noses in the walled vegetable garden, eating up all the weeds, and sleeping at night in the original flint stye. During the day they would also have access to a field and the woods, finding things to do and to eat, including acorns and beech nuts relished in the autumn. If they dug up our grass here and there with their noses, as is their wont, well that just had to be borne, or we must keep them somewhere where this did not matter.

The geese could do the lawn-mowing by eating the grass very short, and because they are mainly grass eaters they rarely ate or even visited the vegetable garden. The turkeys became suicide specialists. Baby turkeys, it seems, are devoted to suicide and nearly always manage to find some piece of string or wire to hang or strangle themselves with, but a few failed and survived to gobble their way around the yard. One Christmas we raised a batch but after having to kill them and pluck them, we decided that as a commercial earner, it was not worth it, even though they had lives of quality. The ducks were free agents since all they do in the garden is stamp on the small plants and eat the slugs and snails which we can do without. They had the river and ponds, and made good use of the lawns to shit on when they flew home for breakfast, which they never missed ... unless a fox had taken a foolish one. It was up to them to choose sensible places to nest and raise their young, and us to find where they were laying in order to steal some of their eggs.

One thing the geese were good at was being guards, squawking and chasing strange people and even biting them. One day, they tried to bite and chase away the bank manager when he turned up to give us a loan. The geese and I rudely asked him what he was doing there. He replied that he had come to see "what he could do for us"! After many cups of tea and subordinate behaviour on my part, he forgave us and gave us far too big a loan, which could never be repaid from the farm income.

The horse stud was small, but we bought a couple of pure-bred fillies at the Arab Society Auction. One sensible grey filly was Crysthannah Royal, a well-bred, well-conformed, small Arab from a very well-known Crabbet stud. The other was Omeya, a very unsensible, thin, excited, light Egyptian-type Arab. She had been left in a field by herself after weaning and was as wild and flighty as they come. But when we saw her glorious head and big, staring, terrified eyes as she whisked past the stable door, and then how she floated along when she was in the auction ring, the die was cast. Chris said he would pay for some of her ... It was only afterwards when trying to pay that I remembered that he did not have any money ... somehow we managed to blag it, and took them home in a trailer. The next morning when I went into the stable, Omeya took one look at me, realised she was trapped, and kicked me hard. I was so angry I swore and roughly led her out into the field while continuing to angrily swear and curse. She was so frightened by my voice and its expression, that she was shaking by the time we arrived at the field, but forever after, she never ever made any motion to kick anyone and she would follow Chris or I to the end of the world.

What had happened, was that the basic moral code and rules of our association, "the social contract", had been established on that first day with that one incident. She never needed to be uncertain how to behave with us again, "don't hurt me and I will not hurt you". If either of us broke the rules. A horse learns what is right as a result of voluntary learning. S/he learns not to bite or kick me, and I learn not to hit her. We both learn the difference between right and wrong. Our code of conduct, or social contract, may vary from that of others but if we learn the rules, we both will know when we have done right or wrong, we are both moral agents.

Of course, a horse learns what is right, not to kick or bite me, as a result of voluntary learning just like I learn not to hit her. We both learn the difference between right and wrong, we are both moral agents, although our code of conduct or social contract may vary from that of others. If she learns the rules we have put in place, she will know when she has not obeyed them: when she has done wrong and consequently she must be a moral agent, a being who knows the difference between right and wrong.

Crystal, who had been handled well by the stud she came from, was not phased by the move, but, I think, at the end of the day, throughout her life, she cared less about us and working with us than Omeya. She was more interested in horse society than a multi-species one. Both these fillies learnt to be ridden, to drive and work on the land, to give lessons to children and adults, to gallop about with us when we both wanted to, do dressage tests, be shown in shows, jump, gymkhanas, flat and endurance races. They also produced a couple of foals each, whom we educated and sold at around three-years-old to selected people.

With our beef cows adopting or fostering second calves, we hardly needed any observational skills to supervise the cow, another "culture" had been put in place. Maybe, one day, now cows are becoming recognised as being individual, feeling "persons", someone will redo this research, and rediscover that it allows more calves to be raised by mothers, live as a herd and learn to be the good sociologists they need to be. Like us, they must be tolerant of bad moods in others, but also ensure that the social contract is upheld. For them this means not hurting youngsters, or bulls attacking the cows, however frustrated and annoyed they might be, as well as respecting all the other social norms.

For economic reasons also it is a good idea to work out a system where calves can be out and about drinking normal milk in a herd with others and be happy and healthy. Making their own choices when they want and being "cared for", that is, liked and desired. The mortality in mother-reared calves is very much less than when they are raised away from their mothers, and they need much less veterinary care, drugs, shelter, and disease control. There has been a lot of research on how to improve the nutrition of calves and suppress illnesses with drugs, but to date little on what is probably of even greater importance: the importance of having a mother. It still seems to be that the calf's combined physical and mental "being", is happier and healthier when mother-raised ... just like humans! The question must therefore be asked, why on earth would it NOT be just as important, given that we are all mammals, and one of the characteristics of mammals is to be raised and suckled by a mother?

One problem that some beef breeds are now facing is that pedigree animals are selected for enormous back muscles (the expensive cuts) and they can hardly walk. Some of them have to have their calves delivered always by caesarean section (e.g. the Belgian Blue). Another problem is that many highly pedigree animals with enormous back ends have very little milk. For many herds of pedigree Hereford cows, for example, the farmer has a Holstein or Friesian milk cow to feed the calves to ensure that they have enough milk and grow well.

Dual purpose cattle, such as South Devons and Welsh Blacks, usually even pedigree ones, produce enough milk for their own calf. The South Devons we had produced sufficient milk for their two calves: that is, five to six gallons per day as opposed to one to two gallons to feed one calf well. Therefore, should we go forward with genetic manipulations, or rather by understanding a little more about evolution, realize that mammals can adapt to a great variety of environmental demands.

At Milton Court after the first year or two, we grew the food for our animals and ourselves, vegetables from the garden, barley and hay in the fields. My mum came up trumps and one birthday gave me one of her Guernsey dairy cows: Ballo. She reared her calf and gave us around ten litres of glorious creamy milk every day and we made different cheeses, yoghurt, cream and clotted cream, butter, and drank the milk by the pint. I worked out that with various volunteers, WWOOFers, students, friends and the family, we saved around £56 a week just in dairy products. We produced our own animal feed, ate some of our own lambs and pork, made hams and sausages, so our food only cost hard work. There were always expenses of course, running a tractor, car, house, taxes and so on and some of this was paid by selling the weaned calves to others to raise outside. The horse liveries and teaching riding also raised some cash.

I took British Horse Society Teaching exams so was a qualified riding teacher. They were a cultural exercise in themselves. I was examined by proper country hunting gentlewomen in twin sets and pearls with posh accents and disapproving frowns. At one of the advanced instructors' exams, I passed all the necessary practical riding and theoretical part, but not the grassland management. At the time I was researching how to improve grassland management for farm animals and horses, but this lady was unable to consider that I might know quite a lot about grassland management for horses and had some different ideas. I had, after all, done the reading and was running experiments on our farm, but instead of discussing the pros and cons of various practices the British Horse Society insisted upon she failed me. It took me a long time to realise if you wanted to pass the exams, you had to say exactly what was expected, not have any idea about why or think about it … I saved up the necessary money (these exams are very expensive), learned what I was supposed to say, and finally passed!

Eventually we acquired a battered old horse trailer and started to take our horses to competitions: showing, dressage, driving and whatever else was near. This gave us goals to work towards and by then, however rude the other horsey people were, I could give as good as I got: no more intimidation.

Endurance riding competitions had started. These were less formal and more fun. It is your horse's fitness that is judged. At that time the rules were being developed, and everyone was very cautious about how far or fast they should go. Riding on a marked trail in an unknown territory looking at the landscape and the wildlife on your fit horse is a very attractive way of spending the weekend, in fact, in my case, not something I will ever give up if I can help it. But, to start with, everyone in the UK was worried that the horses would do too much and there were many horror stories of horses dying in competitions in France, where endurance riding was also starting. As a result, the majority of competitions in the UK were not competitive, no one wins, but every horse is graded according to how fit they are judged to be. To have a fit horse for these rides, it is necessary to spend many hours training in the countryside, preferably up and down hills, and measure carefully the same parameters that are used for humans to see if they are fit enough to do what they have done, and if they are getting fitter. These consist of the time the heart takes to recover to a resting rate, the number of respirations per minute, the temperature of the horse after the event, and whether they are dehydrated or not. They trot up in front of the vet who will judge if they are lame. If any of these measures are not what they ought to be, the horse will be considered "not fit to continue" and be eliminated. As a result, to have a fit horse, many hours must be spent riding, usually fairly fast, over any sort of terrain. This means that much more time is spent with and riding your horse than is usual, several hours per day. This is a great school in being really aware of your horse and what he is feeling both physically and mentally, it is also a good way to stay fit yourself, and an experience that becomes addictive for both horse and rider.

At this time, I was researching the first book on horse behaviour written by a rider and scientist. Eventually after several rejections, it was published by a publisher specialising in horse books: a Mr JA Allan. As a young man he had identified a hole in the publishing market and specialised in books about horses, even though he did not know the front from the back end of a horse. I visited him once in his Dickensian office in the city of London in a narrow street with towering old buildings and big wooden doors, up ancient creaking stairs, to arrive in an office lined with books to the ceiling and papers stacked all over the floor. Mr Allan slowly emerged from his den with cobwebbed windows, bent double, smoking a pipe with long ragged white hair. He crushed up his wrinkled face and asked me direct and very pertinent questions as I perched on the edge of an upright chair crowded with papers. The next time I saw him was ten years later when I was awarded a beautifully engraved wine glass because the book which (after much humming and hawing) he had published, had sold over 20,000 copies.[1] It turned out that people really did want to learn more about their horse's behaviour and mental abilities.

We had some surprising and amusing times on the farm with both animals and humans. I remember driving down the road with the battered horse trailer taking a steer to market when a VW Beetle car passed us, and the driver shouted out "you lost your bl... cow!" It materialised that the bullock had jumped out over the back ramp, landing on the main road. None the worse for wear after landing at around thirty kilometres per hour, he was enthusiastically grazing the verge ... but we still had to take him to market where he would probably be sold to a butcher; should we, could we? This incident re-emphasised the ethical problem that we constantly had to confront. If farm animals have feelings and minds and are in many respects like us humans, should we be raising them for meat, even if they have a life of quality? If one raises thousands, then you do not know the individual well enough to know his personality, but if you have small numbers, you do. Even if this is not a rational argument, it does make it hard to kill or sell the animals you know for meat. On the other hand, if

---

[1] MKW, *Behaviour of horses in relation to management & training*, 1987

we all become vegans and vegetarians, then no animals will be farmed. This will mean that humans will live in an even more anthropocentric world and many races and species would become extinct.

As we know, there are other things that some animals can provide without sacrificing their lives. They can be kept as companions and friends, or work for humans in various other ways, used for transport, milk production and for fibres. But even when kept for any of these reasons, there are always going to be too many, because they have to reproduce to continue to provide these services. It is possible to reduce their reproduction but if you want milk for example, the cow, goat or ewe must produce a young each year. What happens to all these calves, lambs and kids, fifty per cent of whom are male so will not be able to produce milk? Also, there is only limited space to keep all the females. There has to be some population policy.

While pondering this, a hippie turned up driving a couple of Friesian steers. He came from a vegetarian, alternative, meditation centre which was concerned with this problem. They had a dairy herd but no longer enough land to keep all the calves. He had decided to use two of the steers to pull his caravan and wander around the country. His flowing garments and hair, and frequent breaks for "wow man" talk and meditation, temporarily distracted us from his large black and white steers who looked at us dismissively. With reason, it turned out that they were hopping lame ... he had not considered that they would have to be shod to walk on tarmac roads every day, and had no idea how this could be done. But even if it had been done, not everyone wants to drive all the steers that are born around in caravans, say half a million a year in Britain! People don't even want to use them for ploughing anymore. It is a good idea that some are trained to do things of this sort, which allows some mutually beneficial contact, particularly for urban humans, but there are still going to be an awful lot of spare calves. What happens to them if everyone is vegetarian or vegan and no animals are to be raised for meat? How many can be kept in the ever-reducing "nature reserves" and how much contact will any people have with them if they are? A once a year on an "admire the animals excursion" in a theme park? If this is the case then the next generation of humans will never be able to know or learn from any other species as this really does require daily long- term contact and experiences.

The same problem arises whatever animal you keep. You can keep sheep for wool, but they also have lambs. Should they not be able to have lambs? Would castration of males and no sex and no parenting allowed for males and females be a better solution than killing some of the lambs? Consider it carefully: is no life better than some life, provided it is a life of quality?

A common answer to such concerns is to become vegan and not use or eat any animal products. But this causes more problems. In the first place, as we have said, there would then be very few animals around so humans will have even less to do with animals. There would be no introduction to Granny's farm, to help milk the cow or see the lambs, however brief an encounter for most city children. No one will have that unique experience of sitting under the belly of a ruminating, quietly meditating cow to milk her, or galloping flat out along the beach on a mane-whirling, coat-glistening horse, or sitting watching a young lion cub trying to walk and find his mother. Over and above missing out on these experiences and not even realising that they exist, when we know more about how others think and are in the world, it may be that they can show us solutions to some of our problems, but if extinct, these are gone for ever. My experiences with non-humans have changed my life and its aims because I have learnt about how different species may think, and the view of the world they may have has changed mine.

There also would be no domestic or wild predators who eat meat. Wild predators are on the way out today, but they would rapidly become extinct as there would be no place for them in the small nature reserves and farm parks that arise, because they need to eat and they eat meat, hunting and killing it themselves if allowed to. Should we then let them hunt these domestic animals in the farm park or nature reserve? Frequently being hunted and killed would surely cause more distress to the prey animals than being killed at an abattoir, as long as the abattoir is correctly regulated. The only other alternative if we wish to still have some of the predators around is to keep them in captive environments and feed them meat raised for them; here again there are many ecological and welfare problems that need careful investigation.[1]

Not farming animals for meat and consequently having fewer animals, or none, can give rise to environmental problems. There would be no grazing animals maintaining the grasslands to foster plant species' diversity. Even greater environmental problems would arise from producing enough vegetables of a range that vegans demand: they drink soya milk which is causing the Amazon forest to disappear (even if it is organically produced!). They use vegetable oils instead of animal fats and have organic palm oil plantations where there used to be orangutans. More and more land would have to be cultivated to serve ever-growing vegan vegetable needs, where the soil would have to be fed either by human waste, by organic fertilizers, or by using vast tracts of the earth for growing green manuring crops for the soil. The last European forests might disappear, the great prairies and savannahs which are maintained and support cattle, sheep and goats would be dug up.[2] But, the choice should not be between being "vegan" or eating any meat, however it is raised; because there is a third alternative. It is possible to have domestic and wild animals around who are by their presence maintaining the ecology and plant diversity of the area, who are not intensively farmed but out and about in the environment. Such animal management can be both environmentally and ethically sound, even though some animals must be harvested from time to time. It would mean that all humans eat less meat, and have less dairy products, but we could still all have some. Perhaps reducing the consumption of meat to once a week rather than twice a day.[3]

---

[1] We Are All Mammals. Charity number 1184219. "We are all mammals – Welfare and conservation knowledge exchange"

[2] Allan Savory, *Holistic grassland management and The Savory Institute*, https://www.savory.global

There are many things that non-humans can provide not just to serve our physical needs, but to serve our soul, or mind, needs. One of those is that we need to live in multi-species communities where we have evolved. These communities have intrinsic value (value in themselves), not just because we use their products. Multi-species communities are essential for the continuation of the biosphere.[1] So, it would seem evident that what we must do is work towards all animals, whether domestic or wild, having a life of quality (not just one where they do not suffer). We should consume less meat and other animal products so that we can be assured that the animals have lived a life of quality and are not causing environmental problems. We could all become conditional vegetarians. Conditional vegetarianism states that unless you know that the animals have been raised well and lived lives of quality, not just not suffering, you do not buy, eat or use animal products. In some countries, animal welfare standards for organic certification are beginning to approach this position, but not all. Even when raising your own animals, they must be continually monitored to ensure they are not showing distress and that they have a life of quality which fulfils their physical, social, emotional and cognitive needs.

As species become extinct, more and greater environmental problems will be unavoidable. We need to make haste to change this, not by developing more technology, but by realising and acting on the fact that we depend on the living world for our survival and re-learning to live with it rather than change it. But, the next generation, whether meat-eaters, vegetarians or vegans, will all be almost completely alienated from the natural world, even though it controls their lives.

We have a choice in how we keep animals. They do not have to be raised or live in an intensive indoor environment, manipulated

---

[3] MKW, "Ethological, ecologically & ethically sound environments ..." 1989, pp232-247, MKW, "How to measure quality of life ..." 2019
[1] A Tsing, H Swanson, E Gan, N Bubandt (eds) *Arts of living on a damaged planet: Ghosts and monsters of the anthropocene*, 3rd ed, 2017, University of Minnesota

by technology. We can let them live their lives in extensive systems, thereby helping keep the species' diversity. They can be near us and have to do with us, and still have a life of quality, as "free" as in the wild.[1] In this way they can appreciate and be appreciated by humans in different ways: for what they do for us and the world, and for who they are. There will be fewer than there are now, and humans would have to eat less meat, but it could be done, and some of agriculture's escalating environmental problems could be controlled, the soil fed and the ecological holistic cycle continue indefinitely, although humans must radically control their population explosions ... This will not be the case unless we do.

Hobbes, a British philosopher of the sixteenth century, maintained that humans and other animals have a life that is "solitary, nasty, brutish and short". This is necessarily the case with modern developments of intensive livestock rearing (although not usually solitary, rather crowded instead). Today we know more about non-humans and their needs and desires, their emotions and lifestyles than Hobbes did, and we know that many wild and domestic animals as well as humans do not have to live nasty brutish and short lives. It is therefore wrong to mis-reconstruct sixteenth century beliefs and separate humans from animals, whether the reason is to make money, gain social status, or just to have fun killing them. We all recognise that wild, farmed, or captive animals and humans can have a life worth living and have fun, joy and excitement, but it may be short for any of us. Is a life that is short but of quality, better than no life, or a long life which is prolonged by a host of technological inventions? Does it matter if species exist?

These questions raised their ugly head when, as a result of the work I had done with veal calves, the RSPCA asked me to study foxes whom they were convinced were suffering, because of being hunted and poisoned. The man who came to interview me told me, "we, the RSPCA, do not care if foxes exist, we only care that if they do, they do not suffer". But, it matters to me very much whether

---

[1] MKW, "How to measure quality of life in equines ..." 2019

foxes continue to exist. Foxes must be congratulated for moving into cities and living off rubbish (or eating your pet rabbit if he is left outside because you forgot to put him in his hutch for the night), when they have almost nowhere else to live. I turned the job down. Another ethologist from Oxford accepted it and became a world famous wildlife figure.

The usual argument given for the survival of different species is that without diversity of species, we are likely to have an environmentally unstable world which will not benefit humans. Thus, species' diversity has what is called "instrumental value" to humans, that is the continual existence of a great diversity of species will allow humans to survive. This is biologically true as a general rule, but bear in mind that it includes all species: cockroaches and rats, bacteria and viruses, as well as lions, elephants and tigers. Although they may be controlled locally, it may not be a very good idea to try for example to wipe out even the virus for poliomyelitis completely ... it may cause more problems than we know.

The other argument for the continual existence of a diversity of species, and indeed even non-sentient things, is that each of them has "intrinsic value".[1] It is easier to argue for this when the species is another large mammal who is conscious, aware of being in the world and has similar emotions to me. However, very different species, such as pests or disease carriers, if this argument is used, have value in themselves. Even non-living things such as mountains, valleys, rivers and the sea can have intrinsic value as we are beginning to find out from the various environmental problems caused by changes to them. As an ecologist, I was educated to understand that mucking about and changing nature in any way will have consequences since everything is part of "the web of life", even the non-living stuff that is around. Since we do not understand all the interrelations in this great web of existence, a cautious approach would be best: "don't rock the boat". This has always been at the centre of my beliefs: best to respect the natural world, however much we think we know. We can tweak it here

---

[1] Arne Naess, (trans) David Rothenberg, *Ecology community and lifestyle*, 1989, Cambridge University Press

and there perhaps, but not cause large changes or we may find ourselves eliminated from the primordial soup. Now, we are learning that cause and effect is not quite as simple as it seemed when Newton dropped the apple or antibiotics were first invented. Bacteria and viruses can reinvent themselves to cope with our efforts to wipe them out ... humans and other mammals with longer generation times, are not so able to do this.

Quantum physics is pointing to the possibility, that when experimenting with minute particles queer things can happen that previously have not been considered reasonable, for example: the observer can affect the outcome of some experiments.[1] These discoveries point out how little we know about the world we live in, even though we have precipitous technological inventions. Different ways of thinking may indeed help us all in the end.

There are instrumental and intrinsic reasons why every species, and every river, sea, marsh, valley, mountain and plain should continue to exist. It may be that we can make important arguments for changing this or that or trying to wipe out some diseases, but we must not just bulldoze over the world without consideration ... if we wish to continue to exist.

It has become evident that many animals have a "good life" with humans and they also gain from this. This is not only because they are well fed and have veterinary care, but because they have new and different experiences with us and they can enjoy this as much as we do, if it is done well. Take for example your dog who may travel around with you in the car to different places, meet a whole bevy of different dogs, go for walks in new country and learn how to communicate and please. He likes you, enjoys living with you and the range of experiences you have together. My cow likes to come into the stable to be scratched and chatted to, she often chooses this, and she may be harnessed up to do some harrowing or pull a little cart, and provided she is not over worked or frightened, and the work is not monotonous, she willingly participates and apparently enjoys it. Just as we are deprived by having no

---

[1] John Gribbin, *In search of Schrodinger's cat: Quantum physics and reality*, 2011, Random House, p234

intimate contact with different animals, so they can be deprived having no intimate contact with us. Just one example, Shatish, a young horse of our community, became very interested in his reflection in the glass door of the *gite* (cottage), so I opened the door and invited him and his companion, an eighteen-month-old young stallion Larrikin, to look around inside and to walk in if they wished, by talking and showing them. They both looked around very interested, and then Shatish cautiously ducked his head, walked through the door and walked onto the slippery tiles, squeezed around the fridge, carefully bent his long body around the table and chairs and stood observing the wood stove. Larrikan, the junior, followed him and cautiously waited, looking around, smelling the carpet and the tiles. They stood there looking for about five minutes while I climbed on a chair to open the French windows so they could go out. They then walked quietly out, turned around and stared in again. This demonstrates curiosity and confidence in something quite different, and shows an interest in our human doings and livings. They were invited and encouraged by me, but they volunteered, they wanted to come in, I made no demands or even had any physical contact such as touching or leading them. The animal you are associating with can show interest in you too, and enjoy whatever you are doing as much as you do.

It is not only the killing of the animal but even the taking of milk, eggs or fibre from animals in ways where it causes prolonged suffering that matters. Even using them for human therapy where they suffer will not do.[1]

During this period, we had more and more WWOOFers and also belonged to a Quaker organisation invented to foster world peace where people could stay free for a couple of nights. Not everyone who stayed was so idealistic. We found that many WWOOFers and other visitors who had to be fed, warmed, given a room, laundry done for them and so on could be very expensive both in money

---

[1] Dors Li Destri Nicosia, *Equine and human mutual welfare: A whole subject? Critical aspects and possible strategies in equine-assisted activities and therapies*, 2011, PhD, University Bologna

and in our time and energy looking after them and growing enough food for them. Our energy was limited and the energy books must balance so we could continue to run the farm. Forty years on, we still take WWOOFers, and we have had some fun times and made friends all over the world, but we have also had some bad times; one bad apple can have a horrible effect on all. In commercial terms, it has to be said that the vast majority of WWOOFers, students and other volunteers, do not pay their costs with their work, although they may do this in other ways such as by entertaining us, or doing things that they know how to do such as cleaning a house, washing up, but they rarely pay for their costs in the work they do on the farm.

We had some amusing times: one very weak, white toothless young man rode up on his bicycle and declared he ate nothing but carrot juice as he wanted to see in the dark. He even had an electric carrot juicer on his bicycle. One evening when trying to phone in the dark lobby, he roared back into the sitting room deeply distressed, after six months of carrot juice he still could not see the telephone directory in the dark! Then there was Tony who arrived from Lancashire in a home-made smock with big boots and a straw hat. He was enamoured of the life of nineteenth century journeymen and had decided to become one. He had all the right clothes and attitudes of the journeymen of yore. He and Chris would pore for hours over boot catalogues from working-boot companies in the north, where they still made boots with steel toes. Neither of them ever bought any, they were much too expensive, but they dribbled over them with their eyes popping out. Tony came for a fortnight and stayed five years. He married one of our students on the farm, much to her parents' annoyance, and continued for many years after we left with his journeyman life.

Another serious fellow who of course wanted to "share", turned up when we had a roof leak three storeys up. The rainwater had accumulated on the ground floor and was washing around in the sitting room five centimetres deep. Chris and Tony had decided to lift all the carpets and hang them up to drip on lines in the sitting-room, so we had to duck under them gently steaming to huddle near the fire. This very serious alternative youth took one look and rushed away squealing ... thank goodness. Another time a family of orthodox Jews were camping nearby and the father appeared in his little town slippers and Jewish hat to buy some milk. He had to witness the cow being milked so that he knew it was kosher. Chris took him off at 6.30 a.m. each morning, staggering though the mud in his little slippers, to watch Ballo the Jersey cow being milked so he could buy a pint. After about six days of this, Chris took pity on his constantly wet feet and wrecked slippers and suggested that he did not need to come each day as it was the same every day. He replied that no, no he must come and witness it, because, what would happen if someone substituted the milk from pigs for the cow milk? Just imagine milking a pig, never mind drinking the milk!

We had Scout and Guide camps. The Scouts varied enormously in what they got up to and whether they were fun to have or were a terrible nuisance. Since they usually came from the east end of London, they were streetwise lads who had no experience of the country. But if they had an energetic leader with initiative, they did all sorts of exciting things, made ropes and slides to cross the river, went on night walks to see the badgers, played that terrifying game "murder" where one has to find out who the murderer was. But, sometimes "Brown Owl" or whatever he is called, looked on the camp as an excuse to get away from the wife and have an affair with his accompanying secretary and spent the whole week in his tent. The Scouts ran wild, and we had to organise them and try and keep them suitably employed and out of harm's way. After that we were very cautious; only after interviewing the Chief Scout would we allow them to camp. Once the Guides camped; the girls came up to the farm shop we had made in a tent which sold our farm produce and various herbs and remedies that local people had made. They bought a fist full of Valerian, a well-known sedative. After this, the Guides were no trouble at all, they snored away most of the week, although Brown Owl eventually came to ask us why they were asleep all the time! I think the girls might have thought it was pot and they were doing something "ever so naughty and daring", but it was just Valerian, a sedative.

Simba the golden Labrador bitch took over babysitting our son, Jake, who was by then crawling around, and doing his own thing in the walled garden, often falling asleep, guarded by Simba under cabbage or Brussels sprout plants. One day, Simba came rushing up to Chris, rather distressed and worried, and led him back to the barn we had recently built which was full of straw bales. He and Simba clambered about all over the bales looking for Jake, but after about an hour, no sign. I was away, but Chris decided to ring the police in case he had escaped and wandered down the road and someone had picked him up. Before they arrived however, Simba pointedly sniffed in a hole, and Chris scrambled down through the bales to find Jake peacefully asleep at the bottom. He had fallen to the bottom of the pile of about eight bales, and, being in a cosy warm place, rather than howling, he had fallen asleep.

The elder boys learnt to ride, race, swim and gallop around at our camps with other children. One day, an outing to a Pony Club meeting about ten miles away was planned for them and two girls we taught, to meet some other riding children and learn new things. The four children worked hard the day before washing, scrubbing and scraping to have themselves and their ponies looking their very best, and we were proud of their efforts. After a hack there, they arrived at the meeting to find that all the other children had had their ponies boxed to the meeting, and "grooms" who had cleaned them. The spotless ponies and neatly turned out children in regulation riding clothes looked ready for the show ring not a Pony Club meeting in the Downs. Our kids had been clean and tidy when they left but were hot and sweaty on arrival. The woman in charge took one look at the ponies then sniffed and snorted and told them that they could not take part. It turned out that the Pony Club was not for everyday farmers' kids. We went home and tried to do some interesting, exciting things with the ponies, but this experience put my twelve-year-old son right off riding and my ten-year-old in a dither. But, the two young girls seemed to have rather tougher skin, they smiled at the woman, and galloped home relieved, I think.

During this period, I was an honorary fellow in the biology school, University of Sussex (meaning that you have facilities like a desk but no pay, and any publication you write has the name of the university on it). I continued to teach a variety of subjects and research further social organisation and communication between horses. My students and I measured over sixty behaviours, who performed what to whom and what the recipients did. One of the interesting findings was that there was no "dominance hierarchies" in these pastured horses. It turns out that to have competitive hierarchies, horses have to be kept in unnatural conditions where they compete for resources. Our horses had different roles in society and were tolerant and cooperative with each other, generally observing the social contract of the society so that they could remain part of the group. Since the food (and many of the other resources) of large herbivores is distributed all around them. The food of large herbivores is not packaged and does not have to be competed for. Therefore, it would seem unnecessary that large herbivores need to have a "dominance order" which guarantees some individuals priority of access to food and other resources.

Our equines were much more keen on trying to stick together, by showing affection and interest in each than being aggressive and competing. Of course, some were more moody than others, and each had a different role, one was more friendly, another more aggressive, another very socially involved while another spent less time with the group or communicating, there were extroverts and introverts as well. The question then arose, why are they social? What do they gain from this? If they enjoy being with others rather than alone, there has to be an evolutionary reason.[1]

People have suggested that living together helps individuals be protected from predators, but then groups attract more predators than one alone. Perhaps the main reason for living in a group is to enable each member to acquire more information that might help

---

[1] MKW, "Cooperation & competition." 1997

survival:- information about the environment (ecological knowledge) but also about other members in the group (knowledge about each other) and learn by watching and imitating others, which is called "social learning". As a result, they do not have to learn so much by trial and error, which has risks attached. They can learn what to eat, where to find it, who to avoid, who to stay close to and trust, how to find the way around their area from watching, following and imitating. This means they acquire different knowledge in multiple environments, just like humans do and every individual does not have to "invent the wheel"; they can see what others do and learn from this.[1]

The farm prospered and we started having Open Days to show people that organic/ecological farming could work. Although the farm did not make a great profit, it allowed us and our animals to live a life of quality, rich in experiences and fun, if not material goods.[2]

One criticism was that since this farm was in a prosperous, rich part of the country with good markets, a good growing climate, good communications and plenty of money around, it was not a good test of whether ecological agriculture could help by establishing self-sustaining high-net-yielding farms which are run by women, to feed the population in Kenya, my home country. We decided to test this, either we must set up an experimental farm in Kenya, or somewhere where there were similar problems:- a difficult climate, no markets, bad communication, isolated areas, not much money and no state aid. After making enquiries and looking at possible sites we could afford in Africa, it was clear that taking our cattle and horses was going to cost us much more money than we could raise, but more important, there was a host of diseases and without having immunity from life-time exposure, they were likely to contract East Coast fever, rinderpest, sleeping sickness (trypanosomiasis), blackwater, tick fever and African

---

[1] MKW, *Right in front of your mind: Equine and elephantine epistemology*, 2000, MPhil University Lancaster; MKW, *Cooperation & competition*. 1997

[2] MKW & CC Rendle, "Ecological agriculture ..." 1986, pp101-133

horse sickness. We were not going to leave our family of cattle, horses, dogs or humans behind, if they wanted to come, so, Africa was not an option. Was there somewhere in Europe which would have difficult climate and soils, difficult communications and markets?

As luck would have it, just then I was also offered a non-paid honorary fellowship at the University of Edinburgh by a colleague who wanted to work with me on animal welfare problems. Perhaps Scotland would be a good place to develop the next farm. Was there anywhere there with bad communications, a difficult climate, no markets and not much money and that we could afford? One or other of us spent a great many nights going up and down to Scotland on the overnight bus, the cheapest form of transport. Buses have a different clientele than trains because they are usually a lot more uncomfortable and cheaper. One trip I remember well. It was 5 a.m. in the dark and I had asked the driver to drop me off where I would, I hoped, be picked up by the owner of the farm I was going to see on an isolated road in the Borders. The driver stopped and woke me up, I rushed off half asleep, and found myself at the top of a cold, black, bleak moor that stretched for ever. There was, however, a telephone box, but ... I had left my bag, money and all on the bus ... (it was before mobile phones) and I did not even have the ten pence for the phone. I saw the lights of the bus gently rolling over the hill. I sat down and gloomily stared at the mud, but, then, there was a creaking noise of a motor; low and behold, it was the bus coming back ... someone on the bus had found my bag and they had voted to turn around and bring it back to me! All my problems disappeared; since then I have been wedded to the kindly kindred feelings of bus travellers. We finally bought a farm on the Isle of Mull on the west coast of Scotland which more than satisfied all these conditions.

## Druimghigha, Isle of Mull

The wild, wet, windy winter was upon us almost immediately on our arrival on Mull. Horses, cattle, and a newly purchased flock of sheep were installed in non-paddocks, make-shift buildings and under trees against the worst of the rain. But, wisely as it turned out, the humans had a comfortable house with a Rayburn stove to keep the cold out. We had bought 450 acres and rented another 1,500 of moorland off a slightly eccentric older man who sometimes visited. He had had affairs with the wives of most of the local lairds, always had holes in his socks which he wore with gaiters and patched plus fours. He was a sort of "laird imposter" but one of the better things he had done was to employ an enormously tall and strong, soft-spoken local farm labourer Donald MacNellish to make fences over mountains and dales, forests and bogs at Druimghigha (translated as "place of the geese" from Gaelic). As a result, we would eventually when we had improved the soil, have some "in bye" land (cultivatable land which might grow crops). We also had miles and miles of wet moorland. The Laird imposter had said he would leave some vegetables in the garden for us to eat that winter, but when I finally fought my way through a tornado into the little low walled garden, there was nothing to eat. Luckily, we had brought some beetroot and carrots from Sussex, so stored them covered in straw and earth, for future needs, and as it happened, that was soon, since we accumulated WWOOFers and students fast.

A girl student who came north with us, fell in love with a local, drinking, non-DIY hippie who lived in a leaky house, the roof of which was never quite finished even when he died forty years later. The lass was soon pregnant and moved in with him. A lad from Glasgow turned up, very well washed and prim ... he was to study to be a Church of Scotland Minister, so the discussions took a religious turn, learning about Latin masses and protestant puritanism over breakfast, lunch and supper. An enormously fat woman arrived from the Caribbean. Her culture believed that fatness was beauty, and it was only after panting her way up and down the stairs, never mind trying to lift hay bales to feed the cattle, that she found fatness was not all it was trumped up to be. It was not long before she found an excuse to leave. This she managed by complaining to the Tourist Board that we were eating "roadkill". Someone had driven into a red deer on the road that ran through our farm, and we cut it up to eat; why should we leave it to rot? The Tourist Board came to "inspect" and sniffed their way around our house ... we decided to have nothing more to do with them and so ran our own tourist activities for the next eight years, and had more than enough visitors.

That first winter was a problem. Foolishly, I had managed to convince the bank manager by writing a blagging report of all our successes in Sussex, that we deserved a loan of £50,000, and would easily manage the payments. This was very far from the truth even though he may have believed it. The result was that we had to find £10,000 a year just to pay back the interest on the loan. I lay awake every night scheming how we could do this since our income in the first year was likely to be minus £5,000 at least. It was bleak and after all the years of sweat and tears, we could end up losing the farm to the bank. The first thing was to cut expenditure, that meant travel and food. Travel was easy, all we had to do was stay put. But volunteer workers had to be kept fed and we had not grown any food to feed them. So, somehow, we had to keep them motivated to help, as there was a lot of simple hard unskilled labour they could help with. Some "luxuries" had to occasionally be permitted, one of these would be using more wood than necessary to heat the kitchen, or have a very hot bath; it is surprising how when things are tough, such small things become sought after luxuries, and make everyone happy!

Food could not be purchased since there was no money, so we had to glean what we could from the farm. We had bought some barley for the cattle and horses to pull them through, as hay and straw were incredibly expensive. We hoped that the herbivores when they had got to know the farm a bit, like us, would manage to scrounge a living off the bogs and moors. But it was early days and they waited mooing by the building for us to turn up with their rations. On the dietary bright side, we had Ballo the house cow, she kept us in milk, cheese, butter and yoghurt, and the chickens gave us eggs. But we had no bread, flour or potatoes. The only thing we had to replace the carbohydrates was the barley we had bought for the animals. Barley is not usually eaten by humans as it is very prickly, tasteless, and needs the husks taking off and cooking a very long time. It was horrible, but by developing an innovative cuisine and scrounging the moors for flavours, we managed. We had barley with butter and curry powder, barley with creamed nettles and dandelion leaves, salads from dandelions and cow parsley where it grew in sheltered places, mushed up barley with yoghurt, barley with "road kill" red deer, barley for breakfast with rationed jam and cream, barley for lunch with cheese and scrounged salads, barley for supper with anything that could be found. We all lost weight since we only ate what we really needed, but we have never been so healthy either before or since! Not one case of flu, a cold, snuffles, or upset stomach among us although we worked twelve-hour days in all weathers. The priest-to-be had frost bite once because he took off his gloves to try and make the wheelbarrow work, and one of the girls fell off a horse and had bruises. Many came and went that winter in fair and foul weather, but everyone kept their spirits up, even though disasters struck often enough. One of the most tragic was when I went to open the stack of beetroot and carrots to have a treat for Christmas. Although the rounded shapes of the beetroot looked really attractive to the barley-ed workers … when we turned them over, there was nothing inside the skin … the rats and moles had got there first.

The pessimistic, gloomy predictions of the local Mullochs did not help either. When they saw our imported animals, they concluded they "had not a hope" of survival with the wet, long winters, winds and terrible diseases such as red water, black water and goodness knows what, since they had no immunity. They were very wrong, everyone pulled through that long, wet winter except Ballo's daughter who fell down a cliff, hit her head and broke a leg. The South Devons ducked their big brown curly heads into the wind, and produced their calves, and the Arab horses turned their tails to the storms, and produced a foal or two as well as carrying us about in the rain and hail, hunting for the sheep who had gone walk-about. By the spring, everyone had lost weight, but all of us, two and four-legged, were well-muscled and fit, and no one had starved.

We were visited by the Ministry's local Agricultural Advisor. I explained how we wished to farm ecologically, and asked him for suggestions on how this might be achievable on Mull. He scoffed and sneered, telling me that everything we wished to do: grow improved grass, grow wheat and oats, grow vegetables and fruit, keep, breed and give a life of quality to our Arab horses and South Devon cattle, as well as the black-faced sheep, was doomed to failure. On top of this, we had an overdraft of £50,000. He reduced me to tears. I vowed that never again would I ask advice from agricultural advisors; I never have, and although we have made many mistakes in our ecological farming, I doubt whether any of them could have helped ... but they certainly could have hindered with their deeply held beliefs that what we proposed was "impossible".

During this chapter of my life "thinking cow" was a saviour of my sanity, a truly life-enhancing lesson or wake-up call. Cows, bulls and their calves are remarkably tolerant of the most frightful conditions and restrictions they may have to live with. This is not because they are stupid and do not notice. The tests on learning to do new things that we ran later, found that of the five species we tested (filly, heifer, puppy, guanaco, and young elephant) the heifer was the quickest to learn and to respond correctly.[1] So why can cattle cope where other species find it very difficult? It seems to me that their epistemology, or world view, is one where "looking on the bright side of life" is a natural way to live. Even if they are up to their hocks in wet muck, in a miserable shed squashed and

---

[1] MKW & H Randle, "First steps in comparative animal educational psychology," 1996

unable to lie down with ease, they will be standing quietly cudding and looking at the sun shining outside. Almost as if they are saying, "yes things are not very good here, but look, the sun is shining and all is right with the world elsewhere". They are great optimists: positivists who make the most of the most difficult situation and do not care at all about ambition, competing or winning; they just want to live quietly in a cooperative group where everyone knows their roles. Efforts to explain their social organisation in terms of hierarchies and competition is a dogma handed down by competitive primates to each other, rather than the way in which they actually behave when looked at in any detail. So, when disappointed, ignored, frustrated or angry, the thing to do is to "think cow", you would be surprised how it helps. By this method, we overcame the pessimistic gloom of the agricultural advisor.

The sheep we had seen when we viewed the farm were a mangy lot although they were "hefted" (that is they know the area as home and would stay within it), so we decided not to buy them. After a few trips to the markets on the mainland, we bought what we believed to be a much better bunch, but they were not hefted. The result was that they made every effort to travel around on the island. They had travelled 200 miles or so across Scotland in a truck, but would have made their way home eventually, if they had managed to swim across to the mainland. Sometimes the neighbours would ring us and tell us they had some of our ewes, but many of them vanished, becoming the property of the less honest neighbours. We were down about 80 of 300 ewes by the spring gather, but it was our fault; we had not recognised the importance of "hefted" ewes.

The human social round was more difficult than we had predicted. At this time Scottish nationalism was growing and had a nasty side of intolerance of others. In an effort to ingratiate ourselves as common farmers, we gave a very special young working collie to one of our neighbours as he had trouble with his dogs because he was not able to teach them, and could not gather his sheep. It was a pup with a prize-winning trial sheepdog dad, and our very good working bitch, Fuzzipeg. We had bred him to raise and teach ourselves, knowing we would need another young sheepdog on Mull. I took him down to the Scottish nationalist's house, and presented him to the man and asked him to contact me if he had any problems, or needed help teaching him. A month later, he had shot our pup. We may be horrible English people, but what had that to do with the pup?

The nastiness did not stop there, in the following summer, we had been asked to give a demonstration with the horses at the local show which we were delighted to do. To make it relevant to the locals, we performed some Scottish dances with ridden and free horses accompanied by a local piper, and I think it went down well. But, when we came home, all our winter potatoes had been dug up. It turned out, the same Scottish nationalist had put up a notice on our potato crop for the tourists which said "help yourselves to potatoes" which they had duly done very effectively.

Although these incidents were few, they left a nasty taste, and had the effect of ensuring that we concentrated on the social life in our multi-species community on the farm rather than being too involved with the local human one. In retrospect, this was helpful since it was the start of the building and development of our multi-species community ... there is a silver lining to every cloud! One of the differences between the outlook of the human and non-human community was that the puritan idea of suffering while you live in the world in order to make it to eternal paradise is not a canine, equine, bovine or African one. Most human or non-human mammals enjoy their lives even with all its warts, in "the here and now". Even if Africans have terrible accidents and difficulties, the way through this is to tell wild, exaggerated stories of how it happened, and crease up laughing together with all the listeners. Your listeners do not sigh and grimace, nod in sympathy, or groan ... they just crease up laughing at the story, whether it is true or false, but this allows you to enjoy the life you lead, with all its considerable difficulties. We had never needed such a general mammalian and African attitude to life so much, but it really helped; it worked.

Gathering sheep off the hills was a new experience for Fuzzipeg,

Chris's sheepdog whom he had taught in the small fields of Sussex. She found that galloping for miles over bogs and mountains was quite another thing. She was good at cutting out sheep, turning small groups and getting them through gateways ... but finding them on the moors and getting them to run for a few kilometres in the right direction was exhausting and weird for her. We rode the horses to try and help, but neither us nor the horses knew the local botany and could not tell where the bogs were; falling into a bog with your horse is not pleasant, you will be okay because you can jump off and stagger out, but s/he is too heavy and cannot. Soon we all were terrified of going off the paths in case we fell into a bog ... and of course there were the usual terrible local stories of how the neighbours had lost their imported Suffolk punch (a large fat cart horse) or well-loved New Forest ponies in bogs ... "he just sunk right away".

Another discovery about domestic sheep showed how false preconceptions and widely held beliefs could be, and cost us dearly in time and energy. It is generally believed that domestic sheep differ from their wild ancestors in that, unlike their wild undomesticated cousins, they flock together when surprised. This means they can be easily moved in flocks, while their wild cousins scatter and hide. This "fact" turned out to be fiction. We had domestic sheep alright, but they knew very well how to scatter and hide under any odd bit of rock, heather or gorse where we could not find them or get them to move. The last thing they wanted was to be gathered and driven down to the farm in the nice orderly way shown on TV farming programmes! After six days of hollering at each other, falling into bogs, clambering up cliffs on our horses, and running and leaping about all over the 2,000 acres of moor, we had "gathered" at least 150 of them, and had to call it a day, as the rest of the ewes watched us from their multifarious hideouts. Domestic sheep if they do not know each other well, or are in a very strange landscape scatter, they do not gather!

At that time, the government directive was to eliminate scab from the national flock. Scab is a nasty itchy parasitic infection of the skin of sheep, and it was obligatory to dip the sheep in a mixture of "kill all", Roundup, chlorophosphates. In fact, it was even compulsory to inform the police when dipping. So a part-time Mulloch policeman (whose English we could hardly understand) turned up to eat sandwiches, drink tea, and shout at whoever was dunking the sheep that "that one had not been under long enough" (in Mulloch tongue). We flailed around catching and throwing the sheep in the dip while trying not to get it on ourselves. But, three of the geese fell in that night ... they were dead in the morning. When I asked the agricultural department what to do with the waste from this ghastly dip, all they said was, "oh, just let it drain away in the soil". It was several years later that people began to die because of the use of these poisons, and their use was tightened up. Perhaps allowing the sheep to evolve some natural control over scab, and selecting the ewes for this, would have been a better solution and certainly less destructive to the environment. Why is it that without understanding nature, ninety-nine per cent of humans think they can "improve" on it?

Eventually, the dogs, horses, we and the sheep learnt how to gather or be gathered. By the time we left Mull ten years later, we were ready to earn a crust by hiring ourselves out to gather other people's sheep ... The original six days' work has turned into three hours, with two good dogs, two athletic horses who knew the ground well, a CB radio each, and a hefted flock which means they know their way to the farm and stay in that group. Hiring us to gather their sheep would have been an untold benefit to many farmers who, by the time they had gathered one heft walking the moors on foot, sorted and dosed them, had to start to gather the next. By the time they had gathered all the hefts (some farmers had up to fifteen hefts), it was time to start gathering the first heft again to treat for worms or shear or do something else to them. As a result, many of the hill farmers who have hundreds of sheep spend the whole summer walking the hills gathering sheep. With our dogs, horses and CB radios, we could do what they did in two days, in half a day ... but, in retrospect, they probably would not have hired us, because that was their life, and they liked it.

One idea to pay the interest to the bank was to take bed and breakfast tourists during the summer months and feed them from our own produce and another was to take people for rides on the hills on our horses. While Chris struggled with the second-hand machinery we had been able to afford, travelling to auctions on the mainland and driving back over the wet and windy highlands with tractor (no heater) and loaded trailer in the sleet and snow, I attempted to ensure that the horses were going to pull their weight in this money-making lark by being tolerant and thoughtful of the non-riding dupes they would have to carry around. In fact, they became extremely tolerant partners, just as long as riders did not boss them about. But most riders have already learnt that "you must control your horse, you are the boss, he must obey", or "just press the right button and it (the horse) will do it" (the analogy with a motor car is frequent) ... our horses thought otherwise.

We had made a manège over a peat bog by using masses of local murram, gravel and bark chippings from the sawmill in the south of the island. This enclosed space where you can ride around and around is necessary if people are having a go at riding, and also if one wishes to advance in riding skills. The idea is that the horse goes around and around, and then when the rider has the hang of this at different paces, s/he can ask the horse to do simple movements such as turn around and go the other way, change the gait, do figures of eight and anything else you can invent. Our horses had been taught cooperatively, that is, we had always tried to get them to want to do whatever we asked, not have to force them. So, when a couple of elite English country landowners who "hunted with the Quorn" (a very upper crust sort after hunt) turned up on holiday to ride, we expected some trouble. One of them mounted and immediately started bossing Shiraz around by pulling her mouth and kicking her. I told them not to, and that really all they had to do was sit there and ask her by word of mouth. Of course, they did not believe this, so I warned them that it might not be a good idea to continue what they were doing ... they smiled condescendingly, after all they were the elite of the elite riders. Shiraz and Omeya decided enough was enough and took off around and around the school, bucking, rearing, stopping, turning, ignoring their pulling and kicking, lurching and screaming. They frightened the daylight out of these "elite" riders, although they did manage, just, to stay on. After about five minutes, I asked the mares (with a change in body posture) to come towards me in the middle and stop, which they did. The "elite" riders were reduced to jelly. All I said was, "I suggest that you listen to me and are not rude to our horses." They did for the rest of the day, and Shiraz and Omeya behaved immaculately, just as long as they did. The next day we had some small children who crawled around under the mares' legs and sat on top whooping and laughing, flapped about and tried to jump off, get on and all the other games that kids play with horses, and the mares stood rock still, and could not have been nicer to them. Horses are not fools; they recognise arrogance and there are moral codes of behaviour with youngsters, of their own or other species who have a special status in the social contract. This again, illustrates their recognition of rights and wrongs ... they really are moral agents.

We quickly realised that a lesson was always necessary before we would take anyone out on our horses, it did not matter if you were a member of the Olympic riding team or Princess Anne, you would still have to have a lesson. This has worked well for forty-five years, and six generations of horses. It has also been an eye opener on how the horses teach the humans, either the nervous quacking one who they stop for when they are about to fall off, or the one who thinks they know better, lose their temper, or otherwise behave uncivilly.

But after a while, one wonders if a daily lesson with new people in the school followed by a trek about for a few hours every day, was actually included in a "life of quality". For me it often was not, even though I was with the horses, so I wondered how they figured it. Shiraz soon told me. She, being the most stocky of our Arab cross horses had to take the large men. After a few weeks of this, a particularly large journalist turned up. He was pleasant and willing to learn, and I gave him Shiraz to ride. After one round of the manège, she lay down. Immediately thinking she must be ill, I got him off and examined her, took her heart and respiration rate and all the other physiological indicators. She seemed fine, so I tied her outside the manège while the large journalist watched the lesson. She was happily standing staring at the beautiful view, and swishing the odd fly, so I asked his daughter who weighed no more than fifty kilogrammes to get on her ... but she lay down again. This was a wake-up call. I realised that she had had enough and seemed to have invented a way of avoiding the daily drag ... just lie down, then people will get off and you will be spared. If she could go to this length to avoid having to do these rides, then she really must dislike it, and her quality of life was not up to scratch. We sent her to stay with Baksheesh so she would get pregnant and have a break. After having a foal, she never did this again, but we were much more careful how often and how much weight she and all the others carried.

Some years later, we had another example of an invention to avoid work. Crystal, a pure-bred Arab, had done a long-distance ride where she had been over-stretched and was very tired afterwards, due to my ignorance and bad judgement. After months off and re-training, we took her to an event where she was to do a much shorter ride. At these rides, the horses are presented to the veterinary judges before they leave and their heart rates and other physiological measures are taken to ensure they are fit to do the ride. Then they trot up to make sure they are sound. We had trotted her up just before the inspection and she was sound. She passed all the physiological tests well, but, when she trotted up in front of the vet, she was lame in front and was not allowed to start. One is allowed a second opinion, so I got on her and rode her around the site a bit. She did the particular long trot that she never did if she was ever even a little iffy about her feet or legs, and felt as sound as a bell. We also trotted her up in hand, as we have to do for the vet, and she was sound. Then we went back to the vet and asked him to look again ... she trotted up lame again! Presumably she recognised the procedures from her previous experience where she had been too tired at the end; she must have remembered the episode, this is called "episodic memory" and is supposed to be very rare in non-humans. But she had gone one further. Somehow, she had put two and two together: if she limped in front of the vet in a standard presentation, she would not have to do it, but limping in front of us was less important. Perhaps this was just a coincidence, who knows. We know that horses, like all mammals, learn, and make choices and decisions. We know that their past experiences greatly affect how they make those choices. We also know they can be innovative, like us. We need more examples, but such an explanation cannot be ignored.

Jake, our son, went to the small village school on Mull, together with twenty-five other local kids. There was one head mistress and a younger teaching assistant who taught the age range from four to eleven. Mrs Mac, the headmistress, was a strict Glaswegian who would not stand for any nonsense and was committed to teaching the "3 r's": reading, 'riting and 'rithmatic. Mrs Mac's husband was, among other things a poacher (called locally "the fox"). Johnnie Mac, was another fisherman and poacher, who also lived in the village (no relation, they are all Mac-s or Mc-s in Scotland). He often palled up with "the fox" and they would shoot red deer wherever they found them. The carcasses were then put into a large deep freeze that seemed to have escaped the attention of the police, and exported to Germany to be presumably served up with sauerkraut and potatoes in some Hamburg restaurant. Mrs Mac's teaching greatly benefitted Jake, when he went to a school in Devon, he was at least a year ahead of his age group when he started, and his respect for the police did not seem to have suffered. This began my thinking about how to improve the teaching and learning of our four-footed students. Even though they would not be learning the "3 r's", they could be learning the "3 c's": confidence, cooperation and capability!

The years went by and we managed to keep our heads above the water financially, just. If we had any spare money, it was spent on taking part in long distance competitions with our horses on the mainland of Scotland. It is a great way to see the country, appreciate the semi-wild world, plants, animals and landscape while spending quality time with your closest friends with whom you are riding, and meeting like-minded people. One of our home-bred ponies (Aisha Evans), even became the Scottish Long Distance Champion one year, with Shereen another fourth generation home-bred mare, a close runner-up. One year we arranged a weekend long distance competition and the people and their horses came from mainland Scotland and the north of England. It was a fun weekend, although one chap who had come about 350 miles, lost his horse who took off into the hills when he saw the llama Xanape (who had decided to take part in the trot up). We got the horse back but by then the man had freaked out at the boggy terrain, however, I think he enjoyed his weekend, eating, drinking and laughing with the rest of us.

We got to know some of the locals with horses, who were mostly incomers from England and had decided to move to Argyll for the "Good Life". The local farmers were friendly and cooperative lending machinery in exchange for us helping them when required, something that does not happen now farm machinery subsidies and health and safety rules are in place. The horses and I also had some regular riding pupils in the winter months who lived on the island, and the local pony club turned out to be an informal and rather haphazard affair which Jake enjoyed and I sometimes helped at.

One Christmas we had decided to have a mounted treasure hunt in the local laird's forests, we rode around fifteen kilometres in the pouring rain to meet the others in the forest. One young woman, Jill, was to lay the trail using one of the horses we had sold her. She came dressed as a tiger with waving tail and mask and was soon galloping flat out into the forests dropping sawdust for us to follow. It poured and poured, but no spirits were dampened, and we chaotically galloped and shouted, and laughed and twisted around in and out the trees, over the bogs and through the dense undergrowth. The treasure was a bottle of whisky. Towards the end, serious searches began and eventually Fuzzipeg found it. Of course, it had to be drunk in situ since we were all wet and the humans cold as well. The local laird had forbidden us entry to his house or barn (not a very Christmassy message!). We lay under the trees drinking, telling tales and singing until the bottle was finished. Then it was just a question of getting on the horses and riding the fifteen kilometres home in the rain and the dark. Getting on proved a problem but eventually we waved goodbye to the others and set off at a pace. Luckily the horses knew the way well, all we had to do (which in our inebriated state turned out to be much more difficult than imagined) was to stay on as they briskly carted us home. There was one gate that had to be opened by getting off. In my efforts to get on again, I managed to clamber right over the horse and landed in a very wet ditch the other side, but on arriving drunkenly, wet and cold at home we found some of the others, having seen the state of us when we left they had kindly decided to check that we arrived home. They put the horses to bed in their barn and ensured that they were warm and fed. Chris had disappeared indoors to bed, but I collapsed outside the barn in the rain. After an hour or so I woke up very cold indeed, and with a terrible headache, and staggered inside to rant on at Chris for leaving me outside; he did not take it too seriously though since his headache was as bad as mine ... it was a memorable Christmas celebration.

We had our usual range of volunteers and students who came and went. There was a German girl who could not do anything because she had "water on the kidneys". When we suggested she should go to the doctor, she refused and made herself another drink of soaked nettles. A rather pleasant, if ineffective, young Canadian lad came as a horse student and managed to write off our precious only motorcar, a Mini, on one of the narrow roads by driving on the right hand side of the road ... straight into another car which was waiting in a passing place. An American girl turned up with her parents searching for their long lost ancestors who came from Mull. She came back for a year as a student, to be with the horses, after deciding that playing music (she was a student at the Julliard school in New York) was not all it was trumped up to be. We had another American student who had been in the junior athletics Olympic team. She was twenty-one and had to do at least three hours' exercise per day to ensure she could walk and move around as her young body had been so damaged. Then Robin Miller came as a bed and breakfast guest from Western Australia. He came for a night and stayed six months, learning to ride, sort of, and videoed and videoed. He videoed some embarrassing incidents. One, I remember, was when lunging Xanape, the llama. I somehow managed to get the lunge line wound around my legs and toppled over, tied up. Xanape ran off, dragging me along. A few years later we visited West Australia and Robin, and asked to see some of his videos, unfortunately he had recorded his new passion, golf, over them. A couple of gay girls came for a year as students. They wanted eventually to have their own farm. They left to run a tea caravan at the local beach, but I do not think it was very successful. However, I do know that twenty years later, one of them had a small farm in France, and took in WWOOFers. One of her ex WWOOFers came to visit us, he mentioned that she was very, very fierce … perhaps she now realises why I was very fierce with them too! One year we put on our first full horse ballet for the public, we worked away teaching them to understand our rhythms and also to try and reflect in their behaviour the emotions of the music. The evening of the first performance was very beautiful, sitting above the manège with a glorious view over the sea to Rhum as the sun set … but the midges were just terrible. We ended up rushing around giving out a whole variety of anti-midge liquids and creams in order to try keep our audience, but the horses were equally affected, so the performance did not go quite as we had hoped!

One of the first non-vet scientists to become interested in farm animal behaviour and welfare, had invited me to join him in his department at Edinburgh University, and gave me a lab, a desk, and the necessary facilities to do further research. Together, we floated the idea of an MSc in animal welfare, and about ten years later this became a reality and is now one of the most popular degrees in animal welfare worldwide. But my sleeping arrangement in Edinburgh finally got to me. The only place the university would let me have my very small caravan (despite the fact I was voluntarily teaching their students and publishing my research with their department name without pay) was just outside the intensive beef unit at one of the research farms. It was like sleeping outside Belsen, all night long I heard the puffing and groaning of the cattle, accompanied by the rattling of their chains; they were all chained by the neck. It turned one's blood cold; enough was enough. Perhaps the intensive meat-eating public, and the members of the government animal welfare committee should be made to sleep in such a place as necessary experience before making judgements on animal welfare!

By this time, I had become aware that the animal welfare scientists were NOT raising the most important questions. Following Kant's argument, they mostly seemed to believe that being nice to animals was a duty because otherwise we would end up doing similar things to people. They also all believed (and many still do) that non-human mammals do not have many mental events taking place in their heads; they agreed they have some emotions, but not many. But, even then, why would their suffering be less than mine? I found some of the beliefs held by renowned professors irrational and disturbing. As a result, the important questions of whether, when, where, how, if animals should be kept, were being raised by philosophers. One of the first of these was Peter Singer (whom I had corresponded with when we were both doing our doctorates). His thinking about animal welfare resulted in *Animal Liberation*[1] which had been published, helping animal welfare science become an academic growth industry.

One of the markers in my thinking was when I arranged a seminar or workshop at University of Glasgow. This resulted in meeting Mary Midgley, a well-known moral philosopher who had published the first and one of the best texts on animal issues.[2] She subsequently became one of my most important philosophical mentors. This book pointed out simple home truths that some find difficult to accept. This was illustrated at the meeting by a well-known professor of zoology and welfare who mentioned that it

---

[1] Peter Singer, *Animal liberation*, 1976, Jonathan Cape, London
[2] Mary Midgely, *Man and beast*, 1976, Routledge, London

was necessary to "test if ducks need water", not to drink, but water to swim on and be near. She argued that until we had done experiments, it was not possible to conclude that ducks did need water when they were kept and raised for humans to eat. But ... ducks are defined as water birds, they live most of their lives on water, swimming, breeding, raising their young, eating and so on, on water. Why is it necessary to "test if they need it"? This is like "testing whether a child needs to learn to walk". Such beliefs deny evolution. Certain behaviours of any species are in place because the body and mind "being" of that species has evolved to be what it is. If you believe in evolution, or that it seems to be the most likely explanation of how living things are the way they are, then ducks without water will suffer ... if this needs to be tested, then surely it needs also to be tested if a duck (a water bird) is a duck? Where then do you stop with your tests before you can conclude anything? Maybe some individuals could learn to do without it, like some people learn to live without hands or feet, but should we cripple them to provide food or make money? So, if you wish to buy duck to eat, visit the farm where they are raised and fattened, and see if they have any water to swim in.

What struck me most at this meeting was the hypocrisy of some scientists. They put animals through experiments that they personally found emotionally disturbing and unnecessary. For example, the vet lecturer with whom we stayed, taught animal welfare science, yet, she let her land, adjacent to her house outside Glasgow, to the vivisection group of the university. They used it to raise cats for vivisection. She did indeed find many of their experiments with cats distasteful, after all, she had cats of her own, loved them and recognised their emotions, but had convinced herself since she "needed the money", to let her land to allow cats to be tortured in order to try help humans with new drugs or surgical techniques, and argued that this was morally acceptable.

Life continued at Drumghigha, the multi-species community grew and developed, and the farm prospered. Eventually we grew many crops the Agricultural Advisor told me were impossible, wheat, clover, improved grasses, fruits and vegetables, all only from the resources of the farm, using no lime or fertilizer. The natural world was rich, splendid and difficult, but worth the wet, cold and even the summer midges. The farm was glorious, the animals a delight one kept learning from. The visitors were enthusiastic and the volunteers and students hard working.

In 1987, I was contacted by the RSPCA who asked me if I would like to research the welfare of animals in circuses, and I said yes. I knew nothing about circuses except that the people live full-time with their animals and teach them things. The questions I was asked to answer were: Are they cruel to their animals? Should they have the animals they have? Can they be transported without suffering? Should they be restricted to pens and trailers? How do they train them? For the next two years these questions became central as I travelled, camped and studied the animals in all the major circuses in Britain while the multi-species family and farm continued to develop and grow on Mull.

One of the most surprising lessons that I learnt during this period was how adaptable herbivores, canines and humans could be. Placed in a totally alien, wet, cold, windy but beautiful unfamiliar environment, they all rapidly managed to adapt and live happily within it. Another lesson was how important cooperation between species is. Provided you establish that others can have confidence in you, they will cooperate with you and do almost anything if you ask nicely: go off in the rain and wind to find sheep, stand about in the cold and so on. But, only provided you observe the social contract, that is the cooperative treaty. This means that you do not ask another to do things that are really terrifying, or bash them about, or not consider that they might be hungry, thirsty, cold or tired. Cooperation works two ways: neither humans nor non-humans just need enough to eat and drink, and to socialise with their own kind, they have mental needs too which encompass complex thoughts and desires. These cognitive needs vary somewhat between species because they are the result of being that species: that particular body and mind, but they are also moulded by their lifetime experiences; the knowledge they acquire and the experiences they have. Another lesson was that even professional, intellectual humans are not all free thinkers. Scientists follow dogmas and have preconceptions, just like any other mammal.

The natural world can be demanding, difficult, unpredictable and extreme, but it is always beautiful and so often achieves the unexpected, if given half a chance without human restructuring. Allow it to be, respect it and just give it a helping hand to slightly change a direction is all that is required.

## Circuses & zoos; animals & people

Circus people are often not conversant with political developments and are somewhat bemused by attacks on their animal treatment. Previously, their top performers such as lion tamers, presenters of animal acts and trapeze artists have been hailed as Diva-s, and people have flocked to see them perform. Now the same people were being accused of being cruel to animals and considered criminals. Mary Chipperfield from the well-known Chipperfield circus family (who among other things established the first safari park in the UK) had come under particular attack. The Robert Brothers' Circus was another old family circus that was under attack, which the brothers found very confusing.

The circus people knew that the RSPCA was financing me, and they also knew they wanted to close them down. My first challenge, therefore, was to try and convince the circus proprietors to let me into their circuses to study the behaviour of their animals, knowing that I was financed by their enemy, the RSPCA. Somehow, I had to convince them that I was an INDEPENDENT SCIENTIST. Each side must know that I could not be bought, either by the RSPCA or by the circuses. The measurements of their animals' behaviour would tell the story and answer the question whether circuses of their nature were cruel or not. Of course, there are bad circuses, like there are bad parents and pet keepers. This does not mean that all parenting or all pet keeping is bad and consequently must be stopped. Rather, it means that those who are not doing it well, must be given advice and, if necessary, legislation introduced to ensure they improve, and this legislation must be policed.

Independent science has become somewhat rare today, and the RSPCA made no secret what conclusions they required of me, but, for me, it was the science that would talk. A colleague at a conference warned me to keep the copyright of my research findings, and as it turned out, this was very wise advice which, luckily, I followed.

But the first problem was, how was I to obtain permission from the circuses to study all aspects of their animal keeping, training and performing to assess whether the animals suffered?

People, like most mammals, are creatures of habit and become stuck in their ways, convinced that what they have been told or seen their predecessor or someone they admire doing is right and the best way. The result often is that the methods are rarely questioned. This is particularly common where animal welfare is concerned. Nearly everyone is concerned and wants to help, but they are not often well informed, so they join an organisation which has a reputation of doing good for animal welfare, and go along with it without questioning whether what the organisation is doing is really reducing suffering for the animals concerned. This is particularly true of the RSPCA and other welfare organisations' attitude to circuses.

First, it was necessary to make a couple of brief visits to several circuses, to see the animals and talk to the circus people. This, I hoped would give me an insight into how I might convince them that I really was an independent scientist: I might be a threat to their livelihood, but I might also be able to help by showing them where they were falling short, and there was no point in offering me all sorts of perks because I was not for sale. Although some of the circus people liked their animals more than others, and some are liked more by the animals, it usually is the presence of the animals and what they bring to the circus, irrespective of the income, that is important to circus people, whether trapeze artists, clowns, sellers of popcorn, grandmothers, children and acrobats, as well as the animal trainers and presenters. They live in daily close contact with their animals, look after them, teach them and perform with them, and are constantly on the move with them. This means that they have a lot of experiences in common with their animals. Everyone in a travelling circus is expected to help do anything within their physical ability: build the tents and exercise yards for the animals, load them into their vehicles, unload, move them to the ring and so on. Some of the big cats, that is tigers and lions, have been bred and reared either at the winter quarters or as they travelled about. In fact, tigers, one of the most threatened of the large mammals on the endangered species Red List, had bred more easily in circuses than zoos, and they were negotiating selling extra tigers to zoos to exhibit them as the rare species they are.

The strategy to convince the circus people that this scientist would do a disinterested study of their animals, turned out to be quite simple. They needed to know and believe that I lived with my animals in a multi-species community, a bit like them, I cared for them, taught them and worked with them in many ways; in effect, I also had practical experience in all aspects of dealing with them. When not paid by the RSPCA to do research, I made my living from them, in fact as with the circus people, it is the animals that allow me to live the life I lead. There was no way the Circus Proprietors Council would visit our farm to see this for themselves, but if I took along one of our animals whom we had bred, taught and lived with, maybe they would see I did have some practical experience at least. I was called to the north of England to be interviewed by the Circus Proprietors Council and I took Barnabus, my black, white and brown collie-cross dog who went everywhere with me. He lay quietly under the desk while I spouted on about what I intended to do to investigate the welfare of their animals, and then, when asked, Barnabus circulated around all the people saying hello, and did a few tricks. I also told them that I would not visit the circus without him, so, since they liked animals, they had better not deny him entry, like cinemas, most parks, beaches, theatres, some buses and trains, shops and restaurants and almost everywhere else in a town.

They must have realised that I was sincere about animals and knew something about them because I never found any antagonism against me for the research I was doing. They seemed genuinely to welcome me and wanted to learn from me how to improve the lives of their animals, and I received no perks, an occasional cup of coffee in one of their wagons, but not even a meal. I am sure they cleaned things up when they knew I was coming, and that sometimes things happened which should not with their animals but the same is true for any vet in or out of his surgery, pet owner, stable, zoo or even a parent. Sometimes I just turned up, put my tent up and Barnie and I camped with a circus for up to fourteen days while my research assistant (Sue, a young woman sociologist from Mull whom I had managed to persuade to help me) and I spent almost every hour watching and recording the behaviour of the animals.

If circuses were under attack, why weren't zoos? Were zoo animals better or worse off than circus animals and how could we measure this? We decided that we must do some similar studies on zoo animals, but found them a lot less welcoming. Zoos usually have "resident scientists" and the directors or proprietors are often more articulate than circus people. They use arguments that sound scientific for their existence, such as "breeding animals to reintroduce to the wild" or "preserving the genetic pool of different species", which may convince the man on the street it sounds important, but actually they rarely manage to do either of these things! What they do and which should be their strong selling point (but which is almost never mentioned) is to allow anyone who may never have seen or experienced a whole range of animals first hand, to appreciate the enormous diversity and complexity of animals and their great range of abilities; something that is becoming more and more important for the urban human living in a totally homocentric world isolated from other animals. Today, twenty-five years on, animal welfare activists are trying to ban zoos as well, particularly small ones which cannot afford to buy them off with "scientists". As a result, a zoo somewhere is closing almost every week and many animals in good health are being "euthanised".

No one has carefully thought through what the individual animals living in circuses and zoos actually think about it. Nor do the activists think through where the animals will end up when they close and whether the new life is preferred by the individual animal. What usually happens is that many of the animals are killed or put down: "euthanised" on the grounds that they are "too old", "infirm" or "difficult". Some may be sent to other zoos or wildlife parks where, because of their past living experiences, they may suffer traumas because of the unfamiliar lifestyle. Unfortunately, zoos and circuses do not often know that they could consult an animal welfare scientist to give them advice, if they could find one that was disinterested. Meanwhile the animal activists, being much more "street wise", employ scientists who they know will recommend closing that zoo or all circuses, and pay them very well.[1] As a result, some of the last places where the majority of the urban population can have contact with non-human living creatures and experience who they are, are disappearing ... thanks to the animal welfare activists. Ironically, we found that some of the large, important zoos that condemned circuses, were selling their unwanted animals to circuses. They also were calling the circus people in to help with handling or teaching errant animals, when they could not manage!

An example of how it is possible to improve the quality of life for different animals was shown by my research assistant's Angora rabbit farm. She had worried about keeping the rabbits in a group because it was said they would spoil their fleeces by fighting. On my suggestion, she tried letting the rabbits run in a large varied area in groups rather than keeping them in hutches and found that the fleeces were not inferior and, even better, problems such as aggression between individuals were reduced. Of course, a lot of thought must be given to how the environment is designed and managed, but the point is that it can be done: a win-win situation for both the humans and the animals. She was not an animal

---

[1] Stephen Harris, "Foxhunting prosecution professor 'misrepresented science'", 11 August 2018, *The Telegraph*, Hayley Dixon

welfare scientist nor an animal welfare activist, but was open minded and interested in the welfare of non-humans as well as humans.

Sue, Barnie and I drove many thousands of miles up and down the M roads from the north west of Scotland to visit each British circus, whether they were travelling, performing or in their winter quarters. We camped with each of them and watched and recorded the behaviours of their animals for up to twenty-four-hour periods in their living quarters, rehearsals, vehicles, performance and training sessions. Did they show any evidence of "distress", that is suffering either physically, emotionally, socially or cognitively? To do this we needed to know how they spent their time, did they spend more or less time lying, eating, sleeping, running around, etc. than their wild counterparts? Did they show higher levels of aggression? Did they show any stereotypes or purposeless abnormal behaviours? Did they have proper veterinary and nutritional care? Did they have enough exercise? Did they have enough to do and think about? And so on.

We finally came up with a way of measuring "quality of life", using a method similar to that used for humans, which involves measuring their "freedoms".[1] However, instead of measuring "freedoms" we measured how many behaviours were restricted, the reverse of which gives a comparison of how free they are in each environment.[2] Alternate weeks we returned to Mull to computerise and number crunch the results. Luckily my assistant's husband was a wizard with home computers, which had just come on the market. We both had Amstrads (which were impossible to use; one had to have the "Amstrad for idiots" instruction book to do anything), but luckily despite this, Mike was adept.

I learnt a lot about circus people as well as their animals and there was much to admire, although yes, there were also problems for their animals, but these were not insurmountable because they were in circuses. The circus people are a minority who have different priorities from the majority of the population (and

---

[1] http://www.un.org/en/universal-declaration-human-rights/
[2] MKW, *Animals in circuses and zoos*, 1990

consequently are an easy group to attack). Circus peoples' top priority is that "the show must go on", whatever happens, however many trucks have broken down, or big tops have collapsed, or people have become ill or injured, or have even been eaten. Everyone, even the greatest diva, is expected to help with anything they can do to ensure that the show goes on. The famous trapeze artist sells popcorn in the interval, the lion presenter dresses up to become the clown when the clown is sick. The whole family is involved, children begin to learn to perform when they are as young as five and choose what they wish to specialise in. At that time, the state provided a circus school caravan and teacher who moved with the circuses. They started school early in the morning, and were finished by midday in time to practise whatever acts they had selected to learn before the matinee performance, a quick siesta and then the evening performance, where they would help wherever they could. Despite this somewhat rigorous childhood, the vast majority of circus children stay in the circus for all their lives and travel worldwide with their own circus or are hired by others as part of what they call "the worldwide circus family", and, like any family, there were plenty of squabbles.

Compared to theatre companies it is remarkable what the circuses achieve. In theatre, actors are never expected to do anything but act. There is a whole group of other "professionals" hired to do the scenery, clean the theatre, do the costumes, the lighting, the make-up. A play in a ready built static theatre will often take months and months to rehearse and prepare. By contrast, the circuses usually have two performances per day, maybe only a three-day stand, before they take down the auditorium and pack up the props, the backdrops, the lighting, the costumes and pack the hundreds of benches and chairs away in trucks. They have to break down the animals' exercise areas, load them and ensure they have enough and the right kind of food for all the different species, and water and so on, in each species' trailers, then load up the animals and themselves. All this is done after the evening performance. Then during the night, after arrival at the new site, they rebuild the whole thing, including all the animal tents and enclosures, ready for a performance the following matinee. No wonder everyone has to help. One of my strong memories is of stacking the hundreds of chairs after a twelve-hour day watching the animals until midnight, but at least I had not done two performances, sold popcorn and bedded down the animals as well!

Another difference from almost any theatre, concert, or other entertainment is that the animals have not read the script, and there are no script readers hidden behind the scenes. Sometimes some of them know it very well and stick to their roles willingly, but others, if things change or something queer happens, either in their lives or in the auditorium, or maybe they just feel like it, they change the script when in the ring, and might do something quite unusual or different. It is the presenter's job to make this look as if it was in the script, rather than tick off the animal and try to correct him or her. So often the acts changed, an elephant would not lie down, or a camel would not stop running around the ring, or the tigers started an argument, or a lazy lion would do nothing but lie next to the presenter yawning. Improvising and innovation in the performance therefore becomes the name of the game … and of course it must all run smoothly and look as if it was all in the script!

Each circus also had "winter quarters", which was a few acres of usually otherwise unwanted ground, between railways or roads. It would often have a ruin of a house on it, and a stack of broken-down lorries, tents and trailers, and scattered bits of clowns or trapeze artists' sequined costumes gradually losing their lustre as they lay in soaking heaps. There were some animal sheds, often not in good repair, but at least the animals could shelter from wind and rain and lie down on dry bedding, and had wet pastures to wander in. Not one of the circuses we visited actually gave up their trailers in the winter to live in the house, even though one would have thought that with some repair the house would be more comfortable. They were completely uninterested in mortgages, living in houses or earning capital from house refurbishment. Rather, they all seemed to collect circus memorabilia and displayed this and their circus awards in prominent places in their rather cramped trailers, which they painted weird colours, decorated with stars and sequins, and were very proud of.

In today's consumer orientated society, there is a lot to learn from circus people and their animals, but, like some of the animals they bred, they are becoming extinct due to animal welfare activists' often misguided actions.

But, like pet keepers and parents, they did and do still have a lot to learn about how to improve the lives of their animals. There were some unacceptable housing conditions and cramped quarters. Not all the animals were learning new acts or doing things which were interesting to them and would give them some mental stimulation; as a result, some were bored, as in the majority of zoos. But, unlike in most zoos we visited, the people were interested in learning, perhaps because their livelihoods were threatened. Since they had plenty of initiative, they managed with little cost to provide acceptable exercise areas with toys for their big cats, electric-fenced paddocks for their elephants to avoid them being shackled all the time, and made other improvements that we suggested. But the welfare activists and the health and safety laws also shackled them. For example, they were no longer allowed to take the elephants on walks through the towns or down to the seaside to swim in Blackpool as they had done through history. Such activities are good for the elephants and the people, and are at least as safe as riding your bicycle or driving a car around town. They could not take them through the local parks, or even let them wander about in the circus campsite.

However, there were ways that could be invented to amuse and interest the animals as well as the humans, and our job was to try to dream up some of these. For example, one circus had six young stallions who did a liberty act (ran around the ring together or in groups, turned and twisted, and did many other choreographed things as the presenter stood in the middle). Because almost everyone believes that stallions will always fight each other, they are usually kept isolated in individual stables and only allowed to run together in the ring with tight side reins to stop them being able to move their head around and bite each other. But, young stallions do not always fight, particularly when there are no mares around. In fact, they are usually very interested in each other and like adolescent boys will often play rough games; "horse play", but rarely injure each other. On my suggestion, this owner tried keeping the young stallions in a group during the day in their exercise yard and there were no serious injuries, but like boy-play, there were signs on one or two that the play had been rough, with some hair pulled out, or a bruise or nip. The problem was that when the audience saw these things, they immediately thought they were the result of humans hitting them. What the circus people should have done is explain to the audience before they start the act, what and how they come by such scratches. But in the atmosphere of animal welfare activism, this is very difficult. It would have meant video-ing of "before and after" keeping conditions and so on to convince the audiences. So, the result was that the young stallions had to be put back into their isolated individual pens. This, all because of "well-meaning", but ignorant and misinformed, public opinion.

Normally in a circus animal act, there is a presenter (not usually the trainer) who is a young good-looking man or woman in a dramatic costume who stands in the middle with a whip and directs the animals, or looks as if he is. The liberty horses' heads are restricted with side reins so that they have to hold them arched all the time to make them look grander. It became evident that the horses knew the choreography, so they had no need for someone in the middle with a whip looking as if it was all his doing. Why did they need their heads restricted or other pieces of leather or coloured plastic restricting what they did? Such acts build up the ego of the presenter as the guy "in charge", but the public then believe that the animals must be restricted and restrained and chased with whips before they will do their performance. This is not a true or desirable image for circuses. When I pointed out to the presenters and trainers that perhaps they could do away with such restraints and whips, they agreed they were not necessary. The ponies, horses, camelids and elephants and many others all knew their acts' choreography and responded to changes in the music or the position or posture of the presenter. I tried to convince them that if they were to explain this to the public and then perform some acts without whips and restraints, there would be fewer complaints about cruelty in training. We even persuaded one trainer to do his act standing in the middle smiling without a whip, while the liberty horses showed how free they were to do their dances. The audiences expect a sequin-covered elegant presenter to "present", that is be in command of large potentially dangerous animals, and do not recognise that the animals can develop their own dances, with a little help from their human friends. If the audiences were informed and shown how the animals were taught, they might be more interested and sympathetic, and the circus could demonstrate how these animals cannot only learn complex routines, but are also innovative and creative. But, circus people like most others, stick tightly to tradition and find it very difficult to change. This has not helped them latterly, although some circuses (and even some zoos) are beginning to do some demonstrations of this type now, but perhaps it is too late.

There were other areas of education concerning the relationship of humans to animals, and the importance of their conservation, that circuses could demonstrate ... if only they would. For example, they could talk about an individual tiger's personality and how he likes to do this or that or dislikes the other, and develop demonstrations with the tiger doing all sorts of tiger things to show how able and athletic tigers are. They could show how innovative an animal could be by giving them some different things to do, or show one animal doing something the others have never done, and then seeing how the others learn by watching and imitating ... there are a host of ideas to show the different species' particular characters, and their individual differences too. By being shown this with a commentary the audience would become much more interested in how mentally and physically able the different species and individuals are, and consequently more supportive of wildlife conservation.

When people have close contact and get to know animals better and exchange emotions with them, they cannot resist becoming interested in conserving them. This is particularly necessary for the large mammals, lions, tigers, elephants, camels and many others whose survival is very seriously threatened. This could be done by having volunteers from the audience come up and just experience the animals' behaviour, perhaps with the presenter honestly translating what is going on ... They could have acts showing the particular characteristics of the species: leopards are specialist climbers and balancers, lions social hunters and so on.

On visiting various good trainers, who do exist, I learnt how hard they work and how innovative they have to be to better use what the animal offers. This is easier in the circus than in much other animal training, because as we have mentioned, there is nothing special they have to do, they just need to show themselves off, their beauty, athleticism, energy, skills, and how they like or dislike others of their own and other species, including the human trainer. There were two or three trainers I really admired. Years later at the European Circus Competition where I was asked to speak to encourage further improvements in animal welfare, I discovered that in Italy there was a delightful free flying bird act, and recently a special free flying raptor act as well as a forward-thinking lion act, so gradually the message is spreading.

During the study I also saw some unnecessarily rough trainers who just could not get it right. Some of the trainers spend much time outside training time patting and petting or cleaning their animals. Others were honest and fair but did not spend any other time with their animals. The odd thing was that those who were cuddling and telling me how much they loved their animals, did not seem to be better liked by the animals than those who were to the point, although they had to be honest and fair. All of this sparked my interest in how to improve the training of any mammal, which has taken up the last thirty-five years or so.

About one year into the study, the RSPCA got wind that I was doing an independent study. They sent one of their top "wildlife" men to interview me in Oban, our local port, so I steamed over in the ferry and we met at a smart hotel. Over tea and cakes, we discussed at length what animal welfare was, and how it could be bettered. We discussed what I had been doing, and whether or not the RSPCA's position on circus animals was rational and appropriate for the animals, and if not why not. I suggested that their position might be a ploy to attack a minority with different values to gain credit for the RSPCA, rather than being fundamentally to do with animal welfare, because otherwise, why on earth were they not attacking zoos or intensive farm animal keeping where animals suffer in their millions? The fundamental question was, did circuses, of their nature, cause animals to suffer, looked at from the animals' point of view? If not, why could they not improve their animal keeping and training and have legislation passed to that effect?

After eighteen months, with the help of my assistants, I had completed the circus study, and the number crunching, and written the results up. We were ready to address the wildlife committee of the RSPCA. We drove down to their headquarters in Horsham in Sussex and I addressed them with our carefully considered empirical and ethical conclusions. It started badly as ironically, Barnie was not allowed into the lecture room ... the Royal Society for the Protection of Animals did not permit him to come and lie under the lectern!

It continued in the same vein. We showed the amount of distress that the animals in both zoos and circuses showed and that circuses were not much worse than zoos, but both should improve. Things like increasing the amount of time the animals had exercise, improving and enriching their cages and exercise yards, and spending much more time teaching them new things, so that they would have more to think about and do. In other words, reduce the restrictions on their freedoms, socially, physically, emotionally and cognitively.

We also suggested how improvements could be made to increase the circuses' educational as well as entertainment impact, and pointed out where there was a need for improvement in the animals' living conditions and advised them how to achieve this, and advocated increasing teaching without violence. But we also pointed out that the distress we had measured was not endemic to circuses or zoos: neither zoos nor circuses need to cause animals to suffer any more than an urban environment for humans needs to cause humans to suffer. Not all the animals showed evidence of suffering, and there were many ways in which both circuses and zoos could be improved so that no animals did. Consequently, neither could be banned any more than pet keeping, gymkhanas and horse shows, or even parenting because there are bad parents. We had a list of what needed improving and went a step further to suggest how a life of quality, not just the avoidance of suffering, could be easily measured, and how by changing management, this could be greatly increased in both circuses and zoos; something that no one had done before.[1]

When I had finished, there was a stunned silence, and without a word, the whole committee collected their papers and disappeared. I had given them a typed report but had the copyright. The last thing I wanted was for anyone to miss-quote, change, or extract bits and publish it under my name. The full story should be told before decisions were made. I was naïve and bemused by the behaviour of this high profile wildlife committee who were convinced that even though they had been paying me, a scientist, to try to produce the evidence to ban circuses, I had been backhanded by the circuses. The circuses had paid me nothing, nor backhanded me in any way ... but the RSPCA had been paying me. I had believed, in my naïvety, that they wanted an independent scientific study, which is what I had done.

But worse was to come, it was not long before they approached the main tabloid newspapers and provided extracts of my report out of context, despite the fact I held the copyright. Journalists are always interested in a good story and wanted further details on why my report had not argued for the banning of circuses. Luckily, as it turned out, a copy of the report arrived on the desk of a colleague of my eldest son (a journalist for *The Times*). He asked my son if this "Kiley" was anything to do with him. My son ventured that it would be unwise to publish the RSPCA's summary as

---

[1] MKW, *Animals in circuses and zoos*, 1990

I owned the copyright. The story circulated quickly through the Fleet Street pubs, and many papers countrywide wanted a rousing dispute, so they contacted me to interview me. The result snowballed into the face of the RSPCA although papers that had decided circuses with animals should be banned continued to publish anti-circus articles. I was asked to do TV interviews and a forty-minute film to examine circus animals' lives. This involved travelling to Russia as well as visiting circuses around Britain and Switzerland.[1]

I discovered some years later, that one of the members of the RSPCA wildlife committee was active in the Animal Liberation group which, conveniently, had its headquarters in Horsham. He leaked my results, and Animal Liberation activists tried to find me. By that time, we had moved to Devon, luckily, and they did not find our address. It is ironic that I, who have worked all my life to discover how to better the lives of animals in practice and theory, should be threatened by an Animal Liberation group.

Since 1993, circuses with animals have been banned in some countries, and the trend towards homo-centricity has grown with human-only circuses. There is a danger that some of the animal teaching skills are being lost, as today there are practically no circus animal trainers, and by no means all their training was found to be cruel or violent. As it happens, circus trainers throughout history who used violence and fear to teach their animals were often eaten by the large carnivores, or killed with one swipe of a trunk by an elephant they were training. Making such mistakes continues to happen and it is not unusual for an elephant mahout to be killed in Asia, or a circus trainer or presenter to be mauled: even while I was with the circuses this had happened in various countries. So far, every case I have investigated has come down to being caused by that animal having bad experiences and being frightened by humans. Behave badly, break the social contract, and you pay, whether the large mammal is domestic or traditionally wild. All the same through its long history, circus

---

[1] Nature Watch: Captive friends. 2013,
https://www.youtube.com/watch?v=oWqntDuofDY

trainers have taught different species to do some extraordinary things that we would not think were possible for them to learn.[1]

The increasing rarity of circuses with animal acts is of course the result of the "animal welfare activists" who have badgered and bought people of influence and anyone else they could to promote their cause. They have indeed brought the welfare of circus animals to the public's attention, and this has made the circus people pay attention and learn, but they have also more or less killed the circuses in many countries, where, ironically, the welfare of the animals was already often better than in the countries where the circuses with animals continue. The weird thing, as an animal welfare scientist, is how much effort, money and publicity the animal welfare organisations have put into this when even if all the animals in circuses suffer (which they do not), the proportion of the suffering animals which are found in circuses is very, very small: 0.00000001% or thereabouts of animals kept by humans. Think of the billions of animals suffering all their lives in intensive keeping conditions. For example, cattle tied up for six months a year in France, pigs are kept in pens where they cannot turn around, chickens are in batteries for their reproductive lives, or in enormous houses with thousands and thousands of them growing so fast that their legs break, geese are force-fed for *foie gras*, expensive top level stallions kept in stables for years at a time, only coming out for five-minute periods to cover a restrained mare or to be shown off to a potential breeder. What for? Just so some humans can eat too much or earn more money. Then there are also millions of animals used annually for vivisection for medical research, cosmetics and the production of drugs: primates, rats, mice, guinea pigs, rabbits, dogs, horses and so on. In my study the total number of all the animals kept in British circuses was around 350. Yet much of the effort and money given to the RSPCA was devoted to banning them rather than trying to improve the lives of the billions of suffering farm animals.

A little thought, research and legislation, can change the way animals are kept and used and ensure they have a life of quality in

---

[1] "Review of The history of performing animals" in *Animals*, 2015

all environments. Activities such as cock fighting, wild animal trapping, or bull fighting with weapons, do of their nature cause suffering and should be banned. Indeed some terrible things were, and are still, done to make various animals perform, go to war, work in transport, be ridden, and so on. But, even these things can often be done without causing suffering to the animals. There are also extraordinary things that animals have been taught to do without suffering, for example, before the invention of the science of "animal cognition", some circus trainers were already teaching their animals to do things that now the animal cognitive scientists are testing in laboratories! For example, that a horse, dog, elephant or tiger know where they are looking, that a leopard can learn to balance on a very thin branch and walk across to another, that all these animals are aware what mood or emotion the human teacher is in. At enormous economic cost, much of the scientific study of animal cognition is doing little more than re-discovering the different mental abilities of many species; something which any good circus trainer knows because otherwise he would not have been able to teach the animal to do this or that: Xenophon (350BC) pointed this out! We are in danger of losing much understanding of animals that some people in circuses have had for a very long time.[1] It was this research with circuses that alerted me even more than I had previously understood, to how much we can learn from good cooperative animal teaching.

Many animal activists are concerned about a number of terrible practices with keeping animals and are doing good work on bringing them to public attention. There are some practices which involve millions of animals suffering in some countries, but which are not used in others. One of these is mulesing in Australia where sheep have the top of their thigh muscles slashed open to prevent them having flies laying eggs which then hatch into maggots around the anus. The idea is that when the sheep heal, the scar tissue does not attract the flies, but no one takes account of the pain or how many sheep die of this treatment, and it is not at all clear why there are more sheep suffering from this in Australia

---

[1] "Review of The history of performing animals" in *Animals*, 2015

than anywhere else in the world, where sheep are also kept in their thousands but do not have to suffer this. Another practice in New Zealand is to cut the tails off milking cows so that they cannot keep themselves free of flies. The rationale for this is that the tail might deposit muck in the milk ... But how have all the dairy cows in all other countries and through history managed to be milked and have clean milk even though they have tails?

The circuses, despite the fact that I was critical of much of what they did, asked me to address their council again and we discussed at length the standards they would introduce for their members to better the animals' welfare. They also then suggested that my full report should be published, and knew a publisher who would do this, so it was. Chapter 10 in the circus book makes detailed recommendations for the improvement of the lives of the elephants, such as reducing or eliminating shackling, more social contact, more exercise, and more learning of different things. Recommendations for carnivores included larger exercise areas and enriched social environment, and more learning of new and different things. For hoofed stock, access to *ad libitum* high fibre food, suitable places to scratch and roll, and appropriate substrate to walk and run on, and they should have social contact with their own and other species, large exercise yards where they can move at all their gaits, holiday periods and they should not be kept in stalls or tied up for more than five hours in twenty-four. It also argues that many animals should be allowed to breed, as this is part of their natural repertoire of behaviour and allows them many experiences that are valued by all mammals, including humans. It makes suggestions for improved training and suggests a training school to ensure that all animal handlers and trainers are qualified. Finally, it points out the important educational role that circuses could play in learning about and conserving the non-human world. Some of these suggestions have been followed by some circuses in some countries. The book was not a best seller, but most people who are in anyway connected to animal welfare debates in zoos or circuses usually have to refer to it as it remains one of the most detailed studies of circus and zoo animal behaviour.

During this interlude, because we were observing horses, elephants, large cats and buffalo, bears, zebras and pigeons, as well as the circus people at close quarters and for long periods in their home areas, training, transporting and performing, I learnt a great deal about many different species or types that I had not previously had close experience of. They all had remarkable adaptability and abilities to learn an enormous range of things. I was also often impressed by the close animal-human relationships that had formed within circuses, and their mutual dependence, the real beginnings of a symbiotic multi-species community. But, some of the practices that had grown up in circuses on how to keep or train their animals were not conducive to building this symbiosis and needed to be changed. Perhaps the most important message was that it could all be changed so all members of the community had a life of quality. It set me thinking carefully about how to measure a life of quality in any species, anywhere, and to search for how to create and live in a cooperative, symbiotic multi-species community.

A spin off from all the publicity concerning circuses, was that another publisher, Souvenir Press, offered to publish my book on *Ecological Agriculture, Food First Farming* (1993) which includes the results from three of our ecological farms. This was not a best seller, but in 2019, parts of the results are now being widely read, particularly in China and India.

We had been on the Isle of Mull for seven years by this time. Despite the predictions of many, the farm had survived and prospered, and by dint of getting planning permission for a house and selling off a patch of the land near the sea, we had paid off the mortgage, much to our relief. Somehow, I had a yen to return to the mainland, and to have a farm where I could hear the church bells chime on Sundays. We decided to put Druimghigha on the market. The next project would be in a wildlife reserve, and within our budget, no more mortgages, one had been enough for life. Preferably further south in a sunnier clime.

# To the West Country, teaching & measuring quality of life in animals

A new question arose. Could the whole farm become a nature reserve but at the same time increase its net yield and be self-sustaining in energy, food and materials, and allow the animals to have a life of quality? If this were possible, would it be possible to feed the world in the future without restricting wild animals to nature reserves? Could farming and wildlife conservation genuinely be married together on the whole farm? One of the deciding moments of what we were going to do next was a meeting of farmers with wildlife advisors one rainy day at the unfriendly chateau in Mull. We trooped around sodden peat bogs and marshy grassland overlooking a storm drenched sea. The agriculture cum wildlife adviser was maintaining that all we needed to do was have uncultivated pieces around the fields, pour chemicals on the rest, and keep a little piece of woodland. At the end of this day, one of our perceptive students (who went on to become important in wildlife conservation in New Zealand) put it very well: "this man is the real enemy of wildlife" ... Since then we have frequently found this to be the case, it is often the "wildlife agricultural adviser" who is the real enemy of progress in wildlife conservation or in being able to live symbiotically with other animals and plants. It is often those giving advice who believe that the moral right is on their side who obscure the difficult problems and probably caused further species extinctions as opposed to the majority who either do not know or do not care. Our next challenge was to run an ecological farm where the whole farm was a nature reserve which provided a living, fulfilled the tenets of ecological agriculture, and where the animals had a life of quality.

My elder sons were long gone to their own careers, so it was the three of us, Chris, Jake aged seven and I, who went a-wandering in the West Country, after having given up the north as too wet, the south east as too expensive and too many people, and Wales as too culturally different. There only remained the West Country, where Chris had been brought up. We made many trips south in a limping ancient Volvo visiting old Devonians retiring and selling their small dairy farms, and seeing ancient cob buildings built right over the road in little villages. Eventually we came across sixty acres of grass and woodland with no buildings on the edge of Dartmoor National Park. It was within the national park, but since it had no buildings, the question was were we going to be able to put up a barn where the animals and ourselves could live? Commuting to the farm was not an option, since farming is a way of life. Each of us gave each property marks out of ten, on the grounds of what did or did not attract us individually to that property, and the sixty acres called "Little Ash" (because of the young Ash tree where the cart track joined the sunken narrow road) came out top. There were a number of unknowns ... but, we bit the bullet and it took every penny of our money. Luckily, living on a very low income, I had long ago learnt not to mix income with capital, so a little income can be put aside each month. Even if it is only £5, this adds up and is hidden away and comes in handy when stresses pull.

It was February, cold, wet and snowing on Mull. First, we had to move out of the house which was sold. We moved down to the sixty acre nature reserve we were keeping, which was next to the sea. We built a roof over a caravan so the cow, horses, dogs and humans could have a dry space together. Two months later, everything had to be packed into the transporter lorries. The early morning departure in the dark was touch and go. Shiraz, one of the mares who was a very experienced traveller, refused to load and the others went along with this decision. The cows had calmly entered in front, and the sheep had gone upstairs. But the horses refused to load at the last moment. Finally, with a lot of rather careful quiet persuasion including promises of return, Shiraz decided she would go in and of course all followed, but by that time the driver had to rush to catch the ferry. We saw his tyre marks hovering off the road as he roared around the corners and we followed in the ancient Volvo with a caravan stuffed with furniture and anything else left, and a car full of students was to follow later. We all caught the ferry, and the only mishap on this long journey was that the caravan broke down in the highlands, so we abandoned it for later collection, but we never did catch up with the lorry and all the animals.

All the lorry driver had was a sketch map on the back of an envelope to indicate where he was to drop off the cattle, horses and sheep: a field in Devon. When we arrived, to our surprise, he had found the field and dropped off all the animals. They were munching their way contentedly through the long green grass of Devon with glassy eyes ... they had never seen or tasted such luxury before. There was a slight drizzle, but no snow or terrible cutting winds. They reminded me of a bunch of people who have eaten too much good food, drunk too much good wine and collapse replete, gently smiling over their coffee. As we homed down, warm in our sleeping bags in the car, it was clear the move was going to work for all of us, despite the risks we had taken.

Over the next few days, the three of us moved into a caravan parked under some large and generous beech trees in the middle of our patch. It just seemed like heaven after Mull in February, the dandelions were coming out, the weather was sparsely sunny and still, the grass was growing, and the days were lighter and longer. We, like our animals, collapsed in a contented, smiling, somnambulant heap for a week or two before organising planning permission for our barn, plus its delivery and assembly by the family business in the north whom we had bought it from. Jake started school and there was a bus that stopped at our gate to take him. Although none of the other children could understand his west coast Scottish accent, he quickly made friends, particularly with a large family of boys who were our neighbours. Even though we had no sign of any income for a year, life seemed to be wizard, and we could hear the church bells on Sundays. The gentle rolling green country with Dartmoor in the background, and riding around the moor with our dogs and horses in the early spring, just took the biscuit.

The next step was to design the barn as a "multi-species home" which we all could share, humans, horses, dogs, cattle, sheep and llamas, so no one would feel left out or threatened. This involved installing in the barn an ancient mobile home for us with large doors and windows, and organising a multi-species common room where we could all meet. The challenge was that each species must have their own room with their own food, but all the different species' rooms needed to have access to the multi-species' common room to mix and meet when they wanted. We built the cows a low, wide opening off the common room so they could retire to a spacious room where they had twenty-four-hour access to hay and straw. Then the calves needed a small gateway or creep area opening off this, so they could lie down in peace without their mothers munching all their food. Calves like to stay with their peers most of the time, rather than their mothers, so this was popular with them. We gradually acquired llamas and guanacos who had to have a tall, narrow entrance that no one else could get through, to their private suite. The horses, who are always rude to cows, ignoring them and walking so close that they have to get up, had an entrance that all could get through, because we knew that none of the others would be so bold as to invade their private space. But, some of the horses would need individual feeds, depending on their work and physique, so each had to learn where their place was if they were going to get their food. We humans had the mobile home from which we could walk straight into the multi-species' common room and spend time with whoever was there. The cattle or horses could see us through the windows or put their heads in to watch TV or watch us cook or sleep, if they liked.

Because the ground floor rapidly filled up with all these different rooms, plus an open ring area where we could work with any of them, the sheep's winter accommodation had to be upstairs in the barn loft and only they could climb the ramp. Oberlix, my young Arab stallion had a private room next to our bedroom since we did not want all the mares pregnant all the time. He would put his head through the window and butt me with his nose to wake me up if I was late with early morning feeding, and he often kept us awake with his snoring, groaning and nickering in his dreams ... a very noisy sleeper and dreamer.

There were two practical advantages with this arrangement. Firstly, the forage (hay, straw, silage) could be delivered with a tractor and dumped in feeders so that all the animals could be fed in about thirty minutes every third day. Secondly, in the spring the mucking out was done in a day with the tractor, and the muck spread directly on the fields; there was no mucking out by hand with a wheelbarrow, which takes up so much time and energy in most stable yards.

The very real mental advantage was that we were genuinely living with our animals twenty-four hours a day, and after a couple of months, a multi-species community built up where everyone of each species knew where they were, chose who to be with and knew what they should and should not do. We continued daily to learn first hand more about the mental aptitudes of the different species and their individual differences, what they learnt and what they already knew, and how they organised their social lives, in much more depth than previously with all the thousands of hours that I had spent watching and recording their behaviour in the field. They learnt about us, and we learnt about them: how to assess individual intentions and moods; and how to improve communication: visually, auditorily, and with touch and smell and they learnt to watch and listen to our verbal language and interpret it.

Soon, I was invited to be an honorary fellow at the University of Exeter. One particular lecture series I taught was called "animal cognition" ... at last, it was recognised that animals had minds and undergraduates should learn about them. The RSPCA had blocked giving me money to continue any animal welfare research. So, failing any outside financial aid, the obvious way to keep going with research was for our animals to finance it themselves, from the farm and its various activities. We would sell the lambs from our ewes, and the calves raised by the double-suckling South Devons, and an occasional well-educated home-bred young horse.

We also had a top class competing stallion who could court and cover any visiting mares if they wished and be paid for it, and all the horses would give occasional riding lessons to make enough money to pay for their expenses and support our relatively small monetary needs. If there was any spare money, it could be used for researching how to improve the lives of our and other people's animals.

The welfare of rabbits kept for meat or fibre left a lot to be desired in terms of welfare. A student wanted to study them, so we tried out various housing designs with a bunch of male Angora rabbits who lived together. We designed two-storey living accommodation with separate areas for privacy, and plenty of bedding and nest material. Despite all the rabbit "experts" telling us that it would be impossible as they would fight and wreck their expensive fine fibre fleeces, they were undamaged and after clipping their fleeces off twice a year with scissors, they more than paid for themselves. We had shown it was possible to give rabbits an enriched social life where they could make decisions and choice; they were also more profitable and less work than those kept in individual hutches.[1] We had previously more or less cracked how to ensure that animals who lived on a commercial farm could be raised economically without causing them distress and suffering. But, many still believe that as long as animals show no sign of suffering, and have enough to eat, drink, and grow, are fertile and not frequently sick, then they have a "good" life. But, that is not all there is to life for any mammal, including humans. So how could we provide a better life, a "life of quality", for our animals, and how do we know when we have? One way of measuring a life of quality is by looking carefully at the "declaration of human rights", to see what is most important for another mammal: humans. It turns out that it is "freedoms" that defines this.[2] A freedom is mentioned or implied in each of the twenty-eight articles.

If freedoms to perform all the behaviours in the different animals' repertoires are equally important (and today there is no

---

[1] MKW & C Daley, "A Comparison of behaviour ... Angora rabbits," 1998
[2] http://www.un.org/en/universal-declaration-human-rights/

evidence to the contrary), then by measuring their freedoms or lack of them, we can point to the best and worst environment for each species. This is something people often intuit, for example few people would consider that a dog tied up for twenty-four hours had a life of quality even if he was not showing signs of distress, because he is not "free" to perform many of his behaviours. We therefore decided to measure how many behaviours were restricted in the different species in the typical environments in which they are kept. This would point to those who have the most freedom to perform the behaviours in their repertoire.

However, some behaviours within an animal's repertoire may cause suffering to others. For example carnivores hunt, and the hunting will cause suffering to the prey. But, it is probably important to the lion or dog to be able to hunt, but what about the hunted rabbit or bird? He feels too. Suppose one admits that an animal under human jurisdiction should have the possibility of "hunting" (since we have no evidence that it does not matter to him), then perhaps the hunt could be a "pretend" hunt, where the lion, dog, cat, cheetah, leopard or fox can go through the motions of hunting, finding a "prey", stalking, smelling, following scents, rushing, chasing, leaping and so on, but not actually kill or chase a live animal. This, of course, is intuited by people when they know that their dog likes to chase or find a ball. It has now also been taken on board by some zoos who "enrich" the environments for their carnivores by hiding the food, using mechanical antelope that they chase, or whatever one can invent ... and it works because the animals show much less sign of distress and boredom, live longer and are less ill. For herbivores and omnivores (like pigs), there is not usually a problem of hunting others, but they food-hunt, that is they select, learn and eventually know the stage of growth and species to eat. If we keep them in confined areas and provide all their food, we must supply some way in which they can do this finding, selecting and choosing. Large herbivores can also hurt others of another species, particularly when frightened or annoyed, by biting, kicking, standing on them, squashing them and so on, so we should ensure that they are familiar and at ease with other species they have to do with, including humans.

There were some surprising results when we measured which environments were most and least behaviourally restricted. Firstly, the wild did not always come out the most free because often food, water, shelter and freedom from pain or diseases are not freely available and their absence causes suffering. The other observation was that there was quite a lot of variation when animals are kept in the same type of environment, and this depends on how it is managed. For example, when kept in a stall for twelve hours. If they are out and about working or with others for the other twelve hours, they are "freer" and can have a better life than if they are just turned out in a small area alone for those twelve hours.

This means that although there are some environments for animals that are of their nature not able to provide for many freedoms, many others can be greatly improved with a change in management. After all, we cannot all open the stable doors and let the animals gallop off into the sunset, there is not enough space to do this, and anyway the individuals may not enjoy since they often do not know how to live in "the wild" (even if it did exist); but we can with careful management give animals, even in urban environments, a life of quality which has as many freedoms as in some wild environments. By considering freedoms in the same way as we do for another mammal, humans, our attention is drawn to where we need to improve the management to increase the freedoms for that animal. Therefore, it is not necessarily bad news for a pug to be carried around in a handbag, it depends on how much this is done, and what other things he can do the rest of the time. It also depends what other experiences he has had and enjoyed, for example it is unlikely that the posh Hampstead lady who owns and carries the adult pug would like to gallop around in the mud with a bunch of other humans, any more than the pug would probably like to gallop about in the mud with a bunch of other pugs, given his experiences. He is an urban "well-bred" posh dog, and she is an urban, posh, "would like to be considered well-bred", human.

Animal welfare science had not at that time begun to consider what are now called "positive emotions": pleasure, joy, happiness, contentedness. It had concentrated on negative ones such as pain and suffering, and sometimes fear, or "unhappiness", and there was still a reluctance to name and identify emotions that the animal might feel, so measures were restricted to behavioural and physiological changes. But, if they have minds, then they must have a range of emotions and learn about the world they live in. If they can do this, then they have wants, and desires, and they make choices and must show a range of emotions both positive and negative.

Today it is beginning to be recognised by some, that emotions are not add-ons, but are the driving force behind learning, this is often called "motivation", otherwise known as wanting.[1] Every aspect of our mammalian lives are controlled by emotions of one sort and another. These are felt and they, with perceptions, learning, and a host of other mental aptitudes, make up what are called experiences, memories and knowledge. This mental mixing and matching is called "thinking", putting mental things together into patterns.[2] Thinking does not require human language as many have argued, it involves mixing and matching common mental aptitudes (perceiving, feeling, remembering, imagining, dreaming, desiring). It is, therefore, inevitable that non-human mammals think; the leading questions are: what do they think about? And, how similar and different is this from a human? Then again, my thinking may be very different from yours, even though we are the same species, since your experiences, emotions, ideas, and so on may be very different from mine. Your experiences and how you think are therefore private; no one else can have exactly the same experience or feeling as another, although we may talk about it to each other. You can never experience my fear, my red, my love, nor can I experience yours. If this is difficult between members of the same species, then it is doubly difficult to know

---

[1] Masson & McCarthy, *When elephants weep*. 1994; de Waal, *Mamma's last hug*, 2019

[2] MKW & SEG Lea, "Can animals think?" 1997, pp211-40

what another species might be thinking and feeling. One of the important ways we can get at this though is to have common experiences and knowledge, and one way of doing this is to live together. Finding out what and how another is thinking is of course the big challenge, but we can try, and by making mistakes we will learn.

So what is generally considered the most important mental difference between humans and other mammals? The usual answer given today by "thinking people" such as scientists, psychologists, moral philosophers and theologians is that humans have a context independent symbolic language which it is believed opens up "the mind" to develop many other mental abilities. Vocal human language is not often an option for many species because they do not have the right muscles and larynx to make the different noises. But, perhaps they could learn to understand human language or learn to use a language that we could both learn. This could be visual, rhythmic or on a computer for example. If this were the case, then, eventually some other species would be able to tell us something about their feeling and experiences or that of others. Just like Polynesia, Dr Dolittle's parrot who translated her own and others' ideas for Dr Dolittle.[1] The Gardners in Nevada had taught a chimp, Washoe, and several others to use Amslang: American Sign Language, so it might be possible,[2] and Irene Pepperburg taught her grey parrot Alex to actually talk, and mean what he said ... Maybe Dr Doolittle was not all imagination.

It was time I went to visit the Gardners and their chimps in Navada. Sign language could be learnt by primates and maybe elephants, because they had hands or trunks and manipulated things with them so that they could learn to perform various movements used as "words" with these appendages. However, hoofed mammals and even dogs, could push and pull things with their mouths and feet, but they could not make a large range of different movements with their feet or noses to indicate different

---

[1] Dr Doolittle stories by Hugh Lofting (1886-1947)
[2] RA Gardner, 'Teaching sign language to a chimpanzee' in *Science* 165.3894, 1969. pp664-72

words, even if they do learn to understand a lot of human language. We needed to invent some other way that both human and animal can express individual words.

I managed to arrange a lecture tour around the USA sponsored by the Wellcome Foundation and went to see the Gardiners and discussed my problem. They had some interesting ideas, one was to use rhythms for horses and develop that as a language. Our horses, dogs and cows could probably learn to beat a drum with their noses or feet. A computer was the other option, where we would learn common symbols to mean different things (this has now been done with some primates).[1] If I had had access to research funding to work on this with the animals by our firesides or in our stables it would have been an option, but this was very unlikely so not worth the time and effort applying; we had to find another practical way to test their ability to understand human language, without using expensive equipment.

In the days before email, the faculty often met for coffee and talk in university departments so there was plenty of opportunity for cross fertilisation of ideas by random contact. It was a random encounter with a psychologist that started me on the next twenty-five years' research. One of the senior lecturers in the psychology department at the University of Exeter had researched why some children were particularly intelligent or gifted, and how parents could encourage this. I mentioned to him that I was debating how to develop a common language with large herbivores. He pointed out that gifted children were classified "gifted" at nursery school because of their comprehension of language, not because they spoke it particularly well. This set me thinking. Could we use the same techniques that he had found worked with human infants to teach large mammals to learn to listen to and understand human language? We all know that dogs, horses, cats, goats, elephants and cattle will respond to gestures, facial expressions and the expression in the words used. But could they learn to listen and

---

[1] S Savage-Rumbaugh & KE Brakke, "Animal language. Methodological and interpretive issues" in M Bekoff & D Jamieson (eds), *Readings in animal cognition*, 1996, MIT, Cambridge Mass, pp269-88

understand the meaning of the words themselves and gradually build up their comprehension in order to understand the necessary concepts, phrases and sentences? No one had done this in any directed way, although throughout history, there certainly had been people who have lived with and taught non-humans to understand a lot of verbal language, some even in the circuses for example.

Teaching animals has not been seriously studied as a science, either people have a "way with animals" and can teach them, or they cannot. But the Behaviourists of the last century learnt a great deal about how learning works in mammals by doing experiments with them.[1] Today the Behaviourists are having a bad press because what they said is often wrongly interpreted. As a result, the idea of "conditioning" is now understood by most animal training manuals to be some unchangeable/instinctive-like reflex once it has been established. In fact, both Pavlov and Skinner, and many of the others of this school, were very aware that voluntary action is the result of an individual's experiences and therefore responses are very varied, and can be changed. Animal trainers throughout history have always known and used this understanding of how animals learn to do voluntary actions, but today ironically many of the "new developments in animal training" are using a "conditioning" technique (e.g. clicker training) which is based on this faulty understanding.[2] This is not how most learning works. Different species have "instinctive tendencies" because of the way their body/brain, that is their "being", is. They have evolved to tend to do certain things as a result. For example, in humans there is an "instinctive tendency" to walk on two legs once the normal child is old enough. But where, when, how or even if he does this depends on his lifetime experiences, if he has

---

[1] e.g. I Pavlov, *Lectures on conditioned reflexes*. 1928, International Publishers, New York; BF Skinner, "Some responses to the stimulus Pavlov" in *Cumulative record: A selection of papers*, 1972, Appleton-Century Crofts, New York

[2] WT Rockwell, "Beyond determinism & indignity: A reinterpretation of operant conditioning," in *Behavior & Philosophy*, 1994, 22:1, pp53-65

one leg, he will develop another way of moving, or if he has been raised by wolves, he may not even try to walk much but rather gallop around on his arms and legs as some teenage wolf children did.[1] Walking involves a lot more than genes, or "conditioning", it involves the lifetime experiences of the individual being. He has an "instinctive tendency" to do certain things, but this is moulded by his lifetime experiences.

This language comprehension project would require youngsters of several species who had a rich life, emotionally, mentally and physically, like most children, rather than living in a prison-type environment such as a laboratory. It would require a lot of time both talking to the animals, and measuring their responses and the behaviour of the talker/teacher. The major question was, was the ability to learn to talk or comprehend human talk, crucial for advanced mental development? Another question was if they could learn to understand human language in some detail, then what would be the moral implications for their treatment? Thirdly, if this way of teaching some mammals to learn to comprehend human language was successful, could this approach be used to improve their education, teaching or training? Finally, what are the limits of what each species could learn to understand? Maybe, this research could lead to a new science of animal educational psychology.

The first thing to decide was which youngsters of which species it was practical to study. It had to be restricted initially to animals on the farm where we could monitor their lives. We regularly bred sheep, cattle, horses and guanacos, but our dogs were all male, perhaps we could buy a pup. The youngsters who were going to be worked with we would get to know very well, so it would be preferable to use female youngsters as they could thereafter be kept for their lifetimes on the farm, whereas males might have to be sold off after maturity. I had an ex-doctoral student who volunteered to help two mornings a week. We decided to start in the spring and select one female infant of each species. Rosie (our Jersey milk cow), had a daughter Rambling Rose, the mare Druim-

---

[1] DK Candland, *Feral children and clever animals*, 1993, OUP, Oxford

ghigha Shiraz had a filly, Druimghigha Shemal. We chose an infant female guanaco Xanapa, and bought a Springer Spaniel dog pup called Rupert. A couple of years later, Toto, an eight-year-old one-eyed orphan African male elephant in Zimbabwe was included.

The first part of the study, with the youngsters born on the farm when they were twelve hours old (the pup eight weeks old, and the elephant already eight years), was to handle them frequently, talk to them as you would a baby of two to nine months old. They were brought into the schoolroom with their mums (if they had them) for the first five months of the experiment. Once they were not anxious about leaving their mums for short periods, their mums were left outside. As they became accustomed to us and to our talk, we began gradually naming parts of their bodies and ours, pointing out what they were doing and what the speaker was doing. When they were one month old, they learnt to have a head collar put on and to be led on short walks, with their mums in sight. As time progressed, the walks became longer and they saw and experienced different places, situations, animals and people, but again, if they were anxious, and we had their mum, she would come along as well. We gradually named more objects of interest to them, and introduced verbs, pronouns and adjectives for them to begin to associate what they were doing and seeing with the right words, more or less what many mothers often do with their babies. Finally, we had around 300 words which were used.

After the first month, the second part of each half hour session which was conducted twice weekly, was to ask the subject to do simple actions. These were taught as if they were young children, that is by the teacher repeating the word, using gestures, and doing the actions so that they could be imitated. The reinforcement (reward) was laughing and smiling, an enthusiastic "yes", "good girl", or sometimes a food titbit, when they began to get it right. A "directive" was "no", shaking of the head, or frowning was also used when they got it wrong, or were messing about not trying (this is called "negative reinforcement" but it was never violent). They were asked to learn to do simple things that all found equally easy, such as "lift your right leg", "lift your left leg", "shake your head". They were rewarded when they started to do the right action, but later only when they did it better. Finally, they were asked to do it to the words only, without any other clues. The teacher had to stare straight ahead or look away from them and stand still so as not to give any cues where she was looking, or what she was indicating. We continued these sessions for two years and all the results were written up and published.[1] Whatever else it showed, one thing was clear: all these animals learnt fast with the help of a human teacher, much faster than in learning studies in the laboratory where the subject needs thirty to fifty or more trials before they do what is required. Here the subjects started to do the correct responses within seven to ten trials. We had shown that they responded to the same method that is used to teach children to comprehend language very quickly. Perhaps we need to recognise how relevant psychological research and understanding of humans is to improving our understanding of non-human mammals too!

Why had no one ever studied this before? And why did the scientific community not recognise that teaching methods for humans would be applicable to another mammal? Twenty-five years later, some animal cognitive and welfare scientists are at least considering this.[2] There was only one subject of each species, which, in scientific terms, means that the results are usually dismissed. This is an irrational approach, since there is no escaping the conclusion that if one member of a species can learn the task, it is not possible to say that their species cannot do it! Not all humans are thinkers like Aristotle, but humans can produce an Aristotle. Do we need to take fifty humans and see if there is a statistical majority who can come up with ideas and arguments like Aristotle, Plato or Kant, before we conclude that humans can produce such thinkers? Do we need fifty horses to "prove" that a

---

[1] MKW & H Randle, "First steps in comparative animal educational psychology," 1996

[2] https://www.popsci.com/science/article/2010-12/video-smartest-dog-ever-can-pick-out-1022-toys-name

horse can learn to recognise a colour by its name, if one can do it?[1] No we do not, and is it not time that "cognitive ethology/animal behaviour" made an effort to catch up with critically assessed folk knowledge and philosophy from some 2,000 years past?

During this period of my life there were three major lessons learnt. First, one could seriously begin to understand what mental aptitudes all mammals had. Second, we could use what we know about how humans develop and experience the world and apply this to non-human mammals. Third, studying learning theory and knowing it is only one aspect, but putting this together with personal experiences when teaching is quite another.

## Volunteers, students, multi-species living & teaching elephants, buffalo & black rhino

When we came to Little Ash, there was too much skilled work to do to continue to host WWOOFers, so we had a few years' break from welcoming other humans to share our multi-species lives. Although often idealistic and enthusiastic, we had long since learnt that it was not often that these folk paid for themselves with their work because it took much time instructing them in the simple physical tasks we asked them to do and, then, they were often very slow since neither their bodies nor their minds were used to doing them. It was often quicker and a better use of our time to do the job ourselves, particularly if the job was slightly complicated. There are of course positive reasons for having volunteers and students. They may amuse and interest, they may have a skill that we badly need at that time. I remember once on Mull we had just bought a lorry to convert into a horse box and it had an hydraulic lift on its back door. We had no idea how this worked or how to adapt it, but as luck would have it, a young Australian turned up who specialised in hydraulic lifts for lorries!

---

[1] MKW & E Hartmann, "Enhancing learning of verbal cues in horses ..." 2006

... So there are times when WWOOFers are very helpful.

We have shared our twenty-four-hour lives with very many people from a host of different countries, learnt about them, and them about us and our animals, and we have friends or even family all over the world as a result. Living full-time in a multi-species and multi-cultural environment increases the richness of life, however it can also have drawbacks. But, whatever else, it makes one appreciate the importance of tolerance and adaptability. Tolerance and adaptability are the specialities of domestic animals. In the early days we knew less about people I suppose and were less discriminating, so we learnt.

One well-remembered miss-matched couple were a long, thin, gloomy, intense bloke with a ponytail, and a pretty brunette girl who smiled a lot and would have been quite jolly, if he had allowed her. They did our course on ecological agriculture for a year, graduated and were offered a job running one of the farms belonging to John Seymour[1] who had been given several farms to install young people on in Wales by philanthropists. They spent the first six months in Wales, arguing whether to use only horses for traction or have a tractor. By the end of the growing season, the money had run out, they had no crops and the whole thing ground to a halt ... ah well ... it did not say much for our course if this could really happen with our graduates! So we worked away at improving it and selecting the right students. It took about ten years before we realised that the most important thing was to teach the students that nothing was impossible. They had to think how to solve their problems with the resources they had, not go along with the preconceptions of other people. The only dogma we hope we have is, "there is no such thing as can't".

At one time we had seven students, or ex-students (those who stay on to make their living from one aspect of the farm). One was a delightful Irish girl, a new age traveller, who turned up with her skewbald cob pulling her caravan. She realised that although some of the new age travellers were anxious to use horses rather than motors, they had no experience of them and needed it. She

---

[1] The diva of self-sufficiency in the UK Practical Self-Sufficiency

and her enchanting, knowledgeable mare were a delight and graduated to continue on their travelling way. Then a couple of Jake's local pals did the course; one has since developed his own business building chalets out of sustainable wood. Another couple of girls bought two of our young horses and found land and boyfriends locally; some returned to their own countries; some decided that farming was fun but not for them for life. We had fun and tears, boy and girlfriend troubles and delights, babies born, marriages, disasters, divorces and so on, the whole of life seemed to be enacted on the farm. There was no need to go elsewhere for human or animal experiences!

We took to having Open Days once a year and sometimes nearly 1,000 people turned up, organic farming was hitting the headlines. We would put on dances and displays with our cattle and horses, guanacos, dogs and sheep, and some were televised (*Farm Fantasia,* Channel 4). We showed the things our students had made: recycled buckets from the plastic silage covering, macramé pot hangings and harnesses from baler twine, wind-mills from old generators from written-off cars that generated twelve-volt lights, apple juicers with bicycle power, harnesses for oxen and horses made from old tyres, molasses squeezed from our sugar beet and a host of other things that the students dreamt up with a little help.

Workshops on animal welfare on farms, with horses and with pets, continued to be asked for. A blind psychologist (who had written a book on trying to convince Guide Dogs for the Blind to update their training of guide dogs), his dog and I gave a workshop together. Some of the people who later became well-known "animal behaviourists" had their introduction to welfare and teaching animals with us. The animals were always obliging and soon realised that audience applause was worth it. They taught the participants, we just had to make sure they covered the syllabus. But all they got was, admittedly a four-star crust, a lot of admiration and we hope we gave them the particular individual consideration they deserved. Whether a horse, dog, guanaco or cow, they were always perceptive about the students, picking out those who were really interested or who had talent, from those who "knew it all", were just doing it for a laugh or to impress their human friends.

In addition, we were well placed to host visits from scientists, philosophers, PhD students and vets to discuss, learn and watch. A Tanzanian vet working for the UN after doing a course with us, invited me to Malawi to help with the welfare of donkeys. There are over two billion working equines in the world belonging mainly to the poor. They don't all have unpleasant lives although help is needed with some things. For example, how to design and make better harnesses so they do not have harness burns and wounds, and can carry the weight on their backs and not on their necks as bullocks do. But the harness must be made from the materials that they have available since there is no money at all. Teaching them to make it all out of prepared leather, as many have tried, is not going to make any difference to the life of the donkeys, since their owners will never have the money to make or maintain the harnesses. Often their owners are no more violent with their donkeys than they are with their wives, but even then, if they knew how to get along better and teach and live with them without violence, it would be better for all. The poor who own the donkeys do not usually brazenly overwork and underfeed them; they are dependent on them: no donkey, no work, no food for their families. They are only too aware that they must keep their donkey in good health and anxious to do it within their means. Our job was to help them do this.

An international society was started in the 1980s in Edinburgh to advise veterinarians and others involved with working equines, and to research their welfare. One study showed that the major welfare problem for working donkeys worldwide, is not being over worked, under fed, lame or having wounds and illness ... it is worms! Something quite easy and cheap to fix. We were invited to visit a clinic run by the Donkey Sanctuary on the isle of Lamu off the East African coast where they have no motorized transport. Since there is no grass or green materials other than palm trees in Lamu, most of the donkeys who wandered about free, except when they were working, ate whatever turned up on the rubbish dumps, including a lot of cardboard. They also stole what they could from the fruit stalls, being shooed away by the owners like naughty children. But they worked hard, moving all the building materials: sand and concrete blocks as well as tourists' luggage, and everything else around on the island. We hardly saw a donkey that was too thin, abused or seriously ill ... The secret was free worming provided by the Donkey Sanctuary. The donkey queue for worming medicines outside the clinic was long, and when not working, the donkeys were as free as the humans, blocking up the very narrow streets as they meandered around. Their status in Lamu was more or less the same as that of small boys. Some were better off than others of course, but it was a delight to see how they were integrated into the community, together with chickens and their chicks, cattle egrets, cats and dogs. There was no animal apartheid in Lamu: all seemed to get along and were contented ... except for the animal welfare tourists from Europe who thought it all "disgraceful"; ah well, you can't win them all.

We had a Zambian vet visit for a course, and he invited me to Lusaka to teach his vet students at the university. I decided to make a stopover in Kenya and visit the emerging organic agricultural organisations. The British Council gave me travel money, and I was on my way back to Africa.

There were two important developments as a result of that trip. Firstly, while in Kenya, I decided that we should help with the training of their organic agricultural advisors by offering scholarships to Kenyans to come and do our diploma. The second was that while in Malawi helping with donkeys, I heard of a farmer in Zimbabwe who had impressed people by how he lived with many wild animals that had been brought to the farm injured or orphaned. I decided to contact him, and he invited me to spend a day with him. A young white Zimbabwean lad who spoke Shona (it is important to be able to talk in the local language to the people who do most of the work) and I drove down from Harare to visit him.

The owner of Imire Game Park was a delightful old gent, Norman, who was originally English but like many of his generation born and brought up in Asia somewhere. He had had a chequered career (as most young men had as a consequence of the Second World War) and ended up with a large family and a larger tobacco and beef farm, much of which he had recently converted into a nature reserve. This had been restocked with many of the indigenous animals who live on the savannahs in that area (impala, kudu, eland, zebra, buffalo, giraffe, nyala and many others). By clever public relations, he had managed to hold onto the farm after the redistribution of land at the end of the Smith UDI (Unilateral Declaration of Independence) regime when Mugabe took over in Zimbabwe, and he seemed to be the repository for orphaned animals found almost anywhere in that country. There was a young hyaena, a little lioness kitten and a cheeky young warthog running around in the garden, much to the outrage of his wife, Jilly, who grew the most magnificent roses. We chatted, I watched him with his animals and visited some of the enclosures with several adult carnivores living in them. He mentioned that he had previously acquired some black rhinoceros and elephant orphans who were growing up. The black rhinoceroses belonged to the National Parks Department but they needed raising and had been loaned to Imire Game Reserve, who were reputed to be able to do this before they were "rewilded". The elephant orphans had been captured by the National Parks as their mothers had been killed or poached for their ivory. They had been sold to farmers and private nature reserves to raise and were between seven and fifteen years old when I arrived, and getting bigger. A full-grown elephant costs a great deal to keep in food and supervision. Each elephant had to have one and a half handlers (because of days off). Norman had the idea that he might be able to teach them to be ridden, so they could pay their way by greeting tourists and one day giving them rides around the nature reserve to view the animals.

There is a general belief that African elephants cannot be trained, unlike Asian elephants. The origin of this belief has a history. At the end of the nineteenth century, King Leopold of Belgium who had colonised the Congo, hoping to make money from the timber and minerals, had imported some mahouts from

India to train a group of young elephants captured in the forests. The idea was that they would be able to extract the timber as they have been doing for a century or so in Asia, but, both the mahouts and most of the elephants died as a result. The reason for this it seems was that these Asian mahouts were very rough, and frightened the captured elephants who then killed them, and then they were killed. The preconception that African elephants cannot be trained has inhabited the minds of almost everyone ever since and as a result, training African elephants has rarely been attempted, except of course by the circuses. Some circus trainers had managed just fine, but because circuses are circuses, there are training secrets that must be guarded, so how this had been achieved was not widely known or discussed.

Norman could not afford to hire Asian mahouts, so being at a loss how to further their teaching and very aware how dangerous they could be, he took me to see his elephants, having found out that I had practical experience teaching large mammals and five years' research data on how to improve their teaching using ideas from how gifted children were raised. I suggested trying the same techniques with his elephants and rhinoceroses, so the handlers were asked to bring the six young elephants up to us, five bulls and a female, ranging from nine to about sixteen years old. The handlers were using their ankuses (a metal bar with a hook on the end that is stuck into the elephant to make him do something), prodding and poking, but they also were giving them verbal commands, so were not using the ankus too much. As a result, the elephants were reasonably calm. It quickly became evident that both the handlers and the elephants wanted to advance in their relationships and what they were doing, and how, but they were muddled. Norman had a stab at teaching them to be ridden by cooperating with a polo playing farmer, Rori Hensman, who also owned some orphan elephants. The two of them had decided to try teaching them using some of the same techniques used to teach polo ponies. This was a challenge for the handlers (who had never ridden anything but a donkey) because they had no bridles or saddles and consequently no physical way of controlling them.

How could they make them go or stop, turn left or right? If the elephants did not want to do something or did not understand, all they needed to do was shake their heads and the riders were hanging on by their fingernails, and it was a long drop down! So they were at a loss, although they had found that if they used words they knew, they could ask them to go forward, and sometimes they stopped when asked. These lads spent all their days with the elephants ensuring that they did not wander into the farm crops or cause other problems, so they had some idea how to control and relate to them on the ground; when I knew them better, I heard a few stories of their various adventures.

Sitting on the neck or back of an elephant means that the rider can reach their ears with the left or right foot, and tip the edge of the ear forward. They had found that eventually, accompanied by the word left or right, the elephants would learn what this movement of their left or right ear meant, although they did not necessarily want to obey it. The elephants would follow each other on well-known routes but it was not possible to direct one to leave the others or go another way. He would just stop and turn around or start showing signs of impatience and worry, stamping, moving about, shaking his head, lifting and throwing the trunk around; all of this frightened the handlers. I went up to the oldest, largest, fifteen-year-old bull: Nyasha, and started to talk to him in the way we had done with our subjects in Devon, using simple words, gestures and expressions. I repeated the word: "lift your right leg", "cross your legs" and so on, making gestures, doing the movement myself, and praising him when he began to do the thing I had asked, even just a bit ... just like when you teach a young child. He watched me and after a few attempts, he started imitating my action. Nyasha latched on and he began to perform simple movements correctly after only five or six trials. This was a surprise and a delight, and he was well rewarded with food, smiles, claps and laughter. It seemed that the method of animal teaching based on gifted children's teaching would be helpful for Norman and the elephant handlers and might, eventually, help them to earn enough money from tourists to keep the elephants when they were fully grown.

We then went to see the seven young orphaned black rhinoceroses. They were shut into stables at night with armed guards and protective sacks of sand to prevent them being killed by poachers. During the day, they were free to wander around and browse and graze, under the watchful eyes of the armed guard. They had therefore some experience of being herded but not much of being handled after they had been weaned off bottles. However, again within thirty minutes of quietly standing and chatting to them, they came up to me, and I managed to scratch their backs, and one lay down to expose his belly for a more thorough scratch with, what one believed to be, a look of bliss on his face. It looked as if these black rhinoceroses would also be good subjects to teach using our gesture and verbal methods. It was an interesting and fun afternoon, and after cross-questioning the handlers in Shona with the help of my nine-year-old colleague, we drove back to Harare.

Four months later, Norman and Gilly came to visit and interview us in Devon. They had to camp on the top floor of the barn and were woken up when we collected Ramblie to milk in the morning, but they fitted in well, and vetted us to see if we were for real with our relationships and teaching of our animals. They obviously approved as they asked me if I would like to return to Zimbabwe to work with their elephants and in particular to train the handlers, and run courses and examinations for them to make them realise that the welfare of the elephant must come first, so that they had a profession which was recognised and important. This was what I had hoped might happen and I said yes, but stipulated that part of the payment must be that they immobilise and catch one of their wild zebras for me to test whether, if by using the same teaching techniques we had developed, we could train an adult wild zebra to do the normal things that donkeys, mules and horses do reliably and with good temper. Zebra training had had bad publicity. Someone had taught a foursome to pull a carriage at the end of the nineteenth century, but everyone, including most circus trainers, believed that zebras were very difficult to train and always unreliable and bad tempered.

Much of our teaching research has been subjected to the shrugging of the shoulders accompanied by a statement that of course horses, cattle and dogs could learn this and that because they were domestic and their behaviour had changed genetically from the wild. A wild animal could not be taught to do the things that we teach domestic animals to do. If this was the case, then surely it would apply equally to human mammals: tribal people would not be able to learn to live in cities and use the Internet for example. For a time, during the slavery era that was believed, but today we know better and such racism is condemned. The curious thing is that if it is lifetime experiences which allow humans of any genetic background to live in many different environments and acquire different knowledge, why would this not happen among other mammals? Were non-humans and particularly domestic mammals so genetically different behaviourally that they would be unable to acquire other knowledge, were they just pre-programmed robots, and if so why and how, if one believes in evolution, could this have happened to non-humans but not to humans? And anyway, was it in fact true that zebras or African elephants could NOT be trained like their domestic cousins? To test whether or not this was so, we needed to capture, and handle an adult zebra stallion who had been running wild in the nature reserve all his life and never previously been touched or manipulated by humans, and teach him using the method we had developed from studying gifted children. We would measure if he was slower or faster, or unable to learn what we asked, and to cooperate with us without irritation and bad temper. Even if one zebra would respond well, it could no longer be maintained that "all wild equines cannot learn what domestic equines can, and live with humans without trauma". This was an opportunity to do a test that I had been wanting to do.

A few months later, my research assistant (who was a student who had done her PhD with me on the behaviour of our cattle) and I rapidly plunged into a hard-working but fruitful six-week research and teaching programme in Zimbabwe. Early every morning we went to the elephants to teach them to listen, learn and understand more English, and taught the handlers how to do this. I taught both the elephants and their handlers, and she recorded the behaviour of the elephants and the handlers. We began by handling and brushing the elephants and naming parts of their bodies, then progressed to asking them to do simple things to word command. The more confident they became in being handled, and the more skilled the handlers became, the more the elephants learnt, and the more complex were the things we could teach them, with no hitting, poking or shouting. Plenty of praise and a show of delight when they got it right, but also stern voice and facial expressions when they did things that were not permitted, like chasing a rhinoceros or not moving in the direction required when not frightened. The rest of the days were spent following the elephants and the rhinoceroses around as they grazed and browsed, recording what they did, where they went, how they communicated with each other and how their societies were organised.[1]

The evenings we spent studying the rhinoceros in their night-time stables to discover why only one female was breeding although four of them could be.[2] Last thing before bed, all the information had to be put into the computer.

It was hard work but fun, I got to know the handlers very well, and enjoyed their company. They were all farm lads who had volunteered to work with the elephants. They did not know much about elephants or any of the wildlife around, but were anxious to learn and several had visions of becoming guides for tourists, considered a great step up the social ladder that included more pay. As soon as they started to take tourists for rides on the elephants, they were considered authorities on elephants and the local wildlife, so they needed to learn some answers to questions the tourist would ask ... although I quickly learnt that they were

---

[1] MKW, "Communication in a small herd of semi-domestic elephants," 2019

[2] MKW & H Randle, "Implications of semi-intensive management ... Black Rhino," 1996

very good at working out what the person who asked the question wanted to hear and blagging it! But they took our course seriously, and each completed an individual project, such as what plants the elephants were eating, what plants there were in different parts of the nature reserve, how far the elephants walked each day and so on. They were sitting in the bush with the elephants from 7 a.m. to 6 p.m. so had plenty of time to read books, identify plants, learn about how to teach animals, as well as have the odd snooze. In addition, they all took part in collecting data to see if elephant communication resembled human language in any way. This involved recording over ninety different behaviours, who did what to whom, and what was the response. This required identification of the individuals they were watching, and observing them very carefully so that a slight ear, tail or trunk movement or changes in facial expression like slight tightening of the chin were recorded, and what, if any, response there was from another elephant. In this way we could analyse the meaning of different postures and movements, which was eventually written up and published.[1]

We all had a busy, tough month, but everyone seemed to enjoy it. At the end we ran exams for each of the six men and they all passed. It said a lot for the level of literacy in Zimbabwe that they were fluent and could read and write English well, even though some had only been at secondary school a year or two. It turned out that the level of literacy in Zimbabwe in the 1990s was higher than that in the UK, where there was free compulsory schooling up to sixteen years old!

Using elephants for tourist activities could generate jobs and income but it was highly politically sensitive ... if there were political disruptions, as there began to be after my first visit, then tourists did not come and how could the elephants pay for themselves? The indigenous populations often have to live with the problems that elephants can cause. They can destroy crops, push over trees and frighten people. Consequently, they have a bad press among the local small farmers adjacent to national parks

---

[1] MKW, "Communication in a small herd of semi-domestic elephants," 2019

and nature reserves, and most of these people want them removed, not conserved. The indigenous people's ideas about elephants have to be reversed if elephants are going to survive at all in the future. As human populations grow, there is growing conflict between elephants and the local humans, and the local people's interests always trump those of life-threatening interests of the elephants. So how can we get the local people on the side of the elephants and other wildlife? Tourists may come and bring them more money but this had proved erratic. What we needed was for the communities to appreciate elephants for themselves so that they wanted to share the world with them. One possibility was to think how these semi-domestic elephants might help the villagers in one way or another with their work. This might turn the people's bad experiences with elephants into good ones because, like it or not, it is the local human populations who will or will not ensure the survival of their wildlife in the end, not any direction from outside.

Consider for a moment the reason why rhinoceros are threatened and will be extinct, except in large zoos, within another decade. It is, of course, because the local population can enormously profit from helping poachers. Within one night, they can earn enough money from a poacher's back-hander to pay for the schooling and medical care of their children for life, no wonder it is tempting. Elephant ivory can now be sold even legally in many countries, so we expect even more elephants will be killed for their ivory, and young orphaned elephants will die or be caught and sold to be raised and kept in varying situations, from zoos to private wildlife parks. If any elephants and rhinoceroses are going to survive in the future, even in enclosed areas, we will need to interest the local populations in keeping them around, as well as have international legislation to help those who have, or keep, elephants. Of course, both the elephants and the humans who are close to them, must have a life of quality, and we need carefully to consider how to achieve this.[1]

Norman Travers and I agreed that we should see if it was possi-

---

[1] We Are All Mammals. Charity number 1184219.

ble to teach his elephants to do something useful for the local community: farm work, transporting materials or maintaining roads for example. There would be two advantages if this worked. Firstly, the elephants would be helping the farmers do work that would otherwise take them more than treble the time, so interest in the elephants would be generated in the local community. Secondly, the local people, men, women and children could come into contact with elephants, many of whom had never seen, touched or experienced one, and had no idea that they have some mental abilities, emotions and experiences that are similar to humans.

It is only when having close contact and living with them, that one becomes really aware that the other is a living, feeling being who experiences emotions and has desires and needs, like you. Consequently, generally people who have had close contact with another species, become more sympathetic to their survival and conservation because of who they are, and not just as a product to market or an added "aesthetic experience" as one sees them walking off into the sunset through the binoculars. Many years ago I had realised that studying animals in the field was interesting and challenging, but having direct contact when handling or teaching makes one much more aware of their whole body and mind "being", their intrinsic value, and their similarity, in many ways to oneself, particularly when it comes to emotions and intentions. On top of this, when as a result of handling and teaching, the animal does something useful for you, like pulling a harrow or ploughing a field, it combines these experiences in a cooperative activity. If we want the local people to really want to conserve their elephants in the future, we need to use this approach; the elephants must help the local people in one way or another, and in exchange they will help the elephants.

For this reason, Norman and I discussed what we could teach the elephants to do that might be of use to the local community. Yes, they could walk over to the school and let the pupils see them and do something with them, play football for example. We had taught a couple of the elephants to kick a ball which they learnt very fast by imitation, but they liked to kick it back, rather than kick it away to another elephant or human ... demonstrating they were in the same team as their handler!

One thing of value to the rural society of small farmers was to see if they could pull a plough and cultivate the fields, another would be to pull a harrow or tyres to flatten the muddy roads after rain. If we could find a trailer, they might be able to learn to pull it to transport products such as manure, thatching straw or even people. These small farmers had no internal combustion engines and they could not afford them. Even if they miraculously were given one (which had been the case from AID agencies), it was only a few weeks before it broke down or ran out of fuel and no one had the know-how or the money to fix it.

The first thing was to design and make a very simple harness for the elephants, and then see if we could teach them cooperatively to work in harness safely, without frightening or worrying the elephants and with no accidents. Then if they could learn it, would they do it willingly and, we hope, enjoy it? We bought some canvas to make a breast plate to be padded with wool or spongey material and gave it to the local tailor to make up. We made some rope traces and a whipple pole (the pole that the traces are attached to. allowing the implement being pulled to move smoothly forward rather than step by step). An ox plough was borrowed from a neighbouring farm. The first lesson for the elephant was to learn to pull, so some old motor car tyres were tied together for the elephants to pull along the road, and eventually a harrow (an implement that has spikes into the earth and when drawn along flattens it ready for planting) was found rusting away near the farm yard. The old ox cart that had arrived from the South African Cape during the Boers' great trek north was collapsing at the entrance to the house and with a little tender loving care it would rumble forward we hoped, this time to be pulled by a couple of elephants instead of the eighteen oxen it had been designed to be pulled by.

All we needed now was to accustom the elephants to the harness and teach them to pull without being worried. Luckily, I had much experience of harness-training different types of horses, including those believed to be very difficult to train to use safely in harness (such as thoroughbred ex-racehorses, or pure-bred Arab show horses).

It was important that the elephant understood the concept of what he was doing. The idea here comes from understanding the difference between "learning that", learning what is required, and "learning how", that is how to do it in practice. The philosopher Ryle pointed this out using the bicycle as an example: you know "that" a bicycle can be ridden, but you may not know "how" to ride it".[1] The elephants had therefore to learn "that" when they walked forward there would be a pressure on the breast plate and if they continued to pull with this, the implement would follow them. But then "how" were they to do this? They must keep the pressure up by putting one leg in front of the other and continue to pull; when going up hill more pressure would be needed and going downhill care taken so that the cart or implement did not run into their back legs. To start with, any animal being trained for harness work will stop when he feels the pressure on him from the traces, as he thinks that any pressure on his body (like the reins to a bit in the mouth) means "stop". He has therefore to learn the difference between pressure to stop and pressure which means continue. This means we first teach the animals the point of being harnessed to an implement (in this case some tyres strapped together) is to move forward quietly and thus pull the implement, learning: "if you pull quietly and slowly, the thing behind will come along with you". Inevitably the object pulled makes a noise, so it is very important that the elephants do not worry about this or they could take off trumpeting, terrified, with the implement attached making more noise as they run. Then, we need them to be responsive to the voice, understand things like "pull" and particularly "wait" and "stop/ho", and have confidence in the handler.

Because, when riding elephants there is a closer touching bond between the human and the rider, and both the elephants and the riders know each other, we decided that it would be safer to have

---

[1] G Ryle, *The concept of mind*, 1949, Penguin, London, chapter 8

the handlers riding them, talking to them and asking them to do things that they normally did when being ridden. Getting used to the rattle and crash behind is what frightens most horses, so we wheeled noisy wheelbarrows full of tins behind them until they were quiet with this, before we attached them, and finally asked them to pull the vehicle, just one step forward then stop, and then another and stop. If they are okay and quiet with this, then a few more steps until they get the idea and begin to wander along, happily pulling whatever behind. We started with tyres to flatten the road after rain, then, when they were fine with this, progressed to the ox cart, and finally a single elephant ploughing. Ploughing requires a lot of energy and the elephant to walk straight in the previous furrow. It also requires someone to guide the plough behind, and keep it cutting into the next piece of ground. Much to my surprise, our elephants really took to all these things and they were a great deal easier to teach than a normal young horse! They quickly understood how to plough well, and we fantasised about taking them to the ploughing competitions that were sometimes held in Zimbabwe with oxen as well as tractors. Ploughing is difficult to do well, which is why of all the work done on the farm, it is ploughing competitions that demonstrate how able the farmer and his team are.

We would do about one hour's work a day, then they had a good feed. It seemed that the elephants enjoyed it and were amused or at least stimulated by the activity. So, we started to think of other things they could do, for example, a set of harrows had been lost under water in a flooded area. The water was about one metre deep, and Nyasha was asked to wander in and try and find the harrows with his trunk and pull them out. He got the idea that he was looking for something after we had demonstrated this, and wandered about looking and feeling with his trunk encouraged by his rider, and within thirty minutes he had found it and pulled it out all covered in mud and weed! He was encouraged to balance it on his tusks and he walked out of the lake with it. Was there nothing these large strong, willing beings would not do for us? In *Elephant Bill*, JH Williams describes an incredible number of difficult tasks that the elephants normally did in the forest for timber extraction, either learnt or taught themselves, which consequently helped to win the war in Burma for the allied troops.[1] We should not be surprised, but somehow since elephants (and other animals such as horses and dogs) are rarely used for work now, we have forgotten how able and creative they can be, and what willing partners they are if not abused.

The elephants were most curious of the harness. When it was hanging on them they would constantly try and pull it here and there. Although it was tough, they could have ripped it to bits in no time, so to avoid this, we had to make sure that as soon as it was on, they were off pulling something, and then they did not seem to mind it at all.

We started with a single elephant at a time teaching first Nyasha and then Makavooshi to pull the tyres and to plough, which was much harder work. But then we harnessed them as a pair, thinking they could be together and give each other confidence and also pull heavier things. The elephant handlers who had ploughed with oxen, thought this was just wizard as they went so fast and strongly with a three-furrow plough rather than their normal one-furrow oxen plough. Then we put the pair to the ox cart and they pulled laughing wives and children around. Eventually they were asked to do various chores such as carrying compost from one part of the farm to another.

One surprise was how hot the elephants pulling a loaded cart became. We had to stop and let them spray themselves with regurgitated water, and have a rest and a snack at quite frequent intervals. Although they were pulling a much smaller percentage of their body weight than a horse or donkey is expected to pull (around ten per cent as opposed to as much as 100 per cent for a horse or pony), they clearly found it tiring ... They certainly were not very "fit" when compared to horses or donkeys used in harness, even though they wandered around in the nature reserve every day covering ten to fifteen kilometres. This set me to wondering about whether wild animals really are exercise-fit general-

---

[1] JH Williams, *Elephant Bill*, 2001

ly. The zebra we worked on a lunge also showed signs of tiring quickly, and previous reports, too, of using zebras in draught had found that they tired rapidly. In a run-up to the anti-hunting ban in the UK, a colleague had found that the stags that were hunted keeled over from exhaustion before the hounds actually attacked them, because they were not "fit" enough to keep running, nowhere near as fit as the dogs. Perhaps wild animals are not really very fit, after all when you think about it, the same animal is not often hunted by predators (unless it is sick and then it is usually killed); they are rarely chased for long distances or go so fast that they use more oxygen than they are taking in (anaerobic respiration). It is these activities which establish cardio-vascular (physiological) fitness.

Assessing exercise fitness can be done by measuring the speed at which the heart rate decreases after exercise. Today, there are heart rate monitors sold so that runners can check their fitness, and we use these to monitor the fitness of our horses for long distance races. We tried them out on elephants, and found by using the same equipment (which is now constantly being upgraded), it is possible to monitor how fit they are and whether they have done enough or too much and are tired. Perhaps we have selected our working animals, and ourselves (if we exercise), to have a cardio-vascular system that can adapt to longer and faster exercise and to recover faster than some wild animals. It may be that wild, hunting carnivores are fitter but so far no one has looked at their recovery rates, although we are beginning to measure these in working dogs who have remarkable abilities. In this respect, I was recently invited to one of the great sleigh dog races in Canada and the USA, the Quest, to make suggestions for the further development of the assessment of the dogs' welfare by using different physiological and behavioural measures. The dogs are inspected every fifty miles or so throughout the race. These dogs run pulling a sledge for 1,500 kilometres through the Yukon to Alaska in ten days or less and are carefully monitored at vet checks throughout, particularly for exhaustion and lameness. Like almost any other animal and human race, today the organisers and participants are frequently under attack from animal welfare activists. My job was to assess whether or not the inspections were adequate or could be improved to ensure more thoroughly that no dog suffered. There were some improvements that could be made and also further research that needed to be done but one could not help admire the incredible feat of both the dogs and the mushers in doing this race, often with temperatures as low as minus fifty degrees Centigrade!

We can safely say that both the harness and ridden work was enjoyed by the elephants and the handlers and we ended our stay by performing an elephant-ridden quadrille, accompanied by local drummers, and demonstrating all the other things the elephants and handlers had learned. This was in front of an audience from the local schools and residents, some elephant owners and even some government ministers. We also displayed our research results about the behaviour of the elephants and the rhinoceroses which meant many late nights in order to have the data analysed in time. This was a magnificent effort made particularly by my research assistant.

It was a sad reflection of the times, however, that of the six handlers who we trained and who passed the exams, two died of HIV/AIDS within five years, one ended up in prison, one disappeared, and one was promoted to be a guide. Only one remained working with the elephants. A few years later, I trained another group. Of those six, two became guides, one the head elephant handler, and two others were hired at high prices to help with the growing group of elephants used for tourists at Victoria Falls on the Zambezi river. Of the elephants, twenty-five years later, only two of the original elephants remain.

Private nature reserves and tourist businesses throughout southern Africa realised a year or two later that there was a bob or two to be made out of tourists riding elephants, but the problem was shortage of trained elephants and handlers who might be able to train them. Our graduates have built a name for themselves and Zimbabwe, indeed I have met or know of at least ten other Zimbabweans (not all trained by us however), who have been hired in South Africa to help teach and handle the elephants in the fifty or so private nature reserves. Today, further regulation of the elephant tourist business is essential to ensure the elephants' welfare, and safety of the elephants, the handlers and tourists. To this end, there is a growing need for an international association with codes of practice for elephant welfare. Courses must be developed to ensure that all who have to do with the elephants are qualified to look after, handle and teach their elephants without violence. The elephants' needs and desires must come first. It can be done and if it is, it will ensure that more elephants continue to live, breed and have lives of high quality, even if there is no wild left that is not surrounded by human activities and barriers.[1]

The animal welfare activists are against using or keeping elephants in any way. They believe that all elephants should be "in the wild" even though there is very little wild left, in fact practically none in Africa; it all has to be managed in one way or another. Hand-reared elephants or those very familiar with people cannot be placed in nature reserves and ignored, as they seek out humans and often frighten local populations, ending up being shot as "dangerous". The only way in the future will be to keep large mammals in some form of captivity and provide them with a life that is at least as good as living in the wild, if not better. We now know much about what the elephants' physical and psychological needs are. If we wish to provide them with a good life of quality in captivity, we can with a little thought and without spending huge sums of money: although they do cost a lot to keep, that is for sure. This means that they may have to do some sort of "work" to earn money to live and breed. But the advantage of doing this is that it can enrich the life of both the elephants and humans. It also allows the humans to appreciate the elephants' intrinsic value more, rather than just as an aesthetic experience. It is this which will, in the end, be most important for their conservation.

---

[1] 2019 course flyer, Victoria Falls; We Are All Mammals. Charity number 1184219. "We are all mammals – Welfare & conservation knowledge exchange"

Neither the elephants nor the humans must suffer, but they really do not have to. The important advantage is that it gives the semi-confined elephants different things to do and knowledge to acquire so that they can fulfil their cognitive needs. Animals must be provided with a life of quality which is as free of behavioural restrictions as the wild but does not cause others to suffer.[1] If this is achieved (and it can be), the local people and tourists can enjoy the company of non-humans and have them helping each other, building a symbiotic relationship from mutually beneficial activity.

In this homocentric age, this is important for our own as well as their survival. Experiences as a result of these activities can allow animal survival interests to sometimes trump the trivial interests of humans. We need to foster bringing animals and humans together safely to enjoy each others' company and lives, wherever we can, whether these animals are domestic or traditionally wild.[2]

Many object that elephants giving humans rides or pulling things is "unnatural". This depends on what you mean by "natural". It is "natural" for the animals to learn to do different things as they do different things in different wild areas. This is why they can adapt to different environments by acquiring different knowledge. What is "unnatural" about adapting to enjoy being ridden, pull things and do some work, if it means they can live and have a life of quality rather than be dead and finally extinct? I would think evolution would be on the side of the untypical which is not unnatural. Of course, it must all be done without causing stress and distress to the individual and should be a cooperation between the elephant (or other species) and the humans for their mutual benefit. In humans there is now a movement called "convergence", which emphasises trying to feel and understand what another feels and understands in different situations in order to work towards more symbiotic living. This is something that good

---

[1] Course 2019; Development of standards for welfare and conservation of elephants: International Elephant Foundation Conference Pretoria 2019, Jake Rendle-Worthington, Coordinator, ZEWACT

[2] We Are All Mammals. Charity number 1184219

animal teachers and workers have always tried to achieve and we need "convergence" between species as well. Only this will help achieve symbiosis with the rest of the living world. Symbiosis is now understood to be the evolutionary breakthrough in the diversification and stabilisation of life on earth.[1]

One particularly important conclusion of our black rhinoceros research was how one small change in their husbandry helped them to breed. Black rhinoceroses are semi-solitary who have a pointed, almost prehensile, upper, lip. They are mainly browsers and slightly smaller than the square-mouthed white rhinoceros who are much more social, group-living grazers. In the 1970s and 80s it was the white rhinoceros that the world was most anxious about dying out and I had been present in South Africa in the early 70s as some of the first efforts were made to immobilise a few to transport them to different national parks to try to establish new populations. At that time, the black rhinoceros was still prolific and spread throughout East Africa and some of southern Africa, although they were beginning to be poached. Norman Travers at Imire Nature Reserve had been asked by the parks to look after seven orphans, four females and three males, who had been found, and not knowing much about their social world, he had constructed pens where they could be kept at night surrounded by armed guards and sandbags to put off poachers. During the day, they could wander around the park, herded by their armed guards, and join the elephants to be viewed by tourists having lunch at a dam. The problem was that although all the females were of reproductive age when I turned up, only one of them had produced any offspring, one of whom had died. So why were they not reproducing? Reproductive physiologists from London and various German zoos had been to take samples to ascertain if there were any physical problems; the results had been negative. As an applied ethologist, I knew they were semi-solitary animals and only mothers and youngsters up to puberty usually lived

---

[1] MKW, Animal Welfare. Towards symbiosis for the 21st century? Invited Plenary Paper, *International Ethological Congress*, 1991, Kyoto, Abst. 39

together, so I had my suspicions that the conditions under which they were kept might be responsible. As semi-solitary animals, it would not seem appropriate to enclose all the females together in a pen over-night, so we decided we would record their behaviour for a few nights to see what went on. The males were in another large stable and they sparred a bit between themselves, but not to injure each other. One only needed one fertile male to breed with all the females, but each female needed to breed if the population was to increase. After analysing our data and showing how much aggression and fear there was among the females, it turned out that Cuckoo, who was the only breeding female, was a real bully. The other three were terrified of her but they could not get away, because they were all penned together. The obvious thing was to separate the females, give each her own stable where she would have her rations and be able to get on with eating and sleeping in peace. Norman immediately organised the splitting of the stabling into a stable per female, and while he was at it, one for each bull too. I was delighted to hear that all of them bred the following year: applied ethology really could help with conservation, if people would only listen and use their common sense. We wrote all these results up, and published them as one of the papers out of our research centre.[1]

Fifteen years later, at least thirteen youngsters had been bred and put out into the wild, but it is unlikely they survived the poachers. In Zimbabwe's political upheavals, some eight years later, all the adults at Imire were shot one night and their horns taken. Luckily, there were around five weaned youngsters who the poachers did not find, and they have now grown up and reproduced themselves. One very small orphan was raised in Norman's son and daughter's home and became a very large, very demanding family member with a BBC documentary made about him.

Because they were cooperatively taught, reliable, pleasant individuals, two of the six elephants were sold at one of the game auctions in Zimbabwe a few months after we left in 1994. Another

---

[1] MKW & H Randle, "Implications of semi-intensive management ... Black Rhino (*Diceros bicornis*)," 1996

died of an overdose from a tranquillizer which he had quite unnecessarily been given in order to load him into a transporter. Only two of the original elephants remain, but a young female joined them some years later and she produced an infant but alas it died. Maybe she will have another, who knows.

Norman has died, and due to the animal welfare activists, the reserve no longer allows tourists to ride or work with the elephants at all. Another way of interesting people in wildlife conservation is coshed on the head by misguided well-wishers. But, there is a chance that provided elephants' living conditions and training continues to improve, thinking people will begin to question some of the widely-held misguided beliefs for "improving welfare", which stops and prevents activities with humans for semi-captive/captive animals. This in turn could lead to their long-term conservation.

The zebra stallion behaved much like a donkey, and we could handle him all over his body within the first thirty-six hours of capture. Another week and he was leading, and a week later he was being lunged (going around in circles at different paces on a long rein), and was used to blankets and other bits of harness on him. But when we returned to the UK, he was let out and never caught or worked again, as far as I know. He had shown that he was not innately different in his behaviour from a donkey or a wild-caught adult horse and that learning and working with humans was not genetically programmed in domestic horses, because one of a never domesticated line of wild zebras could learn to do normal domestic equine things with humans as quickly as a domestic equine. Thus, "domesticity" does not seem to have affected cognitive or mental abilities and skills genetically any more than in humans. Humans no longer generally believe that races are behaviourally different, so why do they assume that breeds of other mammal species will be genetically programmed to be different behaviourally?

We returned to Imire almost each year for around four years, teaching new handlers and working with the same and some different elephants, and my son, Jake, began working with me. One year I was asked to visit others around Zimbabwe with orphan elephants and teach their handlers, and this spread to consultancies on other hand-reared or orphaned species. Another year the Veterinary Services in Zimbabwe asked me to visit their Cape buffalo foot-and-mouth-free breeding herd that they had had for some thirty years, to see if we could help with teaching them. Trypanosomiasis, sleeping sickness, of cattle is widely spread in Zimbabwe, and the most effective way of eradicating it had been to remove or kill all the carriers ... that means all the domestic and wild mammals. This had been a government policy throughout much of East and southern Africa for some decades, but it had not worked well, partly because the rural people needed their cows, and so they brought them back to have milk and meat when they were not allowed. This re-infected the area, and the government then came again to try and eradicate everything. It now seems an outrageous policy, but such was the interest of colonial governments that this was done in order to benefit the local people, even if it meant killing off the last and rarest of any species. It continues today, in a mutated form; wherever there is a conflict of interests, the humans win and wild animals are obliterated.

One of the far-sighted wildlife vets in Zimbabwe had the idea that if Cape buffalo cows who were immune to Trypanosomiasis could be milked, they could fulfil the role of cattle for the tribal people who depend on them. Because they were free of the "tryp" it would be a great benefit. It would also mean that all the other indigenous species would not have to be wiped out, and the cattle people could continue to live their normal lives. I was asked to visit the vet service's herd of around 200 Cape buffalo and see if it would be possible to teach some of them to be milked, work on the land and be handled like domestic cattle. In fact, about thirty years previously, this had been done by an innovator, an English vet in Kenya and tried out on the Zimbabwe herd of buffalo and eland. These experiments had lapsed, but with the buffalo there remained one old man who looked after the buffalo, who as a boy, had helped with this. We spent a month seeing if we could re-start and develop this further, and most importantly, enthuse those who looked after the buffalo to continue.

   Chris accompanied me this time as he was now a skilled animal teacher. We were given a hut in the staff compound and after carefully looking at all aspects of the buffaloes' management with the old man who had experience, we asked for volunteers among the group of young buffalo herdsmen. Almost all volunteered to learn more about how to handle and teach the buffalo with ease and safety. The first thing was to handle the calves and teach the handlers ways of doing this cooperatively. This would ensure that when they became buffalo cows, they would be easy to handle and could be milked. Then we also found there was one old cow who was still in the herd, who had been handled and even pulled a vehicle when a yearling calf. We started re-handling using the techniques we had researched previously (gentle touching and stroking but with clarity, and as they became familiar, scratching them and lifting off ticks or flies from all over their bodies).The calves were first handled in a crush and the apprentices had to learn not to jump away when they struggled, but to keep calm and keep stroking. After half-hour sessions for four days, each of the six calves could be tied up and handled without leaping about or being frightened. The draught work was even more successful, a couple of cows were brought in and slowly, even though one of them had never had this happen before, a yoke and traces were placed on them and they were harnessed to a cart. The next day, we slowly wandered up the road with them pulling the cart, first with someone leading them, and then without. Buffalo and other bovines do not leap around as much as horses, they put up with more different things without fussing too much, and provided they are not seriously frightened, they are willing to go along with all sorts of things, but they are not fast movers. Their defence is generally, instead of running when scared, to stand and fight. As long as we recognised when they were slightly uncertain about something and let them stand and take it in before asking them to do anything more, it was remarkable how quickly they learnt. The most important thing was to read the animals' moods and intentions well, so as to recognise when they may be feeling slight fear or worry, and not continue with something until they were terrified and all went wrong.

One of the things that really surprised me was how aware these cows were of the size of their enormous horns. While we were harnessing them, just by turning their head fast they could have really hurt us with their horns. The span of the horns is from one to one-and-a-half metres. They carefully moved their head and horns so they never touched the harnesser. Of course, these buffalo had been handled before and were not frightened. It was also invaluable that the older handlers had experience with cattle herding as they came from cattle herding tribes, so they knew how to ensure they were not frightened, and they did it well.

We had them ploughing as a pair and singly within the time we were there, and we led the calves around without trouble, it all looked set that we would be able to teach them and that they would be able to replace the cows in some rural societies. For this we were going to need some research finance as it would take a few years for the taught animals to mature and for the farmers to be taught how to work with them and have them accepted in the communities. I went back to Little Ash, and spent the next six months working out a European Research Grant application with its 5,000 rules, plus the engagement of five universities and the finances down to the last penny. In fact, we traced the vet who had started the teaching in Kenya, and he was all for it, but about to undergo heart surgery; however he signed the contract just before he had the anaesthetic, and we thought all the hard and unpaid work would be rewarded by the project being recognised and financially aided. Months later, I had a response … no … the conclusion, mainly as a result of a junior wildlife manager's decision in Kenya, was that the training of buffalo was "impossible" and could not be financed … despite the fact that on two separate occasions with two groups of people, this had already been achieved! Maybe they were worried about teaching the buffaloes "unnatural acts"; after all humans milking cows or buffaloes is "unnatural" … but curiously, exterminating them by shooting them with guns is not "unnatural", nor is humans learning to read and write, although goodness knows, our primate ancestors never did this.

It had taken me months of work and a great deal of time of a lot of people to submit this proposal. After this, I decided that applying for money for research was a waste of time ... in the months to write it and to await results, we could have had at least a few buffalo heifers and farmers well on the way to success. Maybe we should have been looking for donations to help directly, rather than research money from universities and governments. Most of the buffalo were slaughtered and eaten later during the worst of the troubles in Zimbabwe, and I do not think the veterinary service any longer have a herd or the land to keep them on. Maybe though, as buffalo, like all the other species, gradually become restricted to smaller and smaller areas, and the farmers become more numerous, and Trypanosomiasis continues to be a major problem, someone else will have a go at this, who knows.

## Cattle, elephants, India, Australia & France

By 1998, the thesis on what it was to be an equine or an elephant was gathering momentum and the farm was prospering. We had a herd of ten large, red, curly coated, quiet, cumbersome South Devon bovine persons with a doubly large lumbering bull who we could allow children to sit on, even though because of his girth, their legs would stick out sideways. All the cows had learnt to foster a second calf so that our income was almost doubled. One day, now that cows are becoming recognised as being individual, feeling "persons", someone will redo the double-suckler research, and rediscover that more calves can be raised by mothers, and live as part of a herd and learn to be the good ecologists and sociologists they need to be. Double-suckling is also more efficient in terms of energy and resources than the way today's dairy industry is run; it allows calves to be raised outside in their natural herds and does not need subsidies.

Compassion in World Farming approached me to to do further

research about the welfare of dairy cattle. As mentioned before one of the major welfare problems is that dairy-herd calves are taken away from their mothers around twenty-four hours after birth, and never seen by their mother again. Why is it assumed that this is necessary for the cow to be milked economically? It is standard practice worldwide, even in organic or high welfare herds. We had milked our cows when we needed milk for the house or to sell for over thirty-five years without taking their calves away from their mothers. They were separated at night, we milked them in the morning and then the calves went back with their mothers and suckled for the rest of the day. This worked out very economically viable, was less work, there was milk to drink, and for cheese, cream, butter, and yogurt for a household of up to ten people. It also ensured that the calves were healthier and bigger, and fetched more money when sold. After discussion with Compassion in World Farming, we decided not to put second calves on our South Devon cows for one year, but to leave the cows with their calves during the day, take the calves away at night, and milk the cows in the morning. In this way, we would have some figures on the economics and production of a system in which calves were left to be raised by their mothers in milking herds.

Of course, since the South Devon cattle had not been selected for high milk yields, we did not get a great deal of milk. Their milk is high in butter fat so we had lot of Devonshire cream and butter, and big strapping calves that were worth a great deal more money than those raised away from their mothers. The lower milk production was more than compensated for by the reduction in capital, running and environmental cost. There were savings on labour, buildings, technology, substitute milk for the calves, veterinary care in dairy herds (calf mortalities in the dairy industry are usually around two per cent, even with constant veterinary attention), problems with getting rid of the slurry or manure, and water, and planning of the buildings required.

There has been a great deal of research on how to improve the nutrition of calves and suppress illnesses with drugs, but to date the importance of having a mother and the emotional effect of this on health have not been taken seriously. One has to ask the question: why would motherhood and being mothered have evolved if it did NOT encourage physical, emotional and mental well being? The cow "cares" about the calf which means she misses him and is distressed if he disappears. Given a chance, she mothers him: greets him, calls to him, licks him, protects and suckles him. As a result, mother-reared calves are healthier, socially secure and learn how to fit in with the rules of society. Why do humans continue to "play God" by thinking they can better the results of evolution and nature? They may be able to genetically manipulate the next generation, but the complexities of the natural world always catch them out, and there is ALWAYS a down side. Better to work with what we know than risk causing catastrophic changes which threaten us all. Perhaps before it is too late, a combination of basic ecological understanding and animal and environmental ethics will be taken so seriously that one day concerns for animal welfare and real efficiency of farming practice without reducing species diversity will even trump humans' trivial money-making interests ... but I wonder.

Another ethical cattle problem emerging from human greed is that some beef breeds have been selected for enormous behinds (the expensive cuts) that are so large that they can hardly walk, as they are too heavy and misshapen. Some breeds cannot even give birth normally, they have to be delivered by caesarean section. Belgian Blue cattle are one example, whose meat is sought after particularly by the Belgians. But these artificially selected, sometimes genetically manipulated animals have the same emotional and mental lives as their predecessors. They must suffer profoundly throughout their lives. Why does one never hear any discussion about these farming practices from the animal welfare organisations? Is it just that they are frightened of attacking the agricultural elite? It does not need to be a question of all humans becoming vegan, there are ways in which all livestock can be raised which gives them a life of quality, and which often helps humans in other ways, not only by donating their lives. We must rethink and restructure agriculture and it really can be done.[1] But, there is yet more wrong with the artificial selection of cattle for beef. Many of the top pedigree breeds such as Herefords and Charolais, produce so little milk that they cannot even feed their own calves! In pedigree Hereford herds in the UK, in the case of the "best" show champions who fetch the most money, the farmer buys a Holstein or Friesian dairy-breed cow to feed the calves because their mothers do not have enough milk for them to grow well! Then these calves are used as breeders, thus passing on their very low milk genes to the next generation: even though they may have good fillet steaks, they cannot even raise their own calves.

Dual-purpose cattle, of which there are scarcely any left (a few South Devons, Welsh Blacks and Shorthorns for example) traditionally had strong calves, as well as producing enough milk to keep a group of humans in milk and milk products. Should we continue with extreme selective breeding and genetic manipulation of our domestic animals, with all the risks and ethical problems attached? Or do we finally realise that all we need to do is give evolution a little helping hand here and there by selecting the ones we want for a sustainable breeding population, not by manipulating their genes? We have a duty of care to ensure that our animals do not suffer, and more, *that they have freedom to perform all the behaviours in their repertoires that do not cause suffering to others* and can use their minds to make choices and decisions, learn and acquire knowledge.

We can keep cattle and enjoy their company and their products. They do not have to be kept in conditions of this sort. We do not all have to become vegans to think we are morally correct, in fact there are serious moral questions that have not been addressed by vegans. Rather, we must learn to live with our animals again so we can enjoy each others' company and have lives of quality, even if the lives some live may not be as long as others; which, at least

---

[1] MKW, "Ecological agriculture & improved animal welfare for development in women's agriculture," 1998; MKW, "Integrating food production & wildlife conservation," 2003, Abujan, Nigeria; MKW, "Ecological agriculture. integrating low input, high productive farming ..." 2014

is similar to the wild. But when will we realise this?

In 1999 we decided to take ourselves on a sabbatical trip to India to look at the elephants working there and to meet an extraordinary maharajah who had written a book on training horses in quite a different way. We were also going to Western Australia to visit my sister to see her work with establishing bridle-ways and trying to increase the awareness of the local Australians to the natural world. This trip was to include riding and camping in the indigenous Eucalyptus forests: a pilgrimage to the last remaining natural world in Western Australia.

In the spring we would have to return to Little Ash Eco Farm to do the spring cultivation, but then, with luck and a little help from our friends, we hoped to complete the year by taking a couple of our horses to France to extend our pilgrimage to the remaining natural world in this large and relatively unoccupied (for Europe) country before it all disappeared. We were to have an itinerary which took in places we had been asked to visit and give workshops, earning our keep as we went with our horses, something that used to be common a century or two ago: multi-species journeying.

First, we had to organise the farm. One of our ex-students wanted to stay and run the gardens, delivering organic boxes in the village with a horse and trap for which we had acquired a charming Dartmoor pony: Scarlet. The student, Alex, would keep an eye on all the other things. Various other ex-students lived on the farm or would come in to deal with the horses, riding and doing what they wanted in exchange for keeping an eye that all went well. We lent the pedigree South Devon cattle to a local organic farmer who would have the next batch of calves in exchange for looking after and feeding them.

But, just as we were about to leave, I was called to South Africa. An Italian businessman had been capturing baby wild elephants, training them with the help of Malaysian mahouts and he hoped to sell them at a large profit to safari park owners to give tourists rides. But the South African Society for the Prevention of Cruelty to Animals (SASPCA) were taking him to court for causing wounds and suffering to the young elephants in his care. Despite designing and building an enormous barn with excellent handling facilities, he had made horrible mistakes, mainly the result of ignorance, so the young elephants suffered during their capture, transport, keeping and training. These poor young elephants, between five and ten years old, had been subject to every trauma in the book, including being hit about the head with an ankus (an Indian goad or crook) which caused wounds that had become horribly infected. I already knew about the mahouts' traditional training of Asian elephants, but I had not seen it first hand … until I saw the results on these poor youngsters, isolated from their families, transported, tied and chained, and then beaten about.

I was bundled on arrival in Johannesburg very late at night, into the visiting scientists' annexe, and bounced up early in the morning to take a brief look at the large airy barn. Then I wandered out into the paddocks where some of the young elephants were being taught to walk forward carrying a mahout on their backs. There were also several young elephants just wandering around in the enclosure. I foolishly walked into the area alone without thinking … not a good thing for a hired animal welfare consultant to do … and I paid for it! These animals had had very bad experiences with people, and had recently been poked and hurt by the hired mahouts. Tarzan, a ten year old young bull weighing about a ton was alone. He took one look at me, and presumably took in that a) I was not one of the Malaysians, and b) did not have an ankus, and with a short trumpet and an ear flap (probably of delight), came towards me at an elephant trot. Before I had time to put two and two together, I was bowled over and he was sitting on my back, crunching it with his nascent tusks! Luckily, one of the Malaysians saw what was happening, ran up and chased him off with his ankus. Tarzan leapt up and ran off shaking his head. This incident was entirely my fault, I should have taken on board that all these youngsters had been traumatised by humans, and that I was different, and did not have an ankus and was therefore easy to attack out of defensive threat, or perhaps play with (elephant games are even rougher than horse games). I should not have been in the paddock without a mahout who they already knew. This does not mean that all elephants will always be dangerous, it means that elephants, like children, dogs, bulls, horses, lions or any other mammal who has been traumatised and abused by humans will be frightened and/or aggressive. If they see a chance of attacking or have not learnt the rules of the social contract of what is permitted with humans as well as their own kind, they will investigate, or may attack or play rather too roughly for human taste. Some animals of course, such as bulls (and even cows in Valais, Switzerland) are taught to attack other members of their own species. Some are taught to attack other species when told to, such as police dogs or dogs owned by aggression-loving men. They are taught by humans to behave in this way ... it is not fundamental to their nature and "set in the genes"; it is the result of how they have been kept and treated and what they have learnt. No species of mammal, non-human or human, is "fundamentally aggressive" as a result just of their genes, unless they have neurophysiological problems, but they all can learn to be during their lives by having experiences which encourage this.

I was acutely embarrassed after this incident with Tarzan, it was unforgivably stupid. So, I heaved myself up, and pretended that all was well. That night it took me one hour to get three metres to the toilet, and that with the help of two pleasant British students holding me up and putting up with my groans. The next day I had to march about as if I had a healthy body; it was broken, very painful and creaking horribly, but it had to be a case of "just a little stiff"! Luckily there was a South African nurse, the wife of a cameraman hired to take videos of the elephants, who noticed that I could hardly move. She took me off to the local hospital where after X-rays and much poking and prodding, they discovered that I had five broken ribs on the right side where Tarzan had crunched up and down on my back, but luckily no lung punctures which might have been fatal. There was nothing that could be done other than giving me the most marvellous pain killers. I don't know what they were, but they were great. Drugs may have unsuspected side effects, but they can be a saving grace. In a few hours I was waltzing around as if nothing had happened, modern medicine can be marvellous, do not knock it all!

For the rest of the stay I was more cautious and quickly realised why I had been attacked. The young orphan elephants were being kept tied up for twenty hours a day albeit in a smart, clean, cool barn. They were being given "training" sessions by the mahouts which consisted of being shouted at and prodded with the ankus. I could not understand what the mahouts were trying to teach them, so it was not surprising that the elephants were even more confused. When they did not do whatever it was that was asked, they were poked and prodded so that the skin broke. The skin of an elephant is so thick, that when it is punctured and begins to be infected, the infection cannot drain out and the pus accumulates as a painful ulcer under the skin. Wild elephants often have these ulcers where thorns and stones have punctured the skin, but it was unforgivable that these youngsters should have been so poked in their "training sessions" that this had happened. There was a team of vets trying to open up the wounds so they could drain, not easy on a young hurting elephant recently captured from the wild, where all is alien and terrifying.

The Italian owner was taken to court and rightly convicted of cruelty to elephants, but as usual with such cases, in the end, it was the young elephants who were the losers. They were no longer allowed to be "trained" or learn to do anything with humans as it was assumed that "training" had to be done in this violent way. Ironically, because of this belief that "training" elephants is always cruel, in the end, it was the SASPCA who was responsible for about twenty of these youngsters, spending their lives either in small private wildlife reserves where there were usually no other elephants and they had little opportunity for performing most of the behaviours in their repertoires, or kept in pens untouched and untaught by humans because they were "too dangerous". I visited Tarzan a couple of years later; he had become a bigger and more aggressive elephant as a result of living in a half-hectare pen and having nothing to do, even though he was with two other youngsters.

Elephants who have either pleasant or unpleasant experiences with humans can usually not be "re-wilded" because they will tend to seek out humans in their farms and villages thinking they will get food, or sometimes, they seek security with humans from the large unknown world in which they have been thrust. When they are scared, as they often are by people shouting and chasing them, throwing things and even shooting, then they may attack. The result is that the people and the elephants living where there is pressure on land both become more scared of each other and the frightened elephants attack more often. It is generally the case that if the "re-wilded" elephants have been with humans for a few years, they will not re-join a wild herd for a considerable time, meanwhile they are a threat to the local farmers. As a result, trying to "re-wild" elephants in Africa is difficult. It can be achieved with a lot of thought, but it is more likely to be a death sentence for the elephants who are more or less bound to be shot sooner or later, either by poachers or by those protecting villagers.

The mistakes that had been made with the South African young elephants in their capture, transportation, and particularly their training, were not irreversible. They could have been re-taught that humans could be helpers and friends, and have become valuable members of a multi-species community in enclosed areas, learning and doing many different things. The re-teaching would have taken time and patience, but each of these youngsters had their whole life ahead of them. Instead, the court order (supposedly in place to reduce their suffering), ended in them having to be kept in enclosed places with nothing to do and so they became more neurotic, aggressive, and handicapped individuals and suffered most of their lives. Animal welfare interventions which are done with the best intentions, can often cause a great deal of suffering, sometimes for life. This is often because strongly held unquestioned opinions which are often informed by ignorance are difficult to change, and the situation of each particular individual is not properly considered.

On my way back to the UK, I was treated to a wheelchair at the airport and wheeled through all the usual boring queues for passport control, ticket inspection, luggage weighing and endless waiting rooms (without chairs) characteristic of aeroplane travel. I arrived refreshed at Heathrow after a night on the plane, having taken my wizard pain killers and met Chris to go through the whole process again, but this time without the wheelchair alas. We were flying to Bombay: Mumbai then to Kerala in the south of India to visit the University Federation for Animal Welfare's (UFAW) new programme for training mahouts in how to improve the husbandry and training of Asian elephants.

On final arrival in Kerala, we were met by a young woman mahout, one of the only females, and an experienced elephant veterinarian who treated all the captive elephants in Kerala, about 1,000 in total at that time. The veterinary treatment of elephants' physical diseases was impressive and as good as it can get. However, little serious attention was given to their psychological welfare in their keeping conditions. As for their "training", that was antediluvian. The vets or animal welfare workers see a mahout poking an elephant with his ankus, and because they have no practical experience of teaching elephants, they have to believe when the mahout tells them "that is the only way to teach him". No wonder the animal welfare activists think that any sort of "training" of elephants, anyway, is bound to cause them to suffer, since they are told they can only be trained by being poked and hurt. This is not true as we had shown by using methods developed for teaching pre-verbal children in Europe and in Africa with elephants. But trying to interest the mahouts in another way of teaching or relating to elephants was nigh impossible, we tried and tried, and even did some demonstrations, but to no avail, they just shook their heads and said it was because of some particular day, hour, weather, person, elephant, etc that it was working.

Asian elephants, unlike their African counterparts, just like horses, have through centuries of history been used for transport, ridden for fun, used in wars, worked in forests, and been an emblem of wealth and success. But, curiously, the Asian elephant has rarely been allowed to breed in captivity. If the next generation were born in a domestic or semi-domestic situation, where there are humans around who are pleasant to them from birth, they begin to learn the ins and outs of how to behave appropriately, and their innate behavioural tendencies are moulded by these experiences, just like humans. The lifetime experiences of any mammal are crucial to how an individual will behave in any environment. In Burma, in the 1930-60s, because elephants were the only way of moving through the forests and moving the timber out, this was understood and often put into practice.[1] As a result the elephants performed extraordinary feats of intelligence and skill. Today this has all been lost as they are no longer used and there are no more *oosies* (handlers) to pass on that type of knowledge.

Different races or species may have special abilities, but these are not written in stone. Racism (another type of species-ism) is as misplaced in non-human mammals, as in human mammals. Many of the differences in behaviours between two species, or races, of one type of mammal, are not the result of inflexible instincts written in the genes, but are often the result of preconceptions. It is the expectations of humans that controls how they behave towards the animal and thereafter how the animal behaves towards them. This is because although the human may not be able to read what the animal is feeling and intending, the animal usually can read what the human is feeling. If it is fear and uncertainty, they will also become fearful and uncertain until they have so much experience with different humans that they recognise this as one of many human characteristics. But, the problem is that the naïve human handler's expectations may be fulfilled: "he is dangerous and will run away or attack or hurt me", and he does. "Elephants are wild creatures, they will always be dangerous and unreliable", is a statement too often made without any evidence by scientists as well as lay people. Thus, the human is more often than not confirmed in his/her preconceived beliefs. In addition, the elephant only has to get rid of the human by frightening him a couple of times before the elephant will be considered

---

[1] JH Williams, *Elephant Bill*, 1950, Reprint Society, London

unreliable and difficult and left largely alone, which again confirms him in his beliefs about humans and how to behave towards them. As a result, the human because he is frightened, approaches with an ankus and teaches the elephant to be fearful, who may attack out of defensiveness (fear) or to learn to be aggressive to the human (because then the human goes away and leaves him in peace). Unfortunately, most humans do not even know that this is what they have taught the animal!

During my lifetime researching the life of mammals, there have been two deeply upsetting, sad projects. One, was researching the behaviour and suffering of veal calves, the other was witnessing four- to eight-year-old Asian young elephants in India who had been caught in one way or another in the wild, separated from their families and friends, transported to alien environments by two-legged brutes who hit them with spiked weapons, isolated from any other elephants younger or older, and trussed up with chains before being "trained". That is, they had to learn to respond in mysterious ways to shouts and screams, pokes and hollas which no one had bothered to teach them to understand. The two-legged mammals failed to recognise how traumatised these infants were, nor how incomprehensible their "training methods" were. It was a profoundly horrible experience, and utterly unnecessary.

If it was necessary to capture youngsters in the wild (because otherwise they would be shot), they could have been properly immobilised and captured, carefully transported in groups, and placed together in well-fenced coconut groves (which they had available), with older domestic female elephants to learn from. Then, gradually, slowly and safely they could have entered voluntarily into areas where they could be handled and fed by humans, and gradually learn from and with, the humans without trauma or violence. We had done all this in Africa, why had they never done it in Asia? And who said it could NOT be done? Why should a well brought up elephant familiar with his environment and at home in it, be any more "unreliable" and "unpredictably dangerous" than a human or a horse?

I approached some elephants and tried our methods for a half hour. Here and there even within this time, youngsters became slightly calmer. I worked with a bull elephant in musth. Males come into musth (when they seek out and try and mate with females and attack other males) about every two years for around two months. During this time they have high levels of testosterone (the male hormone) and become more easily irritated and aggressive. As a result, in Asia, males are tethered by a front and hind leg when they are due to come into musth and kept like that until it is over. We had a particularly unpleasant morning when taken to see how they treated the elephants in musth. The chained bull was taunted and jeered at by the mahouts, who threw things at him and waved coconut leaves in his face and tried to infuriate him until he trumpeted, roared and tried his hardest to break the chains and attack them which they thought was funny. After this exhibition, I became increasingly impressed with the remarkable tolerance of all the elephants who had been brought up and treated in such ways.

Since there is very little forest left in Asia, the traditional work of extracting timber is more or less finished for elephants. The result is that the vast majority of elephants are either owned as status symbols by rich people who want to show them off to their friends and clients, or they do tourist work or temple duties. The latter involves living near a temple and then on special days when the elephant god Ganesh is to be revered, they are walked about among throngs of thousands of people who give them money, touch them and revere them. There were a number of elephant bulls outside the temple we were taken to visit, and at least one was in musth, tied up and immobilised. I stayed out of reach (my ribs were healing but still sore), but as near as possible to this bull. I spent about an hour talking and chatting to him. At the beginning, he clearly assumed he was going to be taunted and irritated, shaking his head, stamping about and swinging his trunk. Gradually as I ignored these demonstrations, he began to appear to be more interested in me. I continued to talk in simple English and he appeared to start to listen. Then I asked him to imitate some simple movements I made: lifting a leg, or my arms. After about ten minutes he started to do a similar movement. I smiled, giggled and praised him and threw him a few titbits. In another twenty minutes, he imitated a couple of actions when I performed them and appeared to be vaguely interested. At the very least, he had had something to do and think about and had earned a few tips from half an hour with a human. The question was how could we get the mahouts, who come from their deeply engrained culture to do things differently with elephants, to look and listen in a different way? We took some video and showed them but it was dismissed as "whispering", an occult connection with the elephants.

We saw the best mahouts but even then, we were shocked by their authoritarianism, use of the ankus and their shouting at the elephant. Their only interaction appeared to be to demand the elephant to obey commands. It was a stark contrast to the way we were teaching and interacting with the African elephants in Zimbabwe. Perhaps as a result of the UFAW course, some mahouts were behaving a little less badly with the elephants, but it was not safe or pleasant. We found out that 200 mahouts had been killed in Kerala in the previous ten years, and there were only about 1,000 elephants: a mahout had a two per cent chance of being killed by his elephant! But, when elephants do temple duties, they often had hundreds of unknown humans surrounding, touching, revering and pushing them, yet they had not hurt one member of the public in the same time period. It is the mahouts, who are meant to know and even "love" their elephant, who they kill. Clearly, the mahouts' methods cannot be right.

The next stop was a sawmill to see what the last remaining elephants employed with timber were doing. We accompanied the elephant vet who had to visit an elephant there. The large old cow elephant had a renal disease and had to be treated. For this, she had to lie down and stay very still while she received a blood transfusion in the ear. She did this without any sedatives or tranquillizers, just because she had done it before, and perhaps understood, who knows, that it helped. It was impressive how calm and tolerant of these painful procedures she was.

There was a bull moving and sorting the logs which had just come in on trucks to the sawmill. There was a mahout sitting on his neck who every now and then told him to get another log, or make the stack higher; but it was the elephant who had to lift and carry the logs, balance them between his small tusks and his trunk, place them carefully on the pile and ensure that the pile did not collapse. It was heavy, difficult and dangerous work, he could easily trap his trunk under a badly placed log, or drop one end on his foot, or cause the whole pile to tumble, yet it was he who was carrying, balancing, placing, observing, and calculating how it should be done. There was no way that the rider could either see all that was going on or indeed feel when the position of the log being carried was unbalanced and must change, and give commands for this. It was the elephant who had learnt to make these judgements. The log lifted sometimes weighed tons, and the skills required by the elephants were phenomenal. On one occasion, the pile of logs became unbalanced and started to roll towards some passing humans, but the elephant by skilled and quick decision-making stopped the movement. This required a great number of mental skills that today cognitive scientists still consider unique to humans. This is because none of them have ever witnessed the remarkable feats that non-human mammals can do and have done for generations. Yes, of course, the elephant has a belief about the logs; if they are going to tumble or not, he is making judgements about them, he is calculating if he needs to move the log to the left or right to keep it balanced, and if the pile will hold or tumble when he places the next one on it, and how to place each one and particularly how not to trap his trunk. There is no doubt he must have beliefs and desires, and be able to make choices, calculations and judgements. He has had to teach himself the physical skills how to do all this; since the mahouts do not have a trunk and tusks, or lift and balance the logs, they cannot know. So, just observe an experienced elephant working timber for a few hours; there is no need to do "experiments" to prove whether they can or cannot have these mental aptitudes.

What we find more difficult to understand is what he is feeling while doing these elaborate tasks. We have some idea from how he behaves, but we are still very ignorant about how to know if he is showing, for example, frustration or pleasant anticipation. When he fidgets, throws his trunk about or shakes his head, this could be frustration or pleasant anticipation. Both these feelings will have similar physiological responses, that is a higher heart rate ... but this does not mean that the animal is suffering. Sometimes he may be enjoying what he is doing, sex for example definitely raises heart rates! And what about when he stops a pile of logs from falling on a human? Clearly, he "cares" about not having an accident hurting humans. Of course, he has learnt how to stop the slipping pile, but he was not told to ... he knew that the human would be hurt if he did not; in other words, he had a "theory of mind", he knew the human was a living, breathing, feeling creature with a body and a mind. If the mahouts as well as many concerned with elephant welfare really understood this, they might show the same concern for the elephants as he showed to the human near the tumbling log pile.

After two weeks of travelling around and watching elephants, we were profoundly impressed at how incredibly able, tolerant and easy to teach they were. They can put up with thousands of squawking humans pushing up close to them and touching them in temple celebrations. They can learn a whole series of physical laws concerning moving logs around. The youngsters manage to survive being wild, caught, transported, tied up with chains all over the place, and be bashed and poked by humans without always killing humans on sight.

But one thing was sure, they were being kept and taught in a very different way from our African elephants and it was difficult to see from any of the Indian keeping and training that we had witnessed that they could be considered to have a "life of quality", despite their abilities. We somehow must demonstrate to the world that where people deal with elephants in captive situations, both elephant and human can have lives of quality, indeed lives enriched by contact with each other. Only then, will there be any future for the local people, the tourists, and elephant conservation.

A two-day and night trip on the Indian railway from Kerala to Bihar to meet our Hindu maharajah horse teacher was a delightful experience, with delicious curries ordered at one station and delivered at the next. The maharajah was eighty, and we much appreciated the intellectual exercise of long discussions on many topics that we had, as well as his curious but effective ways of teaching and riding horses without using any violence or any physical restrictions.

The next phase of the trip was a rush to catch the plane, we arrived late and rushed around the airport, it was the last plane before Christmas. We had the wrong boarding passes and were just about to be told to get lost as the plane had taken off when a young Indian man, offered us his boarding passes as a gift ... we clambered on the plane at the last minute and forever afterwards I have had happy memories of India and its people's tolerance and generosity, even if it is not always borne out in their politics.

Western Australia with my sister was a genuine holiday, although the first few days were bed-ridden due to the side effects of the pain killers I had been taking, but thereafter it was learning about the environmental problems, and the fauna and flora. We rode around in some genuinely wild Eucalyptus forests, camping and adventuring together with two delightful horses. Various human relations and friends visited and we all learnt more about mutual reliance between species. We swam with the horses in the sea among dolphins who came to see what was going on, galloped along beaches without any humans for 200 miles, and found our way around with a compass through the enormous open Eucalyptus forests. We learnt how equines are so much more aware of their environments than humans. For a test we played European bird song on a tape recorder one evening around the camp fire. The horses became very anxious indeed, they had never heard anything like it even though their own birds, parrots, often make a constant racket, shouting "28 28" over and over again. How many humans would have recognised the gentle tweets of the blue tits and robins as very very alien and been disturbed? Hardly anyone I suspect.

We returned to Little Ash in the early spring to plough, cultivate and sow the crops for the summer, ride our horses and prepare to ride around France. All was well at the farm, everyone had done what they said they would, and things were growing, so we hitched on our saddles and bags, and bundled ourselves and the horses into the van and drove to Dover. The van was placed on the open deck so both us and the horses had a glorious view of the White Cliffs of Dover as they disappeared into the mist and Oberlix, the pure-bred Arab stallion, and his daughter Shemal put their heads out and looked with interest at the cliffs and the sea, but to our surprise they remained relaxed during the entire journey. We arrived at a friend's stud farm in northern France, and gave a workshop to about thirty people with Obie and Shemal illustrating the points I was making about how to teach and keep horses so that they enjoyed what they were doing. Then, we rode off with all the necessary camping gear for both them and us.

One of the very best aspects of France is their elaborate system of *Randonnees* ... marked rights of way for walkers, riders and bicycles, which weave their way up and down mountains, over rivers, around the coast, and through forests all over France. This means that by following a map, you can ride from Calais to Marseille or anywhere else you fancy without going on major roads or leaving the glorious natural world, although there are brief trips through picturesque villages. We left the cathedral spires of Chartres behind as we cantered through forests and on tracks through arable fields, stopping near rivers for drinks, sandwiches and a quick graze. By about 3 p.m. each day we would begin to look for suitable camping places which had enough grass for the horses, shelter, shade, water and a flat place for us to put our tent. We made an electric-fenced enclosure for the horses over night so they did not wander off, tying the insulators to trees and anything else handy. There were hiccups to start with, but gradually over several days all of us entered into the rhythm. We found that oddly enough it was much more difficult to find food for us humans to buy along the route than for the horses. All the small village shops had closed down, including the bakeries because everyone went to the supermarkets on the edge of the local town. But people were extremely friendly and helpful, even giving us their loaves and croissants that they had just bought, and offering us and our horses meals and accommodation. They brought their babies and infants out to see and touch the horses as we went through the villages and the most difficult thing often was not to stay and talk for too long. Talking is a French forté in which they indulge, usually without restraint! It was that we were just travelling with our horses and living twenty-four hours in close contact with them, mutually dependent, that the people liked ... although they were not about to do it themselves, they liked the idea, so perhaps there was hope yet for humans and animals to live together to their mutual benefit.

During the next four months, we saw a lot of France and met a great number of horses and people, taught about twenty workshops and had great times with our horses. We had been invited to Saumur, the centre of French equitation and spent two weeks there. We rode in their enormous chandelier-decorated indoor manège, and talked even to the *Dieu de le Dieu*: the God of Gods, the chief of all the chiefs, of the French Cavalry School.

Some of the people who had asked us to do workshops were women with a back garden in which they kept their horse, some had stables where they had hundreds (literally) of Shetland ponies. One owner had a field with fifty dogs, was getting divorced, and was worried about her daughter. A plumber on the edge of the Pyrenees parked us in a field dropping off to a precipice. There was a hail storm with stones as big as your head, we grabbed the horses and staggered over to some shacks he had, but the river above had burst its banks, and there was water a foot deep over the floor; nevertheless, we all decided that it was better to stand in the water than to have one's head hit by an enormous hail stone.

Lunch time stops were somewhere the horses could graze, I could identify plants with my flora and Chris could swim and sleep. One hot day, we stopped near a river and stripped off for a delicious cooling swim, well away from prying eyes we thought.

Shemal was tethered to a spike in the ground, and Oberlix was loose as he would never leave her. Half way through our siesta lying in the sun in the nude, someone started a chain saw. Shemal looked up and snorted, then she saw what looked like someone riding a horse on the skyline, she pulled, the tether gave, and she galloped off towards them. There was a major road not more than a quarter of a mile away, so Chris jumped up in the nude and rushed off to try to catch her and Oberlix followed. I watched this galloping brigade and rapidly threw some clothes on. It turned out it was not someone riding another horse, it was someone riding a bike with earphones, quite oblivious to the drama she had caused ... eventually Chris caught Shemal and brought her back, trying hard not to be embarrassed!

About half way through the trip on a very hot day, we decided to go for a swim in a river next to a ford; in we went on the horses and trotted about splashing and larking around. On emerging, Shemal's left hind pastern (the lower joint) was pouring blood, and worse, she would not put it to the ground. She had cut it on a broken bottle that someone had thrown in, and the Achilles tendon was cut right through. We made every effort to stop the blood flow and eventually with the help of a cob web (my sister had long ago told me that cob webs clotted blood, a useful tip) and much deliberation, Chris galloped off on Oberlix to fetch the van about eighty kilometres away. I stayed to nurse Shemal whose leg was swelling as one watched. Chris was back by the last light and by heaving and helping, lifting and holding, we managed to get Shemal into the van on her three legs to join Oberlix. It was a six-hour journey to an old friend of mine from Uganda, Kathy, where we knew we could find a sympathetic lodging for Shemal. Our van was designed and made by us so the horses could participate in the journey, look out the front and side windows, and nudge us when they needed something, so we could keep an eye on three-legged Shemal as we drove carefully. She was a fit, strong six-year-old at the time, and we would know immediately if anything went wrong. Eventually we arrived and unloaded in a farmer's field; by that time her leg had swollen as large as a football all the way up to her hip. However, with encouragement from Oberlix and us, she managed eventually to hop out of the van. She spent the next week hopping about while her wound swelled and swelled and we washed and washed it with a pressure hose, ensuring that despite the flies, it remained clean and could drain. We had several visits from concerned vets and alternative horse therapists. They either advised a host of drugs, stitching and box rest, or swung a pendulum over her, and told us to obtain a whole variety of plants to make homeopathic cures. But they all looked gloomy, shook their heads and said she would never walk again sound. We thought carefully about all these cures, but she was quite cheerful, living with Oberlix, hopping about to graze, and the wound was healing cleanly. If she lay down, she could not get up, so we had to help, but she learnt to call us when she got stuck. We had also taught her to lie down when asked, so when I saw that she really was tired, I asked her to and an hour or so later we could help her up. She remained her cheerful, positive self and her wound gradually healed with nothing but cold water at pressure applied frequently and directed right into what had become a thirty centimetre deep wound.

During the next couple of weeks, Oberlix and one or other of us attended English classes with Kathy's students, visited a host of local villages where Oberlix had to be tied up next to the fountain, while his accompanying rider drank too much wine and ate too-large meals. Kathy absent-mindedly drove into the ditch next to the field so Oberlix and us had to pull her out. In France you can drive a small car without having a licence. *Sans permis* cars are avoided by any sensible driver as they tend to break the rules. Kathy had never managed to pass a driving test, so she had a *sans permis* which often broke the rules and came off the road.

Shemal was getting better and much of the swelling had abated, we had further workshops to give and we needed to continue on our way, but she had to travel in the van, so one of us rode Oberlix, and the other drove her to the next destination. This actually worked out well because we continued to camp in forests and places away from prying eyes. Because we had the van, I had my computer and could continue to write my thesis on equine and elephant epistemology on rest days. The workshops kept Shemal on her feet and thinking, and Obie occupied. The only thing was that the indigenous French population was less friendly and helpful now we had a van. A chateau owner refused even to let the horses have a drink, but low and behold just down the road was a builder who welcomed us with open arms, gave us a field, dinner and a shower but cautioned us not to let the horses get into the chateau owner's field. Of course, in the morning they were nowhere to be found. The had wandered off to find pastures new. By following their spoor we finally located them in the middle of the chateau owner's wheat field, and we very carefully and quietly withdrew with them, as rapidly as possible.

Another evening they were enclosed with plenty of grass but during the night they escaped. This time Shemal had hopped about half a mile to a delicious field of lucerne which the two of them were guzzling, green saliva oozing out of their mouths, the result of infinite amounts of such a delicacy! After this we knew she was really on the way to total recovery.

The pilgrimage continued around Southern France visiting horse retirement centres and competition and teaching stables, and people who did horse "spectacles", dances and displays. The last workshops were at Dieulefit where we borrowed a horse so both of us could ride around in the precipitous mountains of La Drome. Shemal started to come with us on slow, easy rides. It was a beautiful area with plenty of places unoccupied by humans and with marked trails. Sometimes the routes were somewhat extreme. I remember looking quizzically at a track about fifty centimetres wide with a ravine down one side and a precipice up the other, and beginning to shake. How could Obie get along there, he had packs poking out either side and if they came in contact with the side, he would be pushed down the precipice. He began to shake too, but there was nowhere to turn. Luckily, I realised that as Shemal and Chris had wandered on ahead without shaking or incident, it was I who was making Obie scared of heights. I got off and led him along, he recovered his confidence, and all was well ...

but we were careful after that to make sure that we could get along a path before we realised that we could not and could not turn around. Ten years later, our son Jake and Chris with our stallion Oryx and Shemal were attempting a new route over Le Grand Delmas next to our French farm. The path led over a scree which disappeared into a ravine. There was no way they could turn back so by hoping without looking behind, they all managed to negotiate it without sliding off. Luck and good balance on behalf of the horses was on their side, but we have never re-ridden that route!

There were many tales that could be told of our adventures and the people we met and the extraordinary things they did. One workshop was with the Pignons, a family now famous for doing "spectacles"/displays and hired by the largest shows worldwide, including in London. One of the brothers is extremely athletic and can stand up galloping with about five horses abreast and jump jumps; at that time he was just beginning his athletic life, and the horses were certainly interested in participating. The other brother had a beautiful wife who rode "high school" on their Andelusian white stallions. The horses had manes so long they touched the ground and were never out of their stables in case they dirtied them! Yet, they thought they were doing everything they could for their beloved horses. We tried to point out that the horses might not think that their lives were of the best, but like water off a duck's back, such ideas slid away.

Soon it was October and time to return home. We drove back to Calais and Devon. This sabbatical year had taught me a great deal about what fun it was to live together in a mutually dependent way with other large mammals. This of course is something that many used to do with their domestic animals, and even hunter-gatherers surrounded by wild animals, but it is rarely practised today. What insights it can give one both emotionally and cognitively on another's skills and knowledge. By mixing and matching, we can all have a very good life. We, and our horses, learnt a lot about the flora of France, and its geology and landscape, and different people's and horses' attitudes. I had much help from our four-legged family and from discussions with two-legged members to write a thesis concerning their epistemology.

On returning to Little Ash, we took part in many horse events. One test was to see if we could teach by getting the horse to understand "the concept" of what was required, rather than by stimulus–response learning. Would this facilitate the horse to learn movements correctly and consequently, would a six-month period to learn the required movements allow the horse to reach international dressage test-level?

It is also widely believed that, because endurance and dressage require different physical and mental skills, it is not possible for the same horse to compete at international level in both in one summer. Is this true, or does one skill facilitate another? Oberlix, our pure Arab burnished copper stallion, was to test this as we crept up the dressage test scale. We could only attend the tests nearby and we also had to juggle the dates with the endurance races he was doing where he was competing in eighty kilometres and one hundred and twenty kilometre races (at international level). Then at the end of the summer, both he and Shemal took part in the Arab Horse Society marathon, a race over the marathon distance (forty-two kilometres) and they did it at twenty-six kilometres per hour up and down dale, Shemal was first and Obelix was second, a length behind. So, it had been shown that a horse can take part and do reasonably well in international competitions in at least two disciplines in one summer; they can learn to be good in several disciplines, and might even enjoy it more.

The learning from and teaching of the horses, dogs, guanacos and cows in Devon continued and the farm prospered. During our sabbatical while Shemal was lame, I had managed to make some progress with the thesis, which on return to the UK was followed by stays of weeks in Lancaster becoming more involved in what turned out to be, of course, a never-ending study. This was mixed in with trips to help with and study elephants in Zimbabwe, horse work and competitions. But, eventually the thesis was examined, and passed.[1] This was one of the most intellectually challenging

---

[1] *Right in front of your mind, equine and elephant epistemology.* 2000

and interesting parts of my career, at last I was able to study the mind of my equines and elephants as I had promised myself when a six-year-old. There was still much to do and many species to study, but it was a start.

The sabbatical had made us wonder whether France would not be a good place for the next experimental farm and research project; the space that it offered, and its marvellous bridle ways which allowed a real immersion in the living world were very tempting. Yes, there were still wild things and places in France. Perhaps we could find a mountain location to run an efficient self-sustaining farm-cum-nature reserve. First we put Little Ash on the market with the cob house we had built; and a productive organic/ecological farm. But, the night before we had our first potential buyer's visit, we accepted an invitation to stay with friends in Cornwall ... At 11 p.m. the phone rang, our home-built cob and thatch house was on fire. It was the first complete modern house that had been built from cob and all the materials had come off the farm or collected from recycling centres and the dump. We rushed back but the fire services food wagon blocked the drive, so the rest of the night we watched the blaze but could do little. In the morning, the prospective purchasers arrived to find Chris and me shovelling cart loads of ash from the burnt thatched roof and the remains of the cob walls.

A month or two later however, we did sell the farm, and departed. I had been offered a visiting fellowship at Girton College Cambridge to learn more about the experimental approach to animal minds, and they had given me a cottage, a field and a barn for my two horses while at the college. The academic year was fun, spent at Girton among bright people from many nations and disciplines and it helped further to broaden one's approach and make connections in my subject between many disciplines. Shemal and Oberlix were my obliging experimental subjects who confirmed how horses had several of the mental attributes that had previously been restricted to primates. Cycling around the back of the colleges near the river at Cambridge and its surrounds with Rupert (my dog) became another hobby when the flat lands and suburbs of Cambridge had grown somewhat too familiar for the horses and my expeditions. Formal Hall and being all dressed up with good food and interesting conversations with other academics talking about their own subjects, was the icing on the cake.

We needed somewhere to park our horses, dogs and selves quickly and, eventually, bought ten acres of flat, productive, ploughed land and an ugly little cottage near Ely. Over the next six months we converted it into a country residence for horse owners with garden, pond, fenced paddocks, stables and a pretty cottage. During that summer, we rode around in this flat country previously thought to be dull and lacking in species diversity; but, wherever there was a little of the natural world left, it proved to be exciting and of interest, even if it was in little pockets here and there.

## La Combe, Drome, France

It was the end of 2003, and it was becoming crystal clear that something had to be done for the future of wildlife on farms; how could wildlife conservation be integrated with efficient food production for humans? In its modern form, agriculture is the greatest threat to the living world with its almost uncontrolled use of pesticides, herbicides, and veterinary drugs, and deforestation, planting of mono-cultures, over-use of water for irrigation, and pollution of the waterways, never mind the monstrous disregard of the suffering of millions of animals. The problem is that the ever-rising human population has to be fed, but can it be fed by self-sustaining ecological agriculture: a further development of organic agriculture?[1]

If so, how many people could the world support while retaining the great diversity of species that aids a stable, sustainable world? There is now accumulating evidence that eating organic food reduces the carcinogens dramatically in the blood, so people are

---

[1] MKW, "Ecological agriculture and improved animal welfare," 1998

beginning to pay attention, even if only for selfish reasons. But what techniques would be needed to feed the world in this way? And would not wildlife still be threatened? How many people could ecological agriculture feed? And does this mean that human diets will have to change? Can agriculture become more energy efficient? And if so how? Then again, can the whole farm, even if it is in a difficult mountain area, be an efficient human food producing area and at the same time an important wildlife reserve? These were the next questions we needed to ask of an experimental farm.

We bought ruined buildings and 175 hectares at the top of a mountain (750 to 1,200 metre in altitude) in La Drome, France. We reckoned we should be able to produce enough food: wheat, vegetables, fruit, dairy produce, meat, and also wool for around twenty families (of two adults, two children), throughout the year without importing any products from off the farm, and that this could be sustainable indefinitely, once it was correctly structured and farmed. The cultivated area would take up only a fraction of the farm, maximum fifteen hectares, so the rest was to be kept for wildlife. In the nineteenth century, the population of our commune was around 2,000 mostly peasants living in a more or less self-sustaining way, but they had caused terrible erosion of the steep hills by having far too many sheep and goats, which was still very evident. During the twentieth century the human population had declined to under one hundred humans in the commune; our farm and many others had been abandoned as the people left for an easier life earning money in the various cities. The result was that generally it was only hunters who came onto our farm. The French hunters, although they are supposed to be controlled, normally shoot anything that moves including small songbirds and the odd mushroom or lavender picker as well!

The test was to see if by farming ecologically we could feed many more people with far less environmental impact and, at the same time, encourage and integrate all the various forms of wild fauna and flora, and also allow our own animals to have a life of quality. If this was possible, and if other farmers in the commune would try to do this, then not only could the commune feed itself from its own resources, but it could also export food to the neighbouring towns. It was to try to show if more efficient food production integrated with wildlife and with no inputs to the farm, could feed more people at lower energy costs than the current trend in "modern" agriculture. We wanted to live closely with wild animals and plants and our domestic ones. There were hunters on our mountain, but there was no other kind of human within three or four kilometres, and plenty of forests, cliffs and rocks that even rock climbers could not climb.

It was 2003, and foot-and-mouth was restricting the movement of cloven-hoofed animals to the continent, so we could only take our horses and dogs. We arranged transport of all the necessary farm equipment and the horses. The last to arrive were Chris, Jake and me with the two stallions squashed into an over-packed Renault Master. It is a long drive down the motorways but there is generally less traffic in France, and we could have pee and tea stops at the *aires* along the way. In fact the horses were good at telling us when they wanted a break to pee by kicking, and so we would stop.

When we arrived at one of the villages near our new home, the road up the mountain was impassable because of snow. It was Christmas eve. We managed to hire a gîte for the night and the stallions were put in cramped stalls next to each other. They were rather too close for comfort; we knew how they could wind each other up and let all hell loose. However, after the voyage and having given them a short lecture on good behaviour, they minded their manners and behaved remarkably well, so none of us lost face. The next day we had to ride over the mountain in the snow which had inconveniently obliterated all the tracks; the two stallions were delighted at being out and about in new terrain. They decided to squabble all the way as we made our way along paths fifty centimetres wide with a cliff on one side and a scree down the other, ending in a precipice over a small river. They thought not a whit about where they were going, and reduced me to jelly as they snorted, squealed and puffed along these narrow routes instead of watching where their feet were falling and thinking about what would happen if they slipped. When I revisited the path after the snow, I vowed never to ride it again ... but twenty years later, the downhill scree has the odd bush growing on it and we have all got used to extreme paths, so the vow has been broken.

The sun glistened on the snow as we staggered up through the drifts to the ruined house and barn glistening with icicles a foot long dangling from the roof. It looked isolated, enchanting, and even welcoming that lovely winter day. We all stood and stared in wonder as we came over the hill, then the stallions had a whiff of the mares who were already there; the landscape was forgotten as they squawked and squealed, leaping their way towards them.

Little did we know how difficult that winter was to be for all of us. Human accommodation was almost absent. There was no water, electricity or insulation; the draughty stone walls made of limestone had burrows throughout made by the loir (a squirrel like dormouse). The loir made it clear that we were trespassing, and they would come out of the wall and stare down at our shivering selves trying to block up holes, while they chattered and abused us. Eventually however, they came to grips with our occupation as they developed a real skill in thieving food, leaving very large rat-like turds everywhere and making generous nests out of the insulation in the walls. When they could find nothing else, they tucked into the library books, preferring the philosophy section, curiously. The temperature at night was often minus ten to minus fifteen degrees Celsius and all the heating we had was an enormous fireplace which consumed wood by the ton and gave off very little heat which, almost joyfully, escaped up the chimney. We battened down the doors and windows with mattresses to try keep some of the draughts out, but even then we had to huddle close to the fire in our clothes on one mattress.

Since the farm was covered in snow there was no grazing for the animals, and no hay or straw. This is the worst feeling you can have, a number of animals in minus twenty degrees Celsius with snow on the ground and nothing to give them to eat. There was one dark, badly ventilated subterranean barn where they could shelter from the worst of the blizzards if they squashed up, which horses are not keen on doing. We needed help to improve this situation, but in a strange country with a different organisation and a strange language, where were we even going to be able to find the things we needed to buy to build, never mind someone to help build? It took around eighteen months before we knew where we could get all the hundred and one things we needed at some sort of reasonable price ... but the worst nightmare, no forage for the animals, continued because no one would deliver hay or straw up our mountain in the snow at any price. But luckily the young French couple who had joined us, helped locate some forage. They had wisely installed themselves in a mobile home and the ten large, fat, draught-like horses they had brought with them seemed at home in a paddock they had erected, and they had some hay. The French couple managed to steer us through many obstacles and an unbelievable mountain of French bureaucracy, involving dossiers and dossiers (large files with hundreds of pieces of paper in them), and visits to offices all over the department. Finally, I found out, after weeping in the last phase of deep frustration, that, suddenly, things could be arranged even though it had been quite "impossible" before ... French bureaucracy is designed for its own self perpetuation and is quite ridiculously complicated, but, low and behold, it had a human face when the chips were down.

Another education was buying food. It was the first time I had encountered supermarkets as we had always grown all our own. These were a test of strength in trail-finding around acres of shelves and then, decision making, between the 500 types of coffee to buy. Eventually, having discovered that the cheapest was always on the bottom shelf, all one had to do was keep one's head down and scurry around half bent pushing the enormous trolley. The salesmen and women could be rude, but, at the same time, they could be remarkably helpful, pleasant and of course very chatty. Shopping in France takes about twice as long as anywhere else. People are quite happy to wait their turn (and listen in of course) as the salesperson and the buyer engage in long conversations about their families, their homes, their problems. The local pharmacy is where the French hypochondriacs discuss their past and present ailments endlessly with the pharmacist, or anyone else around, and we all wait in queues stretching out through the door ... In our home village Bourdeaux, this has become such a problem that they have just had a new, much larger, building built to house the pharmacy!

We learnt to negotiate these problems, and us humans had enough to eat (at a price), but for the horses, we had to eke out the straw and hay and they were so hungry that they ate every last piece of clean straw from their beds. After which they stood and glared at us. We made sure they had good thick rugs (which they regularly tore on all the thorny bushes), so they could wander around the unfenced mountain in the wind, rain, snow and hail. Actually, they did very little wandering: they too thought it was "extreme living" and stuck close to the barn awaiting another handout and staring at us accusingly for our untimely move. It seemed that from their point of view, we had not sufficiently fulfilled our part of the social contract, "the duty of care" ... They were right of course, and we could hardly meet their accusing stares, and, guilt-ridden, vowed this would never happen again.

Once the horses and dogs had been more or less convinced that life could go on, by our eventual acquisition of some straw and hay and a few fun trips into the mountains, our next task was to try to make our own lives comfortable by insulating the breezy walls, and trying to get some light into the dour, dark ruin by installing windows, glass doors, plumbing and electricity. I remember the first day we had hot water ... it was Christmas all over again and we laughed and jumped about scrubbing and just enjoying the warmth of the shower, what bliss.

Oryx, the younger stallion, was given what later became the dining room as his stable at night. He could push his head through the door into the kitchen whenever he thought he needed attention, while Oberlix, his older half-brother, lived with the mares. On one occasion shortly after they had arrived, the mares tried to go on a home-pilgrimage. One evening, the French couple could not find the mares and foals and spent several hours stamping around in the forests and bush of the mountains in the cold. Eventually they were located right at the top of the farm starting to climb a precipice in the direction of England, but there was a foal missing. The others were brought back to the buildings, but no sign of this three-week-old foal. She was no fool that filly though, the next day when they went back, there she was curled up in the snow just where they had found her mother. She grew up to be a delightful grey mare whom we eventually sold to a friend in Sussex.

Particularly at weekends, guns would be let off around us. We had taken the precaution of learning about the rights of the *chasseurs*: the shooting or hunting fraternity in France. They have political power which seems to go back to the revolution. Previous to that, the peasants or anyone else found trapping or hunting any animals on private land, had had their hands cut off, or were guillotined by the landowners. This annoyed the proletariat who insisted they were only killing for food and, eventually, it was one of the reasons for the revolution. After the revolution therefore, all the land in France was declared free for all for hunting including songbirds. Today, even though there are organisations and some control of hunting, the belief that all should have the right to hunt (shoot) everywhere is still held by almost everyone, and encompassed in the *liberté* of the French motto. Only the minority, those interested in conserving the natural world, feel it should be controlled or stopped. During our trip around France we had heard terrible stories of what happened to people who attempted to stop hunting on their land, from having their cats and dogs shot to a man who had been castrated and left in a ditch to be discovered by some tourists later. Three or four people are shot almost every year by mistake by hunters, and even though a plea not to shoot on Wednesday afternoons so that school children can walk about safely has been made, it is not obeyed. Hunting, that is shooting, is certainly difficult to control when the police, mayors and anyone else in charge of law and order are also hunters. But, gradually things are changing; the mayors and police chiefs are not all hunters anymore, but the hunters, even though now a minority, still retain a great deal of political power and many people are frightened of them.

What were we to do about having the local hunters shooting all over our land? If we stopped them, they would likely shoot our horses and dogs; if we did not we would have a constant invasion of them and their out-of-control dogs and, no wild life. We decided to take the bit between our teeth one Saturday after hearing many shots at the top of the farm. We rode up on the two stallions to find out what was going on. First, we found an adolescent boy carrying a gun, who on questioning told us he was "looking for his dog", the usual excuse. Then on coming around a bend, we found two fierce-looking armed hunters waiting to shoot whatever came out of the bushes. We confronted them and told them we were the new owners and did not want any hunting on our land, that there were clear boundaries etc. The smaller, aggressive fellow waved his gun in our direction, so Oryx approached him, nostrils extended and pushed him into the bushes, gently. We could see how frightened he was, nothing like a big black stallion to sort out a miscreant! The other hunter meanwhile was bending over something: Obie and I went up to him to find that it was a roe deer that had been shot. The fellow knew the new law which says that he needed permission to shoot on people's land. He realised he was in the wrong, so lied that he had shot the animal on the neighbour's land which was only about 300 metres away, and they had come onto our farm on a mission of mercy to dispatch the wounded animal. On examining the roe doe, it was clear that she had received a bullet in her back and would have been paralysed and unable to run. I pointed this out, and they looked a bit hang dog. With Oryx and Obie's help, we ushered them off the property, and the deer had become our property as we all knew. But, knowing what the hunters could do, we decided to try a public relations move, and gave it to them, on condition that they never came to hunt on our land again. Oryx and Chris collected the deer's body, and ritually presented them with the poor thing.

Since then, we have had ups and downs with the hunters, but with the help of welcoming them to English tea and scones, and catching and returning their lost dogs on frequent occasions, they now are beginning to respect our boundaries. We have tried, mostly unsuccessfully, to convince them that we are not against hunting everywhere, just not on our land. Our land is a nature reserve which can act as a breeding ground for their game. We now generally respect each other's territories, and this has been helped along by recent changes in the law. In 2015, the French government at last recognised that "wild animals are sentient", more than a century after the UK, but it has happened. There remain one or two wild cards, one a local lad who several times we have had to accompany off the farm when he was "looking for his dog" ... I have even offered to help them train their dogs so they come when called, and do not run around uncontrolled, get lost or shot by other hunters, but so far there have been no serious takers. All the hunters (and some of the dogs) wear bright red or orange caps, although the rest of the uniform is camouflage, army type. I would have thought "the game" could see their caps as well as humans can, but at least they can recognise each other! Maybe the picnickers and mushroom pickers should wear orange hats too. Riding through the forests in the shooting season is always risky, and we are careful to wear brightly coloured shirts and jackets. Lilka, the "silver bullet" as we call her, is white and she can be picked out miles away in the mountains though, which comes in handy in the shooting season.

The dialogue with the hunters, is not helped by those who are interested in protecting the natural world. The local protection of animals society, are so anti-hunting that they bubble at the mouth when they talk of it. Even at conferences set up to discuss these conflicts, all they do is infuriate the other side. Why is it that neither side can make compromises? After all they are mutually dependent in many ways.

We began to understand some of the cultural differences between the UK and France, some of these we welcomed, some were more difficult. There will always be good and bad parts of any place or culture, concentrating on the bits you like is what we all must do to get along. "Think cow" is now our mantra and has been helpful to many around us too. This is because cows are specialists at optimistic thinking and feeling; however horrible the conditions they are in, they will always look on the bright side and make the best of it all. We should try to do the same wherever we are, but it can be difficult.

Over the next fifteen years we worked hard on developing the ecological farm, but it was tough. There was "extreme fencing" to complete, knocking in posts and putting wire up and down cliffs, and cutting broom, wild roses, and juniper that had invaded every patch of grass, and creating some paddocks to grow much more productive and palatable grasses and clovers for our stock. First it was necessary to improve the soil on stony areas. This can be done by adding humus such as muck, leaves, or any organic fibrous waste and straw. Once humus has been mixed with the existing stones and rocks, the next thing to add are nutrients (NPK as they are called: Nitrogen, Phosphate and Potassium). We needed livestock to do this, so we had barns to build and ruins to reconstruct.

Our animals and ourselves gradually began to feel at home and we had re-built our multi-species community, but we had not participated enormously in the human community. This was largely because we had too much work, not enough energy, and a different language. But I did volunteer to run an English class at the local *mairie* for six years, and through this we became close friends with some of the locals. The class included trips to the south of England and Scotland, and during the classes, we were taught a lot about French culture and had many fun evenings, although I am not sure if anyone learnt any more English than they already knew!

France has an unquestioned belief in anthropocentrism and no religious education or non-materialistic thinking is taught in schools. But there are positive results from this. One was illustrated when our imported South Devon cattle escaped into the mountains. They had arrived from Devon in January when the evenings were dark. They had to be unloaded at the top of the mountain because the driver would not descend our snow- and ice-covered drive. The cattle, when they emerged from the truck, were terrified of the mountains, the snow and ice, the dark and the unfamiliarity of it all after a journey of thirty-six hours from their gentle green Devon fields. They jumped the fences and ran off terrified into the snowy mountains. We spent a couple of nights and days looking for them in the blizzards, then circulated that they had escaped and we could do with help. The mayor of the adjacent commune and several of the residents turned out and helped all day padding through the deep snow, and searching through the forests until we almost dropped. The following day one of the neighbours went out early and found one of the heifers who was so frightened that she had chased him up a tree. Luckily, he had his mobile phone and called for help. This meant we could call the fire brigade because only if human life is in danger will they come out to help. All twenty volunteer *pompiers* lined up with ropes and we managed to direct Emma into an enclosure, and after another couple of days lying alone in the forest in the snow, she followed our Jersey house cow Velu down to the farm. The young bull and a heifer were still lost in the forests. Chris and I managed to trap the bull, again with Velu's help attracting him to her, and we laid a rope trap so that we could quickly twirl the rope around a tree to hold his one-ton weight rather than have him galloping off with two hopeless humans hanging on. There remained Fantastic, a frightened heifer. Every day for two weeks we took Velu up into the mountain to try and find her. Eventually we did, but could not lure her away. However, whenever we asked, the loyal neighbours turned out to help. After two weeks, by chance and with their help, she followed Velu the three kilometres home, warily at first, and on arrival into the barn she had her first feed for a couple of weeks. Thereafter, a well-earned aperitif was served to all the humans.

Anthropocentrism is part of the French make up. It encompasses *liberté, egalité et fraternité*, something most French men and women really do believe in (although their society does not always work that way). I wondered what the British three-word slogan summing up what Britons would like to be known for would be: maybe "Justice, Innovation and Humour" ... well that is what a group of us decided anyway ... even if it is not always true.

The down side of anthropocentrism is that non-humans don't really count except in so far as they can serve human needs, whether this is for food, money making, working for humans in one way or another, human therapy, eco-tourism, being looked at on rambles through the countryside, or appreciated like a work of art, or shot for fun. Non-human sentient beings are not regarded as having intrinsic value. Those interested in animal welfare usually believe that non-humans can feel and suffer, and therefore we must treat them better, but not that other mammals have a mind; someone with a mind is a total "being", and therefore must be equally considered. The denial of this is, of course, not unusual today, take the lifeboat thought-experiment: how many of you when confronted with the choice of throwing out an unknown human or your own personal dog who has just saved your life, will choose to throw out the human? Hardly anyone because even if we might want to, we would not dare; so powerful are cultural beliefs.

There have also been amusing times. After a few years, Chris was asked to be a judge for a new garden competition (he is not very knowledgeable about plants, but smiles a lot). In one of the local gardens entered, the lawn was not as well cut as it could be, nor the beds well weeded. The owner apologised and said he had hurt his arm. We heard from one of the other judges that he had been up a tree stealing his sister's cherries on the family property when his nephew turned up. They started a normal, arm-throwing excited French fight, and he fell out of the tree and broke his arm! Families are very close in France, they holiday together, eat together, meet each other at high days and holidays, but they also squabble a lot!

We had inherited a mighty pile of machines and rubbish from the previous owner whose main hobby had been driving around making roads through the farm. He left his broken-down JCB so we used it to dig holes for planting fruit trees. It used so much oil that we decided to give it to a local farmer who had a couple of hectares of broken-down machines. He liked them, and so did his pigs who made their homes in the various motor cars and old tractor cabs. He was delighted and took it away. Later he turned up with four rather scruffy "pre-alp" ewes (the local breed of skinny sheep with wool like a rough carpet) in exchange. They became the foundation of our sheep flock.

When looking around for some heifers to start our cattle herd, I learnt a lot about why rare breeds are rare; they are unproductive and therefore uneconomic even though believed to be better adapted to local conditions than introduced cattle. As previously on the Isle of Mull, we found the local rare breeds to be less fertile, grow slowly and end up small and with poor conformation, even though they needed the same amount of food as a more productive breed. This is not to say that rare breeds should not be kept going in order to ensure the gene pool keeps its variety, but they are for hobby farmers or those specialising in selling rare breeds for breeding. We found even in our difficult mountain terrain the large Charolais and Limousin were more productive and profitable when given the same lifestyle and food as the indigenous breeds.

We were surprised to see how frightened most farmers were of their cattle. Many of them are horned because they live most of their lives in the mountains and there is no need to de-horn them as when they are crowded in barns. The farmers walk around with big sticks hitting them, shouting, without any dogs to help. As a result, the cattle are frightened too and more likely to attack out of defensive fear. They become difficult to direct and impossible to handle without a crush. It took a family who had bred Aubracs (a local breed) for several generations, all day to run ten heifers through a crush so that we could select two. It is curious that some professional farmers in any country, really do not understand how to move and handle their animals easily even though it would make their lives so much easier. No one teaches how to do this, it is just assumed that the breed or individual is "difficult" or "dangerous". No one realises it is the farmer who has made them so.

We had some exciting moments. When building a dam to make a small lake, the digger, with Jake in it, gently rocked on the ten-metre high dam wall that we had constructed. Luckily, it rocked back rather than crashing over and down, and we eventually managed to pull it off the narrow wall with a steel rope and our tractor. On another occasion, the exhaust of the tractor set the dry bush on fire. This required a call to the *pompier* as the fire spread "like wild fire". Eventually we all managed to control it, and the fire brigade had a large lunch brought in a special wagon for them. When building the barn and securing the three to four metre high roof, there were many moments of holding one's breath as both Jake and Chris had had nasty falls with long-term consequences off roofs before ... but we are all still alive and kicking. In fact, Jake insisted in making a movie pretending to slip off the roof with me as camera man.

Another reason for our new project in France, was to show visitors, students and workers that it is possible to live well in the wild world while learning more about it and enjoying contact with domestic and wild animals. We re-joined Willing Workers on Organic Farms, and during the next eight years or so we had a great number of WWOOFers visiting. They contracted to stay for two weeks before they arrived. Not all were very helpful, but many were entertaining and kept us all going. We had couples and families from all over the world, but the majority were young idealistic students, several with masters qualifications in ecology, although it turned out they could not tell an oak from an ash tree! They learn economics and public relations in their Masters of Ecology, but not how to recognise different species, and then they go off to give ecological advice! Poor old living world, what chance does it have?

Most volunteers had been drawn by the horses, but after being introduced, many began to understand the cows, dogs, poultry or sheep, each having different messages, if we would only listen. It is not unusual to have young enthusiasts who are completely unfamiliar with non-urban living. For example, we had a girl who was amazed to see that carrots grew in the earth and you can pull them up when ready to eat in the summer, another had never done the washing-up by hand; he had always had a machine and spent all morning on a few plates and bowls. A couple came with their horses to study how to live with, and look after, them. Although they "loved them to bits", they were quite thrown by being taught by one of our horses, Shemal, that she has opinions and will make decisions. She must want to cooperate with them, or the human will be the losers. It blew their minds! Another time, a small child crept up towards Moto, one of our collies, who was watching him approach her mat; she snapped at his fingers as he advanced. The child ran off, only to come back with his even smaller brother who was told to go closer and touch her! We suggested that they might like to talk to Kananga, who is an ambassador for dogs, rather than Moto who is not. Kananga allowed the children to play with him. Mother arrived, probably to blast us for Moto's behaviour to her child, but when she saw them with Kananga she simply gave them a lecture about the "instinctive behaviour of dogs"... we did not tell her what her eldest son's idea had been for his small brother!

The students came, studied and helped and left after six months or a year, and often we, like the dogs or horses, became fond of them, but, like the young horses bred here, they needed to experience different aspects of the world and continue with their lives. We keep a visitors book in which they write an honest assessment of their stay, and in general, it seems they did get the message from our animals and the wildness of the farm that it was possible to live well in it. Here is a quote in French: *merci pour ce moment magique, hors du temps et des contraintes de la vie 'moderne'. Ici la nature, les animaux et les humains sont heureux ... que demander plus?* Most comment on the beauty and calmness of the animals and their willing participation in our activities, and all of them state that the experience has been very different and often unique. So let us hope the message that it is possible to live differently in a multi-species community has rubbed off on some of the hundreds whom we have welcomed, with the help of our permanent four-legged family who continue to happily work with us, absorb and give us affection, shape our multi-species community and ensure it survives and thrives.

We had research students from all over the world particularly researching horse behaviour. Once the farm and nature reserve were more or less going, I left most of the work to Chris and the students, and scouted around for scholars in relevant disciplines who would welcome me to their departments for short stays. I was given a fellowship at the University of Bristol in animal welfare to immerse myself in up-to-date ideas from animal welfare science. This time only Oberlix came with me, but he continued to do demonstrations and tests, and best of all from his point of view, he had a delightful pony mare as a companion and eleven months later, another daughter. Another year, I spent a couple of months as a visiting fellow with philosophy scholars at University of California, Berkeley, studying further the philosophy of animal minds.

But aside from that, it has been the Multi-Species University of the Natural World that has been my host and allowed me to summarise what we know concerning the knowledge that other mammal species have. Sometimes with the odd flit to conferences, almost annual visits to elephants in Africa and plenty of action in Southern France teaching, travelling and learning with our horses.

For about twelve years now, we and our horses have been taking part in endurance competitions, which allows us to go on expeditions together, to ride around in new and often lovely places, to meet other horses and humans who are interested in doing the same thing, and of course to enter into a little competition with them. These meetings are our vacations and it is a delight to camp with our horses and dogs among others with their horses and dogs and, for a short time, enjoy a mixed multi-species, multi-cultural weekend, with horses in paddocks and dogs and children running and scampering around among the vans and tents. The horses are given a lot of attention, and groomed to look their best before being presented to the veterinarians who measure their heart rate and other signs of their fitness and soundness. On return after the long ride (and every thirty kilometres during the longer races), the same procedure is performed and the different parameters marked, so that the winner becomes the fastest horse with the best fitness scores. We have taken part in endurance competitions for around forty-five years and followed its evolution, from the lowest to the highest level, and quite often won. Lilka, the "silver bullet" wrote history in 2015 by winning the amateur competition of 160 kilometres at Florac, reputed to be one of the most difficult rides in any country, up and down the mountains and in and out of rivers, plains and forests. She is only pony size, but carried long thin Chris whose legs were almost touching the ground, and galloped past all but five of the international competitors who were going off to do the world championships. Then Shemal, aged twenty-five, recently won a forty kilometre ride against horses half her age, and we have had many other successes, so we must be doing something right although we continue to learn every day, and also, of course, make mistakes.

Both the horses and the riders must be prepared. This entails early morning training rides, riding up and down parts of the mountains at the speed each horse can go, depending on his fitness. The riding competitions range from forty kilometres to 160 kilometres in one day. Some of the horses like to compete, some like just to go along with it all with their rider, or with other horses and riders. But, if they no longer want to do it, they may go on strike and no amount of persuasion during the training sessions will make them run up and down that hill AGAIN, it is a case of "been there done that". Our job, then, is to invent new and interesting things for them to do which will increase their fitness, both physically and mentally. The horses do not have the social rewards of the large cup at the end of the race, or the added social status in the horse world. So, the way in which we try and keep them motivated is by giving them different things to do: working in the garden pulling cultivators or harrows, spreading muck and compost, and weeding between the rows of vegetables for example. They may harrow the fields, or go with one of us to find and check the cattle and sheep and move them to different areas of the mountain. They might pull a buggy up and down our hills teaching people to drive horses, or give riding lessons or demonstrations in workshops that we conduct. Recently they have been taking part in dressage competitions which are conducted online, so no travelling, but there are still proper judges marking the video you send in. Two of our horses have now got to Prix St George level, and this certainly helps them in their endurance work. All of this gives horses and humans a change, helps us develop different skills and means that we can save time and energy because they are doing various jobs more quickly than we can without them. They do dances and displays sometimes (we specialise in multi-species ballets); in fact they more or less take part in every aspect of our lives, although like the dogs and ourselves, they must also have time with their own kind.

Years come and go and we have been at La Combe for fifteen years. The buildings are more or less fixed, at least in the way we like them, the land we wanted to clear is more or less cleared, the fruit trees have grown to give us masses of fruit, the cattle, horses, sheep, dogs and humans have, we hope, lives of some quality and also work for their living, and we believe that they consider that they are family members and colleagues. Perhaps we now know a bit more about the mental attributes of some species, particularly elephants, equines and bovines. The next step is putting this together with a better understanding of each individual's take on the world, his subjectivity ... Maybe in the next lifetime, who knows.

On the ecological front, we also needed to ensure the maintenance of every type of eco-system, the wet and the dry, the open plains and the bushes and forests. For this, lists of as many of the species of flora and fauna must be made and monitored to assess if the numbers or types of species on the farm have changed as a result of our activities. First, we used the animals we had to increase access to parts of the farm that had become thickly colonised by broom, thorns, juniper and wild roses. The cattle are more adventurous than the horses, they will push and shove their way through thick undergrowth, making paths which the horses, chamois, wild boar and roe deer will then use, as well as ourselves. They open up and graze areas, which stops the broom colonising everything, just like the European bison and Aurochs (their ancestors) used to do before humans killed them all off. Cattle living in the mountains and managed well, apart from having a good life, can also be considered "ecological tools".[1] As a result mainly of their efforts, we now have an ever growing number of species of wild orchids: twenty-two to date, and five new areas of natural grassland. The number of the different species of birds and other species was increased, as well as their diversity,[2] although we have not yet identified all the insects. Some progress has been made with the butterflies and they have certainly increased in number and variety in the last years, but this might be due to particular climate changes. Finally, we have shown that twenty families can be fed off this farm if it is farmed ecologically, while at the same time, the whole farm remains a nature reserve.

Most of this limestone calcareous territory is quickly colonised by broom if the grassland is not kept open by grazers, and this is then followed by wild roses and juniper until after about fifty years, beech seedlings start taking over. The major threat to the last remaining oak woods on the west of the farm, are the beech trees marching down the hill, decade by decade. Beech seedlings can grow under the intense shade that beech provide, but oak seedlings cannot. It is a question of cutting back some of the beech and using the wood for building, if we are going to keep the oak forest too. But of course, we need also to keep some of the mature beech forest. The trunks of old beech trees are full of the faces of curious monsters and ghosts which makes their dark forest even more spooky.

Beech forests are generally species-poor, because nothing much likes to grow under them (it is too dark and their litter is slow to break down), except for a few saprophytic fungi who live on decaying matter. Traditionally beech woods have been coppiced (the trees cut down to their roots from which they will sprout again). This happened on our farm probably about every twenty years, and the timber was used in the neighbouring village where there was a thriving silk industry which needed to heat up water

---

[1] Allan Savory & J Butterfield, *Holistic management a new framework for decision making*, 1999, Island Press, Washington DC
[2] MKW, "Ecological agriculture," 2014

(by using charcoal from our beeches) to dye the silk from the silk worms grown on mulberry trees planted in the valley.

Coppicing opens up the forest allowing light to the soil so that many different plants and their accompanying insects and predators can become established, until the surrounding beech trees grow back up and shade it out again. Wild boar like the beech masts and scuffle around for them in the autumn. We also had red squirrels, although they are rare in the beech forests, preferring the easy pickings of the walnuts, and we also planted some almonds. The red squirrels remain very shy, we only caught glimpses of them leaping about or running between the trees, but given another twenty years we could probably have them approaching us and maybe even make assault courses designed for grey squirrels.

One winter, we found a starving wild boar piglet curled up in the barn. She was very weak, so we took her to one of the stables so that we could gradually feed her to restored health. But, Pinociette (because of her long nose) was a little monster! Even after a couple of months living in a stable being hand fed and handled daily, she remained a feisty little thing, rushing at anyone and pretending to bite their boots ... and not everyone would stand still as this squealing, long, black-snouted bundle of black hairs standing on end came rushing at you. She was a specialist at building nests in the straw and it was not easy to see from where she would emerge like a bullet of around five kilograms. When she was surely recovered, we took her away from the buildings and fed her in an open trailer for a few days before she disappeared. But I swear she has now brought her large, snouty, grunty family back and they all stand and stare at us, even when we are only twenty metres away. We have had to erect a low electric fence around the fields we have cultivated to avoid them digging it all up. The idea is that they try and barge through, nose first, but get a good electric kick on the low wire. It usually works but takes a lot of maintenance.

The chamois remain timid but they are beginning to learn to come down into our valley when the shooting season starts, although there are doubts whether they will survive as a population as the neighbouring hunters kill more each year than are born. Let us hope they discover our refuge before they are all extinct. One autumn in the early years, when walking along the top road, a chamois male charged at me at full speed, saw me at the last moment and veered off up the mountain, then another roared past at full speed, a young male being seriously chased by a senior; I don't think in his panic he had even seen me. They are the most excellent mountaineers, they can leap about up and down almost vertical cliffs, graze along paths no wider than ten centimetres, lie down on extraordinarily small ledges, quietly cudding. They and their small fawns gracefully leap over our fences to eat the delicious green grasses early in the year, and they are very welcome. I wish they would live on the farm all year to avoid being shot by the hunters.

The roe deer are semi-solitary, living with their young of the year and maybe the one year old youngster from the year before. The mum chooses a male, who lives usually alone, in the forests in the early autumn. At that time of year, their calls and barks carry all over our valley, and we know there is courtship, choices and matings.

Some vultures have been re-introduced to the Baronnies, a wild area not more than thirty kilometres away by air, and they fly over in droves looking for carcases to scavenge. Golden eagles sometimes nest on the cliffs, but they have recently been displaced because a photographer in a micro-light was too close to their nest. There is usually a raptor or two flying over the valley searching for a stray rodent, but one of the success stories, and the most delightful, are the little song birds who are becoming less cautious. They nest near or in the buildings and feast on all our crops and fruit, waste grain and the insects that feed on these around the homestead. In winter it is the blue tits mainly who will chatter and squabble on the bird table, but many other little birds, wrens, robins, yellowhammers, coal tits, and more are beginning to sing and live around. We don't, I think, have anything particularly rare, but they are all special and can live out their lives here now. It is remarkable to realise that the life of a tiny bird is just as complex as ours, but in different ways. A few swallows and house martins have once or twice nested in the barn, and they are attracted to our ponds, swooping, turning and twisting as they catch and eat the insects living and hatching over them. But, they are dwindling in numbers as they are killed in their flights to and from Africa, and while in Africa. They are becoming scarcer everywhere. Why do so many humans like to kill for fun?

We have put some brown trout in the lake and a few times ducks have flown over to spy out the lake, but none have nested there yet. Badgers live around but there is no risk of tuberculosis (TB) for our cattle who are regularly tested. They like to shuffle about, stealing our mushrooms growing in the big barn. The European beaver lives only around twenty kilometres away as the crow flies, so maybe one day, he will make it up our small river Roland and construct some dams to retain the water for us. Then we also have rats and mice, and snakes that keep them mostly under control in the houses and barns. Sometimes visitors freak out as they see an adder slithering through the bales in the barn; but don't frighten them and they will not frighten you. We have loirs in almost every wall, and bats in the belfry (only no bell), who forage airily over our multi-species bathing pond. Last spring a leverette took refuge for a couple of nights under the milking stool in the barn, he just froze immobile as we lifted the stool and looked at him in his nest. After breakfast he had gone, I hope he managed to find his mother again and continue his life. I don't think we have any species particularly special, but for us, they are all special and have a right to a life that they know how to live.

About six years ago the whole farm was declared a Nature Reserve with an organisation called *Association pour la Protection des Animaux Sauvage* (ASPAS) and we hope it will continue as a nature reserve for ever. One day, maybe someone will re-introduce European bison, and buktan, wild sheep and goats to this last piece of natural mountain.

During this period, I had learnt to listen more carefully to the animals I knew well, and to teach others how to do this, I hope. We had shown that efficient food production can be integrated with wildlife on a whole-farm scale, even in a very difficult climate and topography. We had also shown that ecological agriculture can be more efficient than current non-self-sustaining agriculture, since our farm produces more net than any conventional high input farm.[1] But also, we now realised how hard and long the road is to re-establishing a larger multi-species community that includes the wild things. Even a few years of exposure to humans who hurt or kill them is enough to ensure animals stay well out of the way of all humans for a number of generations. Maybe one day in our nature reserve it will be possible to have some of the "wild" animals unafraid of humans on this mountain in France; it can happen, as we see, in many national parks with tourists in Africa.

## Similarities between gender & animal issues

My childhood was spent predominantly with women, and I had hardly met any boys of my age until I went to university and then I stayed in an all girls' hall. We were three sisters who went to all girl schools, and our mother; father was errant. I had been in a mixed primary school, but, my human friends had always been girls, although I had male friends among horses, dogs, and a duiker (a very small, delicate African antelope who peed for ten minutes once, we timed him). Since we lived in Kenya, we had no male family members to visit or be visited by, no cousins, uncles or grandfathers. Being at boarding school and then busy at home during the holidays, we had little to do with any of the neighbours, although old settler Jack had some sons about the same age as my eldest sister and when she became adolescent, she did show interest in them, and them in her. We had occasional holidays with couples, friends of my parents, so the only males I usually met were my parents' age and I never really got to know even them in the odd weeks we spent together. Our father was distinguished by

---

[1] MKW, "Ecological agriculture," 2014

his absence as I remember, and whenever he was with us it was a special occasion. He was treated with special care, had special crockery, and special meals, and was not to be disturbed ... it was constantly impressed on me how important he was, or that is the impression I had. So, as far as I was concerned, male humans were less familiar than males of other species with whom I lived, and as a result, I had very little interest in them. My experience was that male humans had to be treated differently from normal humans (that is females) and it was required to show them more respect than to females of any age.

This difference in the treatment of the sexes closely resembles the different treatment and consideration of humans versus non-human mammals, resulting in humans' interests always trumping those of other species. Non-humans, like young girls, are expected to respect other humans and "do what they are told" and they are not usually respected for who they really are; they are just sweet creatures who can be fun to have around, now and then.

My impressions as a child were reinforced because from the start, I had been a failure: female instead of male. They confirmed in me the remoteness of human males and strengthened my belief in being closer to non-human friends. The culture supported this, men were still "in charge" and superior; after all, women had only just emerged in law from being possessions, like a dog still is today.

What human males did, saw, thought, and how they behaved remained a not very interesting mystery. To form a deep relationship with any human or non-human, one needs to have contact and build up a "theory of mind" about the other, that is have some idea what their experiences and intentions might be. At that time, I really did not even know if they felt anything much, except occasional amusement and annoyance at our antics. Even if they did feel things, the feelings must be very different from what "we", me and my four-footed friends, might feel. It is very difficult for any of us mammals if someone is remote, superior and alien, to want to understand or become fond of them. Also to recognise similarities in feelings and behaviours, it is useful for individuals, whether humans or non-human, to get to know each other. As a child, there were enough mysteries in my familiar, multi-species world and all its social complexities to keep my brain and emotions seriously occupied, without embarking on understanding these remote male humans.

The distinction was further borne out by what I thought was very unfair, even if males did come from Mars: this was that male humans could have much more exciting lives than females. They could go anywhere, do almost whatever they wanted, and have all sorts of adventures. The army was infinitely exciting to me at one point. I had loved skiing, wandering around in the alps, galloping about having adventures in the bush, surviving, clambering about on mountains and taking risks ... all the sorts of things that soldiers, when being trained, are paid to do. Why not join the army? They even had a mounted section then, maybe I could join the cavalry? I had read stories about how soldiers adventured all over Africa and in the Himalayas, clambering about with donkeys and mules, it sounded great. The fact that they were trained to kill other people did not really worry me that much ... at least they were not being trained to kill elephants, rhinoceroses or horses which would have put me right off them. But when I enquired if I could join, they said yes, by all means but I would have to wear a skirt (a tight one and a tight jacket) and would only be allowed to make the tea! This was well after the war, when the men returning had taken back any interesting jobs that the girls had done, and it was assumed that now women had the vote and the war was over, girls would be going back to the kitchen; after all, the only ambition a female ought to have was to get married and have children. Adventurous outdoor pursuits like The Duke of Edinburgh Awards had not been invented, and even if they had been, girls probably would not have been allowed to do all the tests. It seemed organised adventures with other humans were closed to me.

But, even aged ten, I was thinking about how I was going to structure my life because from the age of six, I had known what I wanted to do ... study the minds of non-human mammals who were sufficiently similar to me that I could really relate to their experiences, intentions and feelings, and also different enough to give me alternative views and understandings of the world. Perhaps therefore I could become a game warden. I had had quite a lot to do with a few game wardens who had come to stay or with whom we had gone on safari, and one nice man called Oliver was sympathetic to my interests and had brought me a young abandoned duiker to look after when he came for Christmas. I had seen the bush houses they lived in, surrounded by so many different species of hoofed mammals and their predators on the African plains. In addition the first ever study on the behaviour of another mammal in Africa had been done by a young game warden employed by my father. Allan Brookes had sat in the bush watching Thomson's gazelles in the early 1950s. He had worked out that by studying how many used the lavatories, and the number of lavatories, one could arrive at an idea of their population. At that time the populations of Tommies as they were called, must have been in the millions, almost everywhere you went in East Africa there would be Tommies, now of course they are not always common, even in the national parks.

I knew my father had had young women working on fishery things in Africa, so, surely I could join the Colonial Service and become a game warden? This would allow me to spend my life where I wanted to be: in the wild, wonderful, natural world with my four-footed friends and maybe make a whole bunch of new friends. I remembered with fondness expeditions into the wild with my father and his band of merry men; nature reserves and national parks were developing and needed game wardens. My father told me in no uncertain terms, the same thing ... I was the wrong sex! So what to do? I could become a kennel maid but looking after sick or lonely dogs was not really for me. I could become a vet maybe, but again that meant only seeing sick animals all the time, and driving around in sick-making motor cars a lot. It also meant five years of study in a city ... there were only five vet schools in Britain, all in large cities.

Thanks to the suffragettes, there was one other option: academia. The suffragettes and their followers had managed to open one college in Cambridge (Girton), so after the First World War, around 100 of them could attend the lectures sitting in the gallery away from the men, and they could even take the same exams ... although their degrees were not recognised for another ten years or so, but they did manage to obtain interesting jobs. Indeed, that was evident from the fact that my father employed young women graduates who were passionate about researching fresh water ecology during the war. Perhaps I could study zoology and botany at a university and maybe thereafter get back to Africa and spend time with animals.

In the fifties, women could go to university and study with more or less equal status to men, but there were very few university places for women, so one had to be better than many of the men to win a place. It seemed that this would be my best option, so I worked hard at my A levels and obtained a place at St Andrew's in Scotland, chosen because it was on the sea, a small windswept town, and with an excellent biology department, and, there was even someone studying the behaviour of an animal, albeit a sea anemone! Despite living in an all-female hostel where men were never allowed, we still met and mingled with young men most of the day, and socialised with them in the evenings, so my education about male humans began. Gradually I became aware that they were not quite as alien as I had thought, nor, as I helped out a particularly dim male medical student with his frog anatomy, were they quite as superior or always brighter than I was.

So the question that had been exercising many women (educated or not) for half a century began to raise its head: why were men thought of as superior and more able than women? Why did young men my age have rights to all these interesting jobs and even earn more money for the same job? Men had more freedoms: drinking in pubs and living where they wanted and with whom they wanted (although I was unaware at that time about the illegality of homosexuality for men). Although women had been given the vote thirty years before in the UK (but not in Switzerland or France at that time), women were not treated equally in everyday life. I can't say this worried me as an undergraduate: it just seemed normal, as I imagine separate treatment and lack of freedoms does for many young Arab women in some Arab states today, as well as many dogs and cats. But, like many women throughout history, I gradually learnt that if determined enough, I could find a way of doing what I wanted and it was often easier to achieve this by working within the system, rather than trying to change it.

During my subsequent career I have often asked myself if I have been discriminated against for being female; the answer is I really don't know. Was it my femaleness that was a top priority in my rejection or selection for various jobs? In the sixties and seventies when female liberation was a driving dogma and almost all women were shouting about the discrimination against them, I could not really believe that it was my femaleness that had caused me not, for example, to be given a tenured university post, or other soft options. Was it just because of who I was that I did not get the post ...? Perhaps I was just too much of a risk? Yes, I probably talked out of turn and was rude to people sometimes. Perhaps it was because it was recognised that, in some way, I would be an un-stabling faculty influence, and unpredictable in terms of beliefs and ideas. Perhaps I would rock the boat. I don't know if this was because I was female, or just because of what I said and did, but all my life I have really desired a university sinecure that comes with a tenured post. I find pretty attractive the idea of taking for granted that one could (a) continue one's research year after year with financial backing provided, (b) continue researching even after sixty with the gross pension they gave you, (c) only have to teach around ten hours at most per week, (d) have conference attendance costs covered and frequently fly to the other side of the world, and stay with all the other delegates in a hotel instead of hitchhiking there and camping down on someone's floor, and (e) on top of all this have a generous salary, more than I have ever earned teaching for university faculties for well over ten hours per week. These sinecures have made my mouth water every year, particularly when my ex-PhD female students took up their tenured posts as if it was nothing but their due! There probably are disadvantages, but they are difficult to see from the alternative academic experiences of my lifetime. Perhaps I was just too dim, not as bright as those with tenured posts? This seemed unlikely as I have encountered some extremely dim people burrowing away in tenured university posts in the most prestigious universities, who have never had anything remotely like an original idea or been able to organise and carry out different research programmes, or teach well and be liked by their students; there were even lecturers with serious stutters who had tenured posts. Another alternative was that maybe they thought I suffered from one of these modern "diseases" such as autism, dyslexia or post-traumatic stress (hopefully not Alzheimer's for a time yet). I remember once one of my colleagues, who eventually became an establishment figure. A fellow of the Royal Society and the rest, invited me into his room one day to have some sort of psychological test. I have no idea what he was testing but I have an idea why. Although I had been in that lab as a post-graduate student, a post-doc and finally an honorary fellow, and had helped them out a great deal in teaching subjects they did not want to teach, demonstrated in practical classes, put the department's address on all my publications, and so on for around twenty years, the other faculty members, whom I knew quite well (all but one were male), were confused. Was I bright and mentally able, or flaffy, unconventional and probably just silly? I don't know what they concluded as a result of those tests either, but it was probably the latter. I applied for a tenured post at that university which I would have loved to have, but, even though I was good enough to be elected an honorary fellow, I was not good enough to even be interviewed for the post ... That smarted, but of course a "jolly good upright fellow" got it, kept it for about three years and then scampered off to another tenured post!

Without a tenured post, the only way to continue with research in academia, was to try to dream up and sell ideas that were recognised as worth researching. This is an insecure way of living and raising a family, so after a while, researchers who take up post-doctoral research, either obtain a tenured post or give up and change careers. Today this usually means obtaining a post working for a company or some sort of NGO (non-governmental organisation) which has already decided on the answer to the research they pay you to do. For me, after finishing my doctorate, trying to understand better the minds of animals meant academia, and that depended on winning post-doctoral fellowships. There may have been discrimination against giving me money for research in areas thought to be "too dangerous" for women, or because there were also men applying who were "good upright chaps", but there was more of a chance of competing on a level playing field when selling ideas that were worth researching and no one else had thought of. A research fellowship usually lasted for two to three years. At the end of this time, not only must all the results be available, but they must be written up and on the way to being published before you could apply for, or be likely to be awarded, another fellowship. As a result, every award depended on your previous performance and even then you were usually competing for the money with those with tenured posts, where it mattered much less if they had the money, because they already had a salary.

Knowing the right people in the right places is one of the most important qualifications for doing almost anything: the "old school tie" network, which is not restricted to English public schools. But, with my background, where I was convinced that my family believed I was inferior and stupid, it seemed to be cheating to use any contacts my biologist father might have, and anyway he would be unlikely to give them to me. Any post needed to be achieved on individual merit. As a result, when selected for an interview for a Churchill Travelling Fellowship, Peter Scott, the bird man, who interviewed me and was a close friend of my father, did a double take when he heard my name, and was shocked to find that I was my father's daughter.

In retrospect, perhaps the way it has panned out has been a lot more interesting, and certainly has opened many different doors to one's thinking, energies and interests than having a tenured university post. Yes, not getting one has probably been, in part, because of being female and consequently considered a bit flaky here and there, but also, perhaps, because of being who I am. People have thought I might rock the boat, which yes, I have done from time to time after finding widely held ideas or beliefs particularly irrational or badly thought through.

I recently re-read Virginia Woolf's *A Room of One's Own* and *Three Guineas*, two rather satirical, wordy, literary works on the role of women, or rather "the daughters of educated men", in society in the 1930-40s. Her summary of the four most important things that women needed to be able to do "to contribute to society" resemble the rules laid out for a monastic life. However, it would be interesting to see if these important attributes apply to my four-footed friends or me; so do we contribute to society? These are:-

(1) "poverty", in other words not pursuing financial betterment if it threatens other principles or species. Our four-footed friends and generally I have had no trouble with this.

(2) "chastity", which she identifies as "a refusal to sell your brain or your ideas for money, or any other reason, such as avoiding conflict". Here I don't think either my friends or I would qualify, we do sell our ideas and brains, not for money but for other things.

(3) "derision", by which she means a refusal to advertise yourself. This is true of most of our four-footed friends, although perhaps it's not always the case, for example a stallion or a dog trying to impress his lady friends and anyone else around does "show off". I have been guilty of this too from time to time.

(4) "unreal loyalties": I read this as doing what you are told because that is what you are told, not because you feel it is right. On the whole, both my friends and I sometimes follow this rule, sometimes we do what we are told when we do not want to, so that we can continue our cooperative relationship together. But then again, sometimes we all do things because it is our culture and tradition and so have beliefs which we accept without question.[1] But, on the whole, I at least, and most of the members of our four-footed family, prefer to have loyalties that they choose,

---

[1] E Avital & E Jablonka, *Animal traditions*, 2000

not because they have been told to have them.

So perhaps, female humans and non-human mammals, can qualify to be "contributors to society". But then again, like it or not, it is inevitable if you are alive and living in any society, you will contribute in one way or another: your presence is felt for good or bad. Which are the "good" contributions and which the "bad"? Each species or individual will contribute in many different ways, and this will only be discovered by looking a lot harder at all aspects of who s/he is, his/her species and his/her particular individuality.

It is more than half a century now since women have had equal consideration in law to men in in some countries. Will the next of society's revolutions focus on the treatment of, at least, non-human mammals by recognising them as "persons", not "possessions" and giving them equal consideration to humans? There is no reason why we should not be able to ensure that their life when living with us is just as good as the "wild"; if not better, if they always have good food and medical care as well as freedoms. We have a duty to provide them with a life of quality, and we can measure the degree to which we have achieved this.[1] If we work together symbiotically so that animals are treated as colleagues or employees, not slaves, then we will enjoy it more. Since the wild is now almost history, if any other large mammals are going to survive, we need to bring them closer to humans in order that they will be better understood and appreciated. Rather than having less to do with them, we should embrace this symbiosis since only in this way can we can enrich each other's lives. We need to work out how to do this, not ask whether we should; it is too late for that. This has to be addressed for gender, race and now age discrimination issues too.

So why were women not allowed the same status as men previously? Firstly, perhaps because humans are primates. Primates have very different roles for males and females and in most species they look very different, so their gender is easily recognised.

---

[1] MKW, "What is it to be a horse?" in *Horse Watch*, 2003, London

Another characteristic of primates is that their food is very rarely equally distributed for everyone to get an equal share, usually it is in patches. The result of this is that most primates go in for power struggles, who should have first go at the food? Because this causes conflict and injury to individuals or even breaks up a society, a way to cut down conflicts has evolved; this is known as "dominance hierarchies" or "the recognition of power". Here, each individual "knows his place" when it comes to acquiring a resource. Because males are bigger, often they are top of the hierarchy, and females may have different hierarchies. This encourages the acquisition of higher status, by moving up the ladder to have more power. To do this, each individual learns to manipulate and deceive others. Cognitive ethologists (those who study the minds and mental events of different species) always assess the other's mind from a human primate perspective, and therefore of course believe that humans are greatly more "intelligent" than non-humans. They even invent experiments to show how deceptive and manipulative members of another species can be and how they consequently qualify as "intelligent" like humans.[1] The enormous range of other mental skills such as communicating emotionally are ignored or downgraded because humans are not good at them.

One unique character of humans is that they have a context independent language. Whatever the situation, the same words mean the same thing: a dog is a dog is a dog, never a cat or a mouse. Human language is also highly symbolic and structured, which has enabled the development of mathematics, the growth of technology and the digital age. This indeed is one type of mental attribute or "intelligence", but there are a great many other ways of acquiring knowledge to live in different ways, and other ways of assessing and sharing the world. But the idea that the human mind and intelligence is superior to any others is hard to give up, even by behavioural scientists in this age of anthropocentrism and despite the constantly growing evidence of the power of humans to grossly manipulate the world (even if such manipula-

---

[1] e.g de Waal, *Are we smart enough to know how smart animals are?*; HS Terrace & J Metcalfe, *The missing link in cognition*, 2005, OUP

tion is going to cause our own destruction). As a result, both the girls and boys engaged in animal cognitive science are striding confidently down the same well-paved path of laboratory tests towards what is believed to be "the truth", dismissing or disregarding any other knowledge about that animal as "anthropomorphic" and therefore a sin ... Perhaps we urgently need more people who rock the boat.

Humans have minds that are very good at what they do, but equally, they can be very bad at doing other things. The human mind and body being is not the only, or superior, one, any more than the historical belief that male humans were considered superior to female is the "truth". In many ways, therefore, animal cognitive scientists are re-enacting genderism. This belief of the superiority of human minds is very close to Descartian dualism, confirming the separation of the mind from the body in the superiority of humans and the greater importance of their needs, beliefs, and technology. Testing rats' ability to learn human concepts such as circles and squares in the lab, when the rats have never seen such symbols before, and showing they are inferior to human infants who have been exposed to such things from birth, is just bad science. When I was a child, I showed some Masai the black and white photos of themselves and their friends I had taken. They had never seen photos before, and they could not understand them ... so they must have been just as "stupid" as the laboratory tested animals! When children or babies of different ages are tested in the laboratories and shown to be superior in such discriminations, they are given very different treatment, they are cuddled, stroked, pampered and given treats, and are often with their mothers. The treatment of rats, horses, elephants or any other species in their keeping conditions and the laboratory is very different. The researchers confidently publish papers in influential scientific journals such as *The Proceedings of the Royal Society*, concluding that the ability of the rat or horse to understand shapes or concepts is equivalent to a one- or two-year-old baby or child. This is then an accepted conclusion and becomes the content of textbooks,[1] confirming our human superiority in all mental abilities!

Apart from this being bad science, it has also lost touch with folk knowledge. Folk knowledge is knowledge which has been accumulated over centuries by people who deal with different species every day of their lives, working, living with, and teaching them, using their products, raising them, loving or hating them, and more familiar with their mental capacities than the urban scientist who is experimenting with them. Yes, of course, there are cultural beliefs or preconceptions concerning other species' mental abilities, but over the centuries, particularly where many animals of a species have been with humans, this has been distilled down to a great deal of real knowledge about such a species; much of which comes from recognising what we have in common in our bodies and minds, our "beings". Frequently, money is spent on experiments by urban-based researchers who have no practical knowledge of the animals they work with (they even have "technicians" to look after the animals). They believe such knowledge must be tested by "science", to prove things that have been known for centuries, such as that elephants or horses recognise each other's calls, and that horses respond to human facial expressions, to name a couple of recent "research" papers published in reputable journals. Another recent paper in an influential journal "proved" that elephants could not imitate a human, yet for centuries elephants as well as many other species have been learning from and with humans by imitating novel acts.

Pertinent to the debates concerning differences between males and females, is a disregard for the importance of emotions. Whether rat, elephant, man or woman, emotions drive what we do, whether it is to learn, unlearn, know something, pay attention, put things together from experience and memory, solve problems, acquire knowledge and so on. To ignore what is going on in the emotional department, omits the driving force behind mental

---

[1] e.g. CDL Wynne, *Do animals think?*, 2004; Clive DL Wynne & Monique AR Udell, *Animal cognition*, 2013
de Waal, *Are we smart enough to know how smart animals are?*

events and how, when, where, what and if things can or will be done. Those who interact with a species or individual of any mammal every day, have to have a fairly accurate idea concerning their emotional states, otherwise they would not be able to work with them, teach them or look after them in the ways they do. They may underestimate a species' or an individual's abilities, because they do not bother to find them out, but they do at least know that a dog can follow a pointing finger, and a horse knows where you are looking even to within a distance of fifty centimetres[1] or that lions will not hunt you when replete. It is as inappropriate to ignore this common sense folk knowledge as it is for a human male to assume that females are "emotional" and cannot be understood. Because of their humanist, anthropocentric background, perhaps research scientists need to "test" such things, but other mammals interpreting human behaviour do not need to do the experiments ... they have already learnt what they need to know. After much investigation of the literature and meeting many people all over the world who work with non-human mammals, the conclusion is that many of the tests done in animal cognition laboratories on species' mental abilities, have not caught up with folk knowledge concerning that species, never mind found out anything new. This is particularly the case with species that humans have interacted with for some thousand years and consequently know quite a lot about.

To progress our understanding of another, we need to drive a middle course, putting what we know about learning (and other mental events) from laboratory studies together with what we know from critically assessed folk knowledge, their feelings, memory, experience, imagination, and consciousness that we experience outside the laboratory. Whether human or non-human, each individual mammal has needs, physical, social, emotional and cognitive or intellectual, and these interweave and overlap, but they must somehow be assessed for each individual if we want to have an idea of his personal subjectivity.

However, the absence of an overall type of power hierarchy,

---

[1] MKW, "What is it to be a horse?" in *Horse Watch*, 2003, London

does not posit that all are equal in every society. Rather, as we have shown, there are hierarchies of aggression, affiliation, showing interest and many other types of behaviour, and certain individuals are decision makers. Each has a different role in the society which is recognised by others.[1] On the whole, however, unless the environment puts great pressure on individual herbivores for one reason or another, herbivores do not go in for power struggles. Yes, they compete for sex and some are more successful than others, but their lives are not dominated by "what the Jones's might think", or how to wield power and be top of the pack. Nor do they go in for the reverse of this, heavy "guilt trips" which seem to occupy the attention of most modern female writers. But this does not make the inferior mentally different and perhaps this is something we might learn in order to live better.

Maybe there will be room in the next decades to disentangle the complexity of non-human mammals' experiences and world views, but doing this will not be achieved by scampering towards worshipping laboratory science, unless it takes on board folk knowledge and carefully assesses the value of anecdotes. As with the anti-sexual discrimination debates, we may have to recognise how much we have in common, as well as the differences and where those are, for each species, and finally, each individual. This might lead to new discoveries and "ways of looking".[2] This might help to reduce some human mistakes and maybe some of the major environmental problems caused by the dogma of anthropocentrism.

I don't regret any of the experiences of my life, nor being born a female. My moral code has been learnt mainly from my four-footed family, and the environments I have been part of, contributed to and lived in. But I know I have not always upheld my moral code.

Humans do not have to cause suffering to other species, they can

---

[1] C Ricci-Bonot & MKW, "The roles of individual and social networking in a small herd of domestic horses," 2017

[2] Henri Bortoft, *The wholeness of nature: Goethe's way of science*. 1996, Floris Books, Edinburgh

enjoy their symbiotic living and enrich each others' lives. In today's anthropocentric world, because of the exponentially growing human population, and ever-increasing consumerism and urban living, the "capitalocene",[1] a species a day becomes extinct, and many fewer people than ever before have to do with any other species but their own. I have been lucky enough to be able to learn from and known intimately some elephants, rhinoceroses, buffalo and lions as well as most domestic animals, but it is very unlikely my grandchildren will ever meet and swap stories with other species. Their children will have no chance at all, unless we act now. Let us go forth and live in symbiosis with them, get to know some of them as well as the living world we all depend on before it is too late. Darwin was right, it is survival of the fittest, but the fittest is not top of the pile, it is those who live symbiotically with others[2] who do this, and we need to get to know all species, both in body and mind.

## Animals stories. What they can teach us

If we can understand other species' or individuals' stories, we can learn something about other ways of living and being which could be a lifeline to the survival of some of us. To do this we must know about different lifestyles, the effect of size and much about another's body and his mind, how he perceives the world, behaves and communicates, what type of social contract he signs up to, and whether he respects and recognises the other as another living, breathing, feeling being. If we can read others' stories, we might learn how to communicate more honestly and with good humour with our own species and avoid conflicts. We might even learn to

---

[1] https://www.resilience.org/stories/2018-01-04/unearthing-the-capitalocene-towards-a-reparations-ecology/ref

[2] L Margulis & D Sagan, *Microcosmos: Four billion years of microbial evolution*, 1997, University California Press; A Tsing, H Swanson, E Gen, N Bubandt (eds) *Arts of living on a dangerous planet.* 2017

live with the natural world instead of playing God and thinking we can improve it, replace it or ignore it.

Stories are one of the things that make the world go around, they sum up histories and experiences, both of the individual and the teller who embroiders them with further ideas, experiences and feelings. One of my pleasures is pondering and pottering around the farm on my young, black, four-legged companion, Shatish, looking at the extraordinary abilities of the living and non-living natural world and reading their stories.

All sentient beings tell a story, whether it is a rat snuffling through your cupboard, or the cow lying cudding in the field outside your window. In fact, there are stories to be told by non-sentient things too: the tree you are sitting under, or even the mountain outside your window, the reprint of the Rembrandt on your wall, or the book of poetry open beside your bed. Stories are to be told by every being, tree, mountain, work of art or building, as well as every book and every publican.

But there is an important difference between the stories told by human arts and artefacts and those of the natural world ... if you dig away at the Mona Lisa, you only make a hole in the canvas. Although some become obsessed with the story she tells and will do anything to save the painting so others can share it, the picture itself is only skin deep, whatever its symbolic significance. You can find out more about the experiences of the painter, the history of the painting, who the Mona Lisa was and so on, but the story is of the painting, how and why it was painted, not what it is. So, what you read into it is your stuff, not hers; she is just colours and canvas. My artistic sister taught me this many years ago when I asked her what her obscure patterns and paintings were saying ... it is up to you!

But the everyday stories that have a constant meaning are stories of the natural world and they can be told only by digging and digging. The more we delve, the more the story and the mysteries are illuminated. As we dig, further mysteries of detail, chance and chaos confront us, until if you are one of those who digs smaller and smaller, your head whirls with the circles within circles of quantum physics or microbiology.[1] If you dig around time, as we consider it, you have geological stories. If you dig around a whole living being in all its aspects, both physical and mental, then again your head will whirl with the complexities of the way it is structured, works, and grows, as well as the extraordinary interrelatedness of all things and its physical, social, emotional and cognitive, and how all these can adapt and change.

Normally we see, or take into account, two or even three dimensions of the story being told, but within the sentient world, there are more dimensions that must be added. If this is not done, then, like art, the story of another animal becomes only the story of its interpretation, not that of the individual who is telling it.

So, to hear the story of the teller, we must do serious homework. First, we must learn as much as we can about the species the teller belongs to, their body, size, lifestyle, anatomy, physiology, senses and how they work, their brain and neurophysiology, the total range of their behaviours, their feelings and what and how they acquire experiences and knowledge, their life-style and ecology.

Today, after a century of research we have a considerable amount of knowledge about different species' bodies, although not as much as about our own. So, we have some idea about our similarities and differences from other mammals, much of it gained from vivisection. We have knowledge about neurophysiology and brain similarities and differences, and know something about what the different parts do. Ironically, we know this because of the tacit reasoning underlying the use of all laboratory mammals in medical research and experimental psychology: they are similar to humans and consequently will react similarly to different experimental treatments. Therefore, they are useful to test drugs, surgery and other things that might help or hinder humans. The results from experiments where the animals often suffer are then applied to reduce human suffering. The added advantage from the homocentric point of view is that the animals do not shout about the injustice or sue. Rats, primates, dogs, mice,

---

[1] Margulis & Sagan, *Microcosmos*, 1987; Tsing, Swanson, Gen, Bubandt (editors), *Art of living on a dangerous planet*, 2017

guinea pigs and other easy-to-keep animals have, since Pavlov, been used to learn about learning and how it works. It is from them that the science of human educational psychology was developed and continues to be nurtured.

Despite this, we constantly invent humans as the only rational, thinking, conscious, cooperative being ... a superior species with a range of mental skills and emotions far greater than any other mammal. At the same time however, since Linneaus in the eighteenth century, who began the classification of animals and plants, and later, Darwin and his contemporaries, in the nineteenth century, who developed evolutionary theory in the twentieth century, it is the similarities between species which lead to their classification as relatives. One group are mammals which includes humans are defined as mammals because they have so many things both in body and mind in common. They have body and mind similarities with similar functions, these are called homologies. This has of course been tacitly recognised, which is why non-humans have been used to better understand the body or mind of humans. But it is time we stated this clearly: we have a great deal in common with other mammals in our whole "being", which means, if we are rational, that we should give them at least equal consideration to humans.[1] Today, we use animals to learn about humans where it is convenient. Where it is not, we consider ourselves unique ... maybe we are not so rational after all! One begins to wonder sometimes, when someone turns around and sneers "don't talk to your dog, he can never understand what you say", whether humans really have "minds" or are they already robots? After all, today, humans are almost desperately building the hardware and playing computer games in order to be a type of power-hungry robot.

Most of the animals I have known have much shorter lifespans than humans. Consequently, I have often known them for their entire lives, and even several generations, so the stories recounted by our animals relate to their whole lives; birth to death. There are many individual stories to tell, some of which have already

---

[1] MKW, "The mental homologies of mammals ..." 2017, p87

been told, from Konyok's and my puppy and childhood stories, to today, Velu's story as she sternly limps about, eating, cudding, remembering and maybe wondering. In the interim there are stories woven by Baksheesh, Aderin, and a host of other equines, or the elephants, Nyasha and Macovooshi, or Cuckoo and her clan of black rhinoceroses. Just a few will suffice to hand-stitch the tapestry of the life of a two-footed learner about quadrupeds' life experiences.

## Sneefus

Sneefus came from the dog pound as a young male, black, retriever-cross-everything, he was about six months old when we picked him up. He fitted in with my two small boys and took part in many expeditions on foot and horseback, he travelled to South Africa on the ship with us and spent six months in solitary confinement on returning to the UK in case he was carrying some diabolical disease. Up to this point he was the whole family's colleague and friend, involved and interested in everything we did and happy to take part with joy ... but after his confinement, he was a different person. He had been shut up for no fault of his own in a kennel where he saw me about once a week, and other humans rushed around, throwing in his food, mucking out and cleaning his kennel, but with hardly a word of encouragement. He must have been very confused, it was so unlike his lifetime experiences so far with humans, and it did not make sense to me either. There were vaccinations available which would have replaced the six months' quarantine and because vaccinations did not count when bringing a dog to the UK, there were many dogs being smuggled in without any vaccinations. It seemed more likely that the dreaded rabies would be brought in because of the quarantine.

I suppose what happened to him is what occurred to me when I thought I had made lifelong friends at school and university who, I assumed, were real loyal friends. But it did not work that way, close human friends disappeared, were disloyal, or did not bother to come when called with anguish. I learnt about the fickleness of human loyalty, while Sneefus and the others of my childhood had taught me about their steadfastness. Having placed all his love and loyalty at our feet, particularly mine, I was disloyal and left him to fester in that alien quarantine kennel. He came home a different dog; a dog's dog, he was polite and pleasant to humans, but that was it. His attention and loyalties were to other dogs, both male and female. He made some human part-time friends and learnt to exploit them, stole food put out for cats, crept in to sleep in comfortable places made for others, and he certainly left some bitches pregnant. One lonely old lady was so pleased with his visits that he would contently eat a large dish of food put out for him and then lie by her fire while she talked to him on winter days. She even had a special dog flap made for him to come and go as he pleased, so he was often absent from home. Another time, we received a call from a farmer about five miles away across the downs. He had taken up residence in their car, went for drives with them and to and fro with the children to school ... and of course they could not bear not to feed him. The reason it turned out for his change of location, was that the farmer had what Sneefus thought was a stunning bitch on heat ... There were some follow-up puppies of course!

After his incarceration in quarantine, he followed his own agenda. What a lesson this was in broken loyalties. I missed his close, loyal attachment and swapping of the stories of our common experiences. Finally aged about fifteen, he was hit by a car on one of his expeditions, so at least he did not have to grow old groaning by a fire when he had so many other plans he would have liked to put into effect.

## Oberlix

Because his pure-bred Arab horse mother, Omeya, whom we had bought as an errant three-year-old, had won a series of races, we were offered a free covering by Aboud (a three-year-old pure-bred stallion who subsequently became a show champion and was exported to Bahrain to improve the Arab horses there). Omeya seemed to agree with our choice, and became pregnant in the first three weeks with him. Eleven months later, Oberlix was born out in a field with no trouble, slipped out all slimy and when dried turned out to be this brilliant reflective chestnut all over. He was everything one had hoped for, brilliantly beautiful, with a very correct body, lovely flowing movement, and a proud Arab head with enormous watching eyes, and he developed a character to match. At that time, I was just beginning to wonder how to teach young animals to listen to and learn to understand human language. So, I treated him like he was a nephew or a grandchild I was fond of. He lived in our multi-species home, mainly with his mum of course, but he also spent a lot of time with me, learning about humans. In turn, I was learning about this wilful, and delightful but proud, young chap who was developing a strong code of conduct. He grew fast and it was not long before our son Jake sat on him for the first time, and he came on walks and trips with us, and generally took part in our family life and we took part in his.

As he grew up, like all young stallions, he started to assert himself, grabbing things with his teeth, "horse play". But, being grabbed and bitten on the arm was not my idea of fun. So he had to learn how to behave differently than with the other horses with us. An early lesson in respect for others came when one of the male guanacos attacked him because the guanaco thought he had run off with his girl. Oberlix was tied up at the time and could not out-fight the guanaco. He learnt in one rough lesson never to be disrespectful to the small camelids; it was a serious lesson in inter-species respect and understanding, the guanaco bit his forearm and tore some of the muscles.

He grew in body and mind, and since he was bigger and stronger than us, we had to develop and learn a mutually agreed social contract. His part of this was that when with humans, whatever the situation, he must not act according to his strong desire, by for example, touching up or mounting a mare someone else was riding, or squalling and squealing while striking out with his front legs in excitement. He must not leap about and pull a human off his/her feet, or bite or kick a human ever. He must also watch out for young humans who may be wandering loose among the horses, and not tread on them, even if they did not know the rules. All of this meant that he had to learn what is called "desire-independent reason", that is, in some conditions he must control his desires and obey the social contract and, in others, continue to live in the way he knew. This is something that is supposed only to be possible for humans, but he and many others of different species have also shown me that they can control themselves when it is necessary, and consequently continue to have their freedoms. Of course, the human also has a part to play in this bargain or "social contract". They must make allowances and know the rules too.

So Oberlix would be allowed to do whatever he wanted at times when that did not conflict with us or another species' interests. When he was with humans, he had to learn to do whatever he was asked, provided it was reasonable, and first he had to understand what that was. He also had to learn the social contract of the other horses so that he behaved appropriately when he was free with them. We humans must know and respect the equine social contract, too. For example, if we did not want the stallions to fight when they were loose, then we must not put them together close to the mares. But, when they were with us or being ridden, even if they were with the mares in season, it was against the social contract for them to either try and cover the mares or to fight with each other. Oberlix's overriding wish, like all yearlings, was to be accepted by the other horses and by other species he was living with. He wanted above all to do the right thing so he could be "one of us". It was both the other horses' and our duty to show him what that was. It involved things such as, do not hurt youngsters, do not try to rape a mare and be nice to everyone as much as possible so they will be nice to you. Charles Kingsley's *The Water Babies*, which is a homily on how children must learn to behave and why, is very relevant here. Kingsley pointed to the moral code or social contract that children must learn: "do as you would be done by" and if you break the rules: "be done by as you did".

Gradually, the social contract between us developed and became more elaborate. I soon realised that we could ask Oberlix to do almost anything and he would do it, so it became my responsibility not to ask him to do anything so dangerous that he might really hurt himself. At the same time, I realised that he had great skills of movement and knowledge of the environment so he should be able to make many choices himself. Once, walking across a very narrow plank over a rather deep ditch, I turned around to find him following, carefully weaving his front feet one in front of the other so he could balance on the plank, rather than placing them side by side as is typical of a horse walking.

But this social contract is not one-sided, I also had to respect his needs and desires to ensure that we could work cooperatively together, he gave in to my wishes here and I gave in to his there. Perhaps because he was who he was, he taught me a great deal about what was reasonable to ask and when, and when it was reasonable to leave well alone, and we had twenty-two years of a good life together, adventuring, enjoying each other and trusting each other implicitly. We won a raft of events of all kinds, and I hope he felt I listened to him as he listened to me. Once, we won a jumping event and then went on to win a cross country event at an Arab performance show. We entered the jumping just for fun because we had never done any formal jumping in an arena. I was so delighted when he jumped the first fence that I told him what a great guy he was and continued to do this after every jump. I could feel him grow with confidence, and we finished clear and in the fastest time. The next event was a cross country where you gallop about up and down, jumping over rather large fences fixed in queer places. My heart was in my mouth as we slithered down a muddy slope to be confronted by an enormous couple of fixed bars on the top of a bank, jumped them and landed way down below. I closed my eyes, went with him; he cleared it all and won that, too, even though he had not had any practice at all. It was his confidence that had won and by telling him he was brilliant I had helped to build it. But what "having confidence" means needs to be considered. It is an awareness of what you can or cannot do, it therefore involves an acute awareness of yourself, or "self-awareness". Since confidence can be dented or changed, grow or disappear, it also requires some "thought" about it, in other words, some reflections ... Therefore Oberlix's confidence is an illustration of "reflective self-consciousness", something that nonhumans are not recognised as having by scientists today. Yes, the individual has learnt that he can do this or that, and that gives him an expectation of himself being able to do it again; yes, of course it is learnt, and can be further developed or not, but the point is that it is in all us mammals. It can be developed and it often matters because it results in different individuals being able to achieve different goals.

After Oberlix and I won a number of eighty kilometre races, I decided we should spend a summer doing both endurance races and dressage competitions. Everyone in the horse world believes that it is not possible for the same horse to do well at a high level in two such different disciplines. The nearest they get to doing different things is combined training: jumping, dressage and cross country (galloping over more fences). Each type of race and dressage requires different muscles, bodies, and mentalities, as well as different equipment and different riding techniques. Humans can ride horses in these different disciplines, but it is very difficult for the same horse to adapt and do this at an advanced level. It was time we tested this. At that time in dressage, there were practically no pure-bred Arab horses competing and no stallions, so the judges were inclined not to favour Arab horses who have a different way of moving than a dressage horse. It is true, Arabs find it more difficult to play by the same dressage rules, but they should be allowed to try. So first, we had to teach him to move differently in the dressage arena than when doing endurance. The idea was not to cover ground quickly with a long low stride as in endurance, but to almost hover above the ground with all legs in suspension from time to time. He must keep his head in a different place from its normal position. It must be higher and with the chin in, and still. Both he and I also had to think carefully not only about what he was doing, but also what he might be doing next, as there is a series of exercises to be completed in a particular order and in a very particular way; it included things like going sideways, or in tiny circles, changing lead legs in a canter frequently, going backwards, turning around on the hind legs only, doing trots on the spot, and very elevated trots throwing the front legs forwards and so on. We decided to start with the easiest competition and when he obtained fifty-five per cent, which was a pass or "satisfactory" at that level, we would enter the next level. The idea was to see how far up the scale we could go in the six-month summer season when he was also competing in advanced endurance races of over eighty kilometres and more. The end of the season would be rounded off with the Arab horse marathon, a forty-two kilometre fast race across country that horses from our stud had previously won.

With the assistance of a dressage judge/instructor friend, we practised the movements, competing first in the novice level and then went up through the levels: elementary, medium, advanced, Prix de St George, finally reaching intermediary, the one before the Grand Prix (the top Olympic level) by August. We did not win any of them, but we competed, and in all of them obtained more than fifty-five per cent. We could have continued, as he knew most of the movements by then, but time and money ran out. Meanwhile, on alternative weekends, we won a couple of endurance races and finally, he came second to his daughter, Shemal, in the marathon. It proved that a horse can compete at a high level in both endurance and dressage in the same summer, and that it did not need years to teach the horse these movements if they were taught by a teacher who understands that the horse is able to learn the concept of what was required and then work himself towards that.

Oberlix told his story in his actions, his enjoyment of life, his willing participation in everything, and in our mutual delight in our contact: never was collective intentionality more deeply felt than when I rode him; he taught me real cooperation and the importance of confidence for both horse and rider in order to achieve whatever goal you set. He was one of the great mentors of my life and daily told stories, and also read them from humans and other species around him. He had absolutely no difficulty reading my stories, feelings, desires, ambitions, loves and hates, and did his very best to help them along, usually, but without being a push-over, as it was important that I respect his needs, desires and wishes.

## Shemal

Shemal is Oberlix's daughter, born to Shiraz, daughter of Baksheesh, the foundation stallion. Just before she was born, we had prepared the research on the comprehension of language and she was to be one of the subjects. She and the others would all be exposed to gentle handling by a human soon after birth, and to simple English words within the first week of their lives. All went well with her birth, and as with Oberlix, she was talked to and handled by us shortly after birth, then left to allow her mum to mother her and for her to learn who her mum was. Once the various subjects were a week or so old, they were taken on expeditions with their mums. Once they were around two months old, they would leave their mums behind for just a few minutes to start with. If any of them were anxious, we would stop until they had calmed down and wanted to go on. During these expeditions, they were introduced to different things, places, people, and animals and we spoke the words to describe all of these.

By two months, we started to teach Shemal and the other "cog" subjects to do simple movements when they were asked in their two hours a week school time. The first few tasks were to "lift your right (or left) leg", "shake your head", or "paw with your front leg", or "put your head up or down". The way the teacher gave the instructions and how she behaved with each individual was recorded, as well as how the student behaved and how many times the words and gestures had to be repeated before each did the movement, how well s/he did the movement, and what her/his reward was, a titbit or a "good girl, well done". Shemal was very quick to learn these things, but not as quick as Ramblie, the heifer calf, who usually only needed three repeats of the word, with gestures or a mime by the teacher before she did it, but Shemal took part willingly in everything and made the most of it enthusiastically.[1]

Occasionally, we took all the school mammals on expeditions in the horse van, like a school outing. We took them to the seaside; they had never seen the sea before of course. Each of them approached slowly, staring, puffing, snorting, advancing and retreating. It was low tide and the waves were very small, but they were enough to cause much leaping about. Shemal was by no means the least active here, dancing, snorting and leaping. Eventually, they all paddled in about ten inches of water, walking around rather carefully, watching the waves, then we all had a picnic lunch on the beach, and a couple more hours of playing before loading up to go home. The next expedition was to a forest where they all followed along freely. Someone came along on a horse, Shemal took one look at this other horse, and cantered off after the horse and rider. After about 300 yards, when out of sight, she realised she was alone without any of us, and she did not know the vanishing horse or rider, so she rushed back. But we had had time to hide behind trees and in the bushes. When she arrived back where she had left us, there was no one there! She panicked, rushing up and down and calling ... we could not let this go on, so Ramblie soon emerged from the trees, burping and setting about eating the grass around, and gradually we all joined her; the relieved Shemal relaxed and grazed too.

Shemal was learning more and more words, and how to read humans and their emotional states, intentions and desires better and better, as we were hers. It was not long before she was also doing the normal horse things, pulling carts, being ridden, being taken to competitions to meet other horses, going off on camping expeditions and so on, and of course she started to win. Her confidence could hardly be dinted, even if ridden by an unconfi-

---

[1] MKW & H Randle, "First steps in comparative animal educational psychology," 1996

dent beginner. She was quick to learn the social contract with us as well as with the other horses, and as she became older, she taught many others. In fact, it was with Oberlix and Shemal that the sub-culture of the multi-species community was founded. Today she is the grand matriarch, aged twenty-five, of our stud, she has won a whole range of different types of competitions, and has had two foals who are also competing. Her role now is chief decision-maker and older lady and she is very much in charge of the younger herd of which she is a part.[1] They listen to her stories as we must too, and she will ask with different amounts of urgency for whatever she requires. One of the most important things, even in her older age is that she is not left doing nothing for too long; she must go places, do things. Yes of course, eating well and keeping warm and remaining comfortable are important and perhaps taken for granted much of the time by her, but she comes and stands at the fence demanding to go out and do things, even when all the others have disappeared to talk horse among themselves. She has very strong opinions which everyone of any species must respect. She teaches people to ride but if any rider thinks they "know it all" and starts bossing her about, she will leap about and scare them out of their wits. If she is asked to do something inappropriate, or in a disrespectful or thoughtless way, by for example being threatened, cajoled or clouted, she will just not do it. Yet, she is always willing to over-extend herself for someone she likes and respects, and will try to do anything for them. Cooperation works both ways; it never involves domination.

There is no doubt that living with these horses and many others, has changed my attitude to life and living, my ambitions and goals, my beliefs and ideas. Perhaps one of the things that has attracted humans to horses is that they are in many ways very like us: they get into panics and stresses, they are easily frustrated and annoyed and show it, they are what people call "sensitive"; although they have different skills from ours. One thing they crave most, is a calm, comfortable life where they are quite confident in what they do, and are given half a chance to develop and go forward to

---

[1] Ricci-Bonot & MKW, "Roles of individual & social networking ..." 2017

achieve their goals. They must have confidence in themselves, and if they have to do with humans, it is vital that we recognise this and build their confidence, not smash it as we hit and terrify them into doing things with no explanation, which with a little encouragement they would be only too glad to do.

They love to do different things, go to different places, they love to be praised and regarded with respect and they love to be with people who help them achieve these aims. Perhaps they have too big a cerebral cortex to be content with the life that their wild ancestors lived, they want to be challenged and, like us, do more than wander around in a social group, finding food and breeding. If they are not given a good life, as is often the case, then why do they put up with the way they are taught and treated without always becoming psychopaths? After a lifetime of trying to sort out the "behavioural problems" a horse is said to have, by changing the way the humans behave with them, I am convinced that horses need stable, emotional attachments that they can rely on, but in addition they must have mental stimulation. If this is not so, then why have so many of them, today and throughout history, still done what they do with us? They really do not have to, they are bigger and stronger than us, and could frequently avoid doing the things they do daily. What they do develop towards humans is a real and deep loyalty and in exchange, they require mental and physical stimulation, and a unique blend of attributes for any friend. These things I have learnt from horses.

Over the years I have travelled about in many parts of the world looking at cattle and how they are kept: dairy cows, beef cows, working bullocks and cattle ploughing or transporting things. One thing that never ceases to amaze me is, whatever the conditions they are living in – they may be up to their hocks in muck, have nowhere clean to lie down, be very cold, wet, hungry, or thirsty, squashed up too close, or tied up for months and months – for them the glass is always half full. They always look on the positive side of things, almost as if they are saying: "yes I am very hungry, lonely, cold and up to my hocks in shit ... but the sun is still shining and the world goes on". Unlike horses, they don't really have ambitions, they just want to "be" and to enjoy quiet, comfortable lives. Indeed, they have to be kept in very lousy conditions and be hit about to become aggressive or irritated (like many bulls), they rarely show their frustration, and if they want something, they usually just go on working away at the barriers with their weight and patience, until they get it! One of my students gave our cattle a box that they were supposed to learn to push on one side with their noses to obtain a reward. It took them thirty seconds to solve the problem, quite different from what was expected, they found that there was no need to push one side, they just smashed it up ... and there was the reward!

Even if a calf is separated from his mother, both mother and calf will give the odd howl but continue with what they are doing and after an hour or two find each other again; there is no rush, "everything will be okay in the end" seems to be their motto. My life was changed as a result of the four years I spent studying them. Why not indeed enjoy the moment? Why have human ambitions become irritated and frustrated so often, why not "think cow"? I don't always manage of course, but I do try and it has helped me through all parts of my subsequent life, and maybe helped some other people who have learnt to "think cow" too. Cows would be very good human therapists and also teach many ordinary people different ways of living that would enrich their lives.

## Rambling Rose

Ramblie was one of the cognitive subjects, and she learnt to understand around 300 words. She was born to our house cow whom we milked and she came into the school room twice a week for courses in human language comprehension. She quickly learnt to do all the things we were teaching the others, in fact when we worked out the results, she was the quickest to learn, and considerably quicker than the puppy Rupert. She came on expeditions with us and the others, and took part in all the various demonstrations that we did, but being a bovine, everything she did, she did with contemplation and more slowly than the others. She had trouble learning to lie down, it is not something cows do easily, they are not given to leaping about, rolling, and moving fast like the other subjects, but eventually she did it when we asked. The story she told me was how aware bovines are of their environment and how quickly they can learn things when one thinks they are just standing in some sort of absent haze. She showed us clearly that she had often worked out what was required and was just taking her time to do it in a proper, considered way, with no flaps or panics, no obvious excitements and frustrations; she was just coming to terms with it all and willingly taking part in everything, but at her pace. Most older cattle people know to never hurry cattle, and they repeat this piece of folk knowledge to others.

## Effendi

Effendi is a young South Devon heifer, born in France and she has learnt to pull the harrows, and to pull a cart slowly. She is of course easier to teach than her companion, a young Arab stallion of the same age, because when things go wrong, she just stops and will not move rather than leaping around and causing an accident if one is not careful. She never shows she is panicking, but that does not mean she is not afraid, and so we must take things just as slowly as we do with the young stallion. People are wooed into believing that bovines are not scared and are stupid, but both of these ideas are often wrong; they can be just as scared and are by no means stupid. They are just considering and waiting, something that we could all learn to do better.

# Velu

Velu is a Jersey member of the bovine clan, she has been with us now for sixteen years, a long life for a cow, and today she hops about with a bad hip, enjoys her eating, being with her calf and her grown up daughters, no doubt teaching them the social contract. She can be milked by every sort of person who "wants a go", and gives us glorious cream-rich milk. We never take the calves away until the mother weans them so they live with their mothers; if we want some milk, we separate the calf at night and milk her in the morning, thereafter the calf rejoins mum, has the rest of the milk and stays with her for the day. This system has many advantages. It seems to suit the cows once they have learnt that milking by a human is not so bad; after all, the human does not bite the teats like a large calf often does, and with luck will give her a bucket of food while he milks. All the cows close to the farm are handled daily and they very much enjoy having their heads and backs scratched (which they cannot reach with their rough tongues). Velu often thanks us with a lick on the face, her serrated tongue almost taking your skin off. She will queue up with the horses to have the ticks taken off in the summer, as well as allow the hens to jump up under her legs and catch the biting flies.

She always tells us to take time, to sigh, to contemplate, to observe the great living world around us, to enjoy the moment whatever is happening. Sitting on a stool milking the cow for twenty minutes each morning is a time for relaxation and contemplation, for close contact with a warm furry side, and for listening to the soporific regular squish of the milk into the pale. It is "cow meditation time"; an important part of the day that has probably avoided us making many rash decisions and really does encourage unhurried contemplation of the world around.

# Macavooshi

I first met Macavooshi when he was a late teenager on my first visit to Zimbabwe. He had already been mounted and ridden a little by the not very skilled handlers who were also learning. On the repeated visits to Zimbabwe and Imire where he lived, I increasingly admired him. He seemed to be able to put up with the handlers not grasping how to teach or handle him without irritation, he was always pleasantly mannered to everyone, and showed interest in them and did what he was asked, even when it came to being put in a plough or pulling a cart. He took it all in his stride, despite being by then a big strong bull elephant. He learnt a lot from his elder colleague Nyasha I think, and then he continued to use this knowledge to live in some symbiotic way with the humans he had to deal with, including tourists, and the other elephants. Mac learned fast and then often put his own innovations into what he has learned and was doing. He learnt for example to pick up a log and lift it, and also how to balance it and place it where required, and for him it was something that he immediately grasped. When he was asked to find a watch in a pile of hay, it took him only half a minute to understand what he was supposed to do and then do it. Recently he has been learning to follow the smell of individual humans and trail them, the idea eventually being that he will be able to help with anti-poaching work. Today, they have decided no longer to ask him to work by ploughing or even giving tourists rides. Prior to that, he was particularly aware of the need to stay very still while a person with an intellectual disability or a very young child was handed up onto his back for him to carry around.

One of the mysteries is that, although currently treated with hormones to prevent this, he has never come into full musth when the bulls often become aggressive and unpredictable, and yet he has fathered a daughter. Perhaps this is because he is the eldest and his early experiences, even when he was orphaned and brought as a small suckling elephant to Imire, have always been reasonably acceptable.

The story he tells is how possible it is to cooperate with and live symbiotically with an adult bull elephant, long may he continue to live his life, which he seems to do, whether working with tourists or helping humans in different ways, as well as keep elephant time. It is possible to have elephants in semi-domestic situations and for them to have a life of quality; perhaps others will follow Macavooshi's example and realise it can be done.

## Cuckoo

Cuckoo was a mature, matronly, large, one-and-a-half-ton black rhinoceros when I first knew her. She lived with a group of six other black rhinoceroses, all of whom had been rescued as orphans when their mothers had been shot. Black rhinoceroses are not very social. Mum and infant usually live together for a year or two, then mum may get pregnant again by a visiting, usually solitary, male, and some eleven months on, the next baby is born, when often, the older brother or sister may wander off to make his or her own way. These rhinoceroses were adult at the time I first met them, and were living in a group where the females were all shut up in a *boma* together at night, and the males were in another one. This meant that the four adult females could not get away from each other during the night. They could escape each other during the day as they wandered browsing through the nature reserve, but even then the armed guards kept them reasonably close to each other so that they could survey them. We had found that Cuckoo (the only female who had become pregnant and produced an infant) was a bully, and the others had to spend their nights in fear of being attacked by her.

Cuckoo's bullying had clearly been learnt during her upbringing at some stage, and one day she separated from the herd and disappeared into the bush. I volunteered to go and find her. Somewhat cautiously I walked into the bush, keeping a good look out for any moving rhinoceros-like object. When not wanting to be found, like all mammals, although large, and normally noisy, rhinoceroses make not a sound. They can move as stealthily as a cat through the undergrowth, and freeze totally still for minutes at a time. Peeking out from behind a tree, I eventually found her. She was around thirty metres from me, standing looking in the direction in which I was advancing. It did not take her long to see me, and she made a couple of rather half-hearted charges as I ducked behind different trees. But now the trouble was, how was I going to get her to move towards the night-time accommodation, there was no way of contacting anyone else to help; I did not have a mobile phone or a walkie talkie radio with me. After about half an hour, one of the scouts turned up, and he had a long whip made of wire, which was one of the ways they made the rhinoceros go where they wanted them; but swishing this around only made her charge more often and her charges began to have some determination in them. I eventually managed to persuade the scout that the only way might be to give her time, and that very quietly, after a while, we should try to drive her without using the whip in the right direction. After about three hours, we had her out of the bush and walking towards the *boma* where her evening feed awaited her, but it had been a touch-and-go situation. What I realised was that she had learnt that by being a bully and charging either people or other rhinoceroses, she could get her way and be left alone. This had been the result of how the handlers had behaved towards her when she was younger. When she started to do this, they were frightened and would clear off, so she had learnt to do this. Years later, one of her sons who had been weaned early so she could get pregnant again, was causing the same problem for his handler and I was asked to visit him. What I found there was that the young male, on seeing the handler, would put his head down and do a mock charge, whereupon the handler would give him some food to distract him, because he was frightened. Of course, it had not taken long before the youngster had been taught to charge and had become a problem. We worked away at changing his mind on this, but I could see that although the situation was progressing while I was there, it was very unlikely that the re-teaching would continue when I left. Luckily, this youngster was of an age to be repatriated to an isolated nature reserve in another part of Zimbabwe and was transported shortly after I left. I do not know what happened to him, but they do say that of the thirteen young rhinoceroses they eventually managed to repatriate, half of them are still alive … unlike Cuckoo and her entire clan who were shot by what is reputed to be a military brigade one night in their custom-built protected *boma*.

One of the astonishing things about the rhinoceros is how, in many ways, they resemble equines in their behaviour. For example their ears are conspicuous and they have developed more or less the same ritualised ear movements that equines and camels have, flattening them against their head when irritated. They roll on their backs like horses do, which none of the bovines do. This they do by lying down on their sides, and moving their legs in the air to roll onto their back in order to scratch it. Sometimes they can roll right over to their other side, for more scratching and moving about as their robust short legs kick about in the air, before getting up, front first (not back first, like cattle) and shaking themselves. They are also extremely cued into visual communication, responding to slight visual movements or facial expressions from other rhinoceroses, humans, elephants or others. Hence Cuckoo could read her handlers' stories quickly: they were scared and therefore would run away quickly, but in her *boma* where she was never chased with whips, she let one handler scratch her back and eventually collapsed in a heap on the ground, in what appeared to be a pleasant ecstasy of being relieved of itches she could not scratch herself. Rhinoceroses are odd-toed mammals, like equines, and their behaviour in these ways shows how closely related they are. How is it that humans can almost wipe out a whole species who do them no harm, before they even know them, and for trivial and fictitious reasons?

There are many other individual stories I could tell, but let us now consider some things in order to be sure one is reading their story, and they are reading ours, rather than inventing things.

Non-humans can "read" the stories of other individuals and species, but they may read them in different ways and interpret things differently from a human. Dogs with their acute sense of smell can read scents, find them, follow them, learn from them, understand who and what has been there, where they were going, and maybe even why sometimes. When you let your dog loose in a new place, for him it is like being in a library full of the books he wants to read. Then again, what about an elephant? He has this amazing combined hand and nose that he can smell with, feel with, manipulate with and communicate with, and he uses it constantly. Over fifty per cent of all conversations between elephants involve trunk movements.[1] Imagine what it might be like to have a hand combined with a nose, and one not two; he reads smells and often tastes as well as being able to caress, push, pull, twist around or find something with his trunk. Also, it is even more difficult to understand, if the trunk is not moving right in front of him, he is doing it all blind. Then again, horses and cattle, antelope and deer, for example, are reading smells day and night throughout their lives. But to verbally-orientated humans, losing your sense of smell is hardly considered a handicap: to most other mammals it is a disaster. Considering what and how smells are read by non-humans moves us away from our "comfort zone". The smell is in the air, it is travelling here and there, it may tell the recipient something about where it is coming from and when, but most of its information is here and there, then and now at the same time ... This opens up very different world views to us; we are forced to think differently about space and time. But, this does not mean that non-humans do not also have, or can learn, our idea of space and time: far and near, before and after also exist for them. It means that they have a new dimension to how things are considered, similar to how quantum physics forces us to think.[2] When the mathematics are translated into ordinary words, the

---

[1] MKW, "Communication in a small herd of semi-domestic elephants ..." 2019

[2] J Gribbin, *In search of Schrodinger's cat: Quantum physics and reality.* Black Swan, 1984, Penguin Random House, London

ideas are so queer they make us shake, but they are also real!

One thing that most equines, bovines, elephants and rhinoceroses excel at, is reading body language.[1] They do not have to know you or any other species or individual to be able to read what you are feeling. This can be a major problem when trying to handle, "tame" and teach them ... they already know your intentions because they have read your body language. Just a slight movement here, or a tightening of a muscle there, is enough to tell them that you are afraid, angry, pleased or happy, like or dislike them, are having fun or are bored or irritated, and whether you are showing off to other humans, or just not paying attention to what you are doing, or are depressed or worried. They have read the emotional story of your day so far and after a few instalments, probably have a fair idea about your emotional life ... whether or not you have read and understood theirs. In fact, it is this reading that allows us to teach them so many things, because they know when they have it right, and when they have it wrong. They frequently can tell when you believe they can learn something, and when you really do not, so they do not need to even try! Teaching non-humans is intellectually challenging. For example, how do you convince them that YOU believe they can understand it and will eventually do it? Not only do you need to read their story as they tell it, but you have to tell yours honestly, in the way you wish them to understand. If you lie, they do not need lie detectors, they see it clearly, and it will put a damper on their learning with you. Perhaps some animals could be taught to detect lies for us too!

Some people and animals throughout history have understood this and managed to do remarkable things together, but today much "animal training" is publicised as an exercise in "behaviourist" understanding, where it becomes a kind of robotic learning, and "new techniques" such as the fashionable "clicker training" or "protected contact conditioning" are expounded at length using a rhetoric and jargon. The rules people have developed may help the urban, complete beginner who has never encountered a thinking being of another species before, but it misses the whole point

---

[1] Allen, "What is it to be a horse?" In *Horse watch*, 2003

of the animal's ability to tell, to read and to understand stories. Believers in behaviourist approaches remain beginners, and animal training goes nowhere new or interesting.

Humans have found out quite a bit about how to improve the teaching of children through what is now called "educational psychology". By applying some of the ideas to the other mammals we are teaching, it is surprising how and what we can achieve with the animals we work with; "animal educational psychology" should become the buzz words, not "animal training", if we want to get anywhere interesting or discover anything new.

We stumble about in our beliefs about non-human mammals, and find it difficult to recognise our mental similarities because we pay no attention to reading other species' and individuals' stories, floundering around convinced ours are the only, the best, and superior. We also have a raft of "beliefs of convenience", for example, how awful it would be if we really did think those cattle chained up in the University of Edinburgh's agricultural barns were feeling and suffering similar things to Jews being held in concentration camps. If that is the case, then how awful it is to leave your dog chained up for twenty-four hours because you are going out and he chews the carpets? Or your horse shut up all day every day in a loose box alone? Why would any mammal who does not characteristically live in such a way, NOT find such treatment as difficult as any human would? Why should it not be as emotionally and physically trying for them as it is for a human? We know that they feel, learn, think, remember, make decisions and choices and have experiences, and if we open our eyes, we see that they can both tell and read stories. Although we may not always correctly understand them, at least we should try.

The majority of humans today live in cities and have very little contact with other species or plants, but we have plenty with people, night and day. In fact city lives are almost exclusively anthropocentric. One of our students admitted that the only time he was ever alone was when in the toilet. As a result, for humans who hardly ever interact or even see or touch a dog, the dog is as strange as if it comes from Mars. In addition, the dog is usually portrayed as an instinctively reacting robot with a few feelings thrown in. No wonder they cannot believe that a dog has a story to tell, and sometimes one with lessons for those who understand it. Rather, "talking to dogs" is believed to be something that autistic, handicapped, dumb or mad people do. Although today, people may have gained plenty of knowledge from the Internet, and can acquire still more, they cannot read what Fido is saying. In fact, they do not even know that he is telling a story, never mind know that he is reading yours.

On the other hand, because of our homocentric lives, we often exaggerate what we think we can read in the expressions of other people. For example, in many novels the author will mention an individual's story which has been read in the expression of their eyes. The number and type of expressions in the eyes that are described are mind boggling. As someone who has spent a lot of her life carefully observing others, humans and non-humans, I have often tried to read such things in the expressions of other humans' eyes, but the only thing I can usually tell just from the eyes, without looking at the expression of the rest of the face, is where they are looking, and the intensity with which they are looking. Their intentions and any further details about their feelings has to be read from the rest of their facial expressions. There is much research on how we, and now other species, recognise other individuals from their faces. But, we can discover much more about other human feelings from what to another mammal are usually very obvious facial expressions, body postures and movements. So these "eye stories" appear to be the result of the writer's imagination, but they are accepted because it is something the reader aspires to be able to do. We all like to believe that we can tell another human's true story, but can we? The problem is that many other mammals who are familiar with humans can do this because they can read very subtle cues from slight muscular changes which may or may not be intended to communicate emotions; we humans are not good at this at all, because we concentrate on talking. As a result, only very obvious facial expressions, which would be considered exaggerated and ritualised from a horse or dog's point of view, such as smiling or frowning, tell other humans how we are feeling. The subtlety is in our verbal messages, the words, the order in which they are produced and the emphasis put on them; this is what we pay attention to and have a facility for understanding, or some of us do anyway.

It is this context-independent language which takes us on long symbolic journeys and allows us to learn from and understand other humans. But, the subtleties of the immediate meaning in a specific context, particularly related to feelings are often lost. As a result, it is not uncommon for humans watching another's expressions or listening to their talk, to misinterpret what that individual is feeling. This may result in unnecessary squabbles and even divorces. We can easily lie to each other about our feelings, or mistrust each other because of an imagined visual or tonal change in a vocal message on the part of the communicant. This is far less likely among other mammals, not because they are stupid, but because they have read a whole series of subtle messages from the body, face and actions of the communicant that humans do not see. Of course mis-communication can also be the result of the recipient's expectations, rather than the intentions of the communicant. So even sophisticated non-human visual communicators can also make mistakes in interpreting what another is feeling and his intentions, but usually they are more tolerant of these mistakes perhaps because of their rather different social contract which rarely involves power struggles; often these mistakes are just ignored. We humans could learn how to communicate our feelings better between ourselves and avoid many misunderstandings if we learn ways of communicating from other species of mammals. I am often really surprised how even people close to me misinterpret a message I am trying to put across whereas with my quadrupedal friends, they see it all, even when I am trying to hide something.

If we take the trouble to understand non-humans' individual stories and their reading of ours, and take account of their similarities and differences to us, and learn from the often un-judgmental mutual respect they show for each other, we might learn more about how to communicate honestly and with good humour with our own species. There are people who recognise this potential and use various animals to try help humans with problems, in "animal assisted therapy", but alas if the "animal assisted therapist" is an unquestioning humanist, they often miss the whole point. S/he believes that the only really important thing in his/her work is that s/he helps humans overcome their problems whether they are physical, mental or social. Of course, the therapist will argue, the animals used must be "loved" and "well cared for" but they are used as mere tools to achieve human cures. One would think that using animals in this way would be an ideal opportunity for pointing out what we can learn from the animals used, whether they are dogs, dolphins, horses, elephants or whatever. But the therapist, who may be very skilled at talking to and understanding the human's problems, is rarely able or willing to read what the animals are telling him or her, which one would have thought was the whole object of the exercise! Let me give a recent example. A woman therapist who runs a successful business using horses to help people overcome social problems invited me to give a workshop on how to begin to understand the stories of horses and how to read what they are feeling. The course was to be held at her friend's stables outside Paris. Many of the horses lived in architecturally elaborate nineteenth-century stables, except when they were giving lessons outside with humans. This in itself is not necessarily unacceptable to horses, it depends how it is done. But some were kept permanently in muddy little paddocks of around twenty-square metres, sometimes in groups, sometimes alone. The weather was wet and the mud ankle deep. I had injured my knees and walked rather slowly observing these horses, thinking how I could suggest to the stable owner that improvements were needed: "horse therapy". The human therapist and her patients strode out ahead of me, talking about their problems. I turned up late at a squalid paddock and read the story of the ponies standing fetlock deep in the mud, gloomily looking into the middle distance. I was supposed to use these horses to help flatter the human students with their often very trivial problems, and to disregard the horses' problems. My moral obligation was to declare their story, which none of the others, including the "animal therapist", had stumbled to. A little thought would have improved things for both horses and humans without costing money. I refused to work with the horses and told them why. Shortly after, an improved arrangement was made which allowed the horses to be out of the mud and brought into a ring to work with us ... briefly, for all of thirty minutes! The woman who ran the course was furious that I had pointed out that the horses also have a point of view (even though one would have thought that was one of the central ideas she might be putting over to the human patients). She refused to pay me in full and told me I had been "rude" and therefore would never be asked again. I heaved a sigh of relief, but lost a lot of sleep thinking about the humans who had paid a lot of money to attend and were so involved with their own "problems" that they could not recognise the stories and suffering of others organised to help them.

Another time, a doctoral student of mine was working on the welfare of horses kept to help mainly physically challenged humans. She found they were kept in stables all day and night, and never did anything different, and that most of them could not associate freely with any of the others. Also, that the people running the stable and all the animal therapists helping were only interested in the humans and their problems. The animals suffered, and there was enormous wastage as many of the horses could not continue to tolerate their lives. Money was wasted and there was no furtherance in the understanding of the horses, who were meant to be there to "help" the humans. Another doctoral student of mine did some research at some Guide Dogs for the Blind kennels, when she pointed out how the young dogs were showing signs of distress in their kennels. Instead of management trying to improve their lives, she was thrown out and had to re-start her doctorate elsewhere. There are some people who are beginning to understand that therapy for humans will come from a better understanding of the animals' point of view, which may teach the humans to be more sensitive to looking and learning from others of any species including their own, but they are rare.

Maybe it is the humans without problems who need to take a step sideways, and understand and believe that others of all species have stories to tell, even if we find the language initially difficult to understand. A stallion story illustrates this. We had two stallions who were half brothers and the older and smaller one, Oberlix, had known his younger larger brother, Oryx, from birth. Stallions' "instinctive tendency" is not to let another stallion near the resident stallion's mares and if the visitor insists, to chase him, and fight him until the intruder gives up, or the resident stallion is killed or injured. The mares look on but do not take part, although they may attack the visitor if he wins the fight. Both of these stallions had lived with our mares at separate times and knew them well. As a result, both believed that they were the resident stallion, and the mares did not seem to favour one or the other particularly. We like to ride our stallions out with each other and any of the mares and without a great deal of physical restraints in the sense of bridles with bits, or pieces of leather tying their head down and so on. They have all learnt that when with us, they do what they are asked without being physically restrained, and do not attack each other. But there comes a time when the two stallions are close together with a mare, perhaps in oestrus, and they cannot manage to control themselves, and a fight might break out.

Both of these stallions found a way out of this situation, so they did not have to fight, risk injuries and be harshly ticked off by us. What they learnt to do was when things got tricky, they would pay attention to something fascinating over the hill and turn their head away from the other with head up and ears pricked, orientating towards the often imagined or very slight change elsewhere. This redirected the attention of the other stallion and the mare towards things that were potentially more important: something that might threaten them and that they should be aware of, even though it may be imagined. This gave enough time for the touchy moment to pass, provided we were careful to ensure that they did not dwell on the situation and kept their attention directed at doing other things: changing the gait, performing a new movement and so on. Then, all was well and no fights were needed and no one lost face either. This was a solution that they had found themselves. In the end, the two stallions managed to live together without fighting, provided there were not mares anywhere near them. They had not entirely cracked it ... but they had got further than most with avoiding trouble with each other.

A more testing story-reading comes when you are asked to relate to, deal with or teach a large and potentially dangerous individual who has an angry or fearful story to tell. This is usually because their stories have not been listened to previously by humans. This might be a horse, dog, bull, tiger, lion, leopard, elephant, buffalo, rhinoceros, grizzly bear or any other large mammal. Unless they are physically restrained, you are in what has euphemistically been called "unprotected contact", and it is a case of get it right, read the story they are shouting out, or die. But the amazing thing is that usually people who have free contact with these frightened and angry individuals live to tell the tale without using drugs, technical aids, or excessive restrictions. Perhaps this was something to do with those who dare read the story, or perhaps the frightened or angry large mammal recognises that the human is trying to help, even if not very effectively. I, like most people who do this, have made painful mistakes, but the more one learns, the more forgiving the animals seem to be.

## A Cape Buffalo

Let me end by telling a Cape buffalo story, where with the help of some experienced cattle keepers we got it right. Cape buffalo, among hunters and anyone else who listens to humans' stories, are believed to be the most dangerous of all the African animals. This is because they do not hesitate to attack when wounded, and are known to have killed more people than any other species. However, when tales of our success with the elephants began to be circulated in Zimbabwe, we were asked to help teach some Cape buffalo cows to be safely handled and taught so that they could replace the small farmers' domestic cows in areas where there is trypnosomiasis which kills cattle but to which Cape buffalo are immune. The buffalo were extraordinarily receptive to our methods, and in no time we had a couple of cows harnessed to a cart, pulling it along slowly, and we had started handling the very wild calves, who, within a couple of days, were much quieter. We finally had the cows pulling the cart without being led, directed by words and gestures from the driver in the cart.

One of the most astonishing things was how careful the buffalo adults were that their enormous horns did not touch us when we were harnessing them, even though they could swing their heads about and, even without intention, catch us with them. It was strange and probably a bit scary for them, but they must have taken great care not to catch us with a horn and consequently hurt or scare us. But all the humans we were teaching were not able to accept this as they constantly used sticks and shouted, and leaped about when the buffalo scared them. After demonstrating how dangerous this was, and gradually building the confidence of the handlers, the buffalo calves became calm and behaved just like domestic, well-handled cattle calves. Bovines certainly learn fast and are able to take on board, with relaxed good humour, what others might find horrifying. The project never went further alas, but if only more people would really try to listen to the animals' stories, helped along by recognising our similarities with them, and also recognising that they, like each of us, have private lives and experiences, it would help both the animals and the people a lot. It is the experiences and the accumulated knowledge of the world that each individual of each mammalian species has that we need to try to grasp. To do this, we need to take a step beyond the anthropocene.

We do not know how to do this for humans, we do not even know what another consciousness is, it is a metaphysical problem "beyond physics" with many answers. How do the mushy white matter of the brain and the electrical connections of the neurons become things like thoughts, ideas, beliefs, choices, decisions and feelings? There is mental stuff that, so far, is not materially measurable, which some say does not even exist,[1] yet the common sense of it is that we all know it does: we feel things and are aware of things; we are conscious of being alive and in the world (unless we are unconscious or dead). Our Newtonian physics has been very helpful in explaining the physical world up to a point, so we more or less have an idea about light and radio waves, electricity, neurons and the structure of a brain, and we can more or less measure the messages travelling through this or that neuron and synapse ... but there remain things like knowledge and feelings, thinking and decisions, choices and desires which are not measurable or even sometimes identified in different people. We may measure where and how they travel in the brain but we cannot experience them, yours and mine are different. There is no ultimate "truth" concerning mental and emotional activity: what I experience and feel is true for me, what you do is true for you, this is otherwise known as recognising the subjectivity of another, or the "personhood" of another. It is private stuff which is not composed of measurable stuff. Every sentient being is an individual with a slightly different body and brain, and on top of that their lifetime experiences will always be different. What individuals learn, understand and experience is unique, even in the case of identical twins, although if twins are raised together, they may get closer than most to experiencing similar things and better understanding of each other. But when all is said and done, the fear of one individual is not the same as the fear of another: you may see and know that he is frightened, but you can never experience his fear, that is private. Even seeing red may have a different significance or feeling or "beingness" for each of us.[2] This is still considered, after three decades of debate, to be the "hard problem of consciousness",[3] which we have made little progress in understanding.

---

[1] e.g. D Dennett: https://plato.stanford.edu/entries/consciousness-animal/

[2] Nicholas Humphrey, *Seeing red: A study in consciousness*, 2006, Harvard UP

[3] David Chalmers, "The hard problem of consciousness" in *Journal of Consciousness Studies*, 1995, 2:3, pp200-219

Nagel in 1976 threw up his hands at the insolubility of this problem when he discussed what it might be like to be a bat, another mammal, with different special senses. But he did point out it was "something like" being a bat and set the world thinking how we could begin to understand this. It is the "big problem" of consciousness,[1] understanding another's epistemology, knowledge and experience of the world, that is, his "world view", whether he is a man, monkey, spider or horse. There have been tomes and tomes written about all this of course, but today there remains some form of non-materialistic, un-measurable stuff (sort of "spiritual stuff" if you like) which only the individual has access to. It cannot all be explained by giving up God or by selfish genes. Scientific materialism has not even begun to crack it.[2]

If this is a problem for a member of one's own species, how much more of a problem will it be to try to understand the world view of another mammal? Is it ever possible to experience the same thing and have some similarity in the view of the world as another mammal? Yes, it is, because there are overlaps between closely related species. We recognise that we have many characteristics of both body and brain in common and there are many similarities in our sensations, behaviours and feelings. If we recognise and carefully outline what these similarities are, we will have taken the first step in understanding another's world and feelings. After all, you know what your dog is doing when he sees that you are furious, because he has stolen the cake. He knows you are furious and probably knows why (if you have taught him not to steal cakes), so he will either quickly get out of the way or lie down, show he is sorry by squirming up to you and trying to lick your fingers, and perhaps he may even believe, for a time, that he will not do it again. He knows that you are furious with him, and

---

[1] T Nagel, "What is it like to be a bat?" in *Philosophical Review*, 1974, 83: 435-456; Jeffrey Alan Gray, *Consciousness: Creeping up on the hard problem*, 2004, Oxford UP, Max Velmans, *Understanding consciousness*, 2009, Routledge, London

[2] e.g. Richard Dawkins, critiqued by Mary Midgeley, *Evolution as a religion: Strange hopes and stranger fears*, 1985, Routledge

why you are. All of us humans, dogs, horses, elephants, lions or rats recognise many of the emotions being felt by the other:[1] fear, friendliness, disinterest, pleasure, joy, arousal, aggression, fury, attention, terror and so on because we behave in the same sort of way when feeling that emotion. Sometimes we make mistakes and wrongly interpret the other's feelings; but, if we try harder, get to know the other well and know something about their lifetime experiences, we really can begin to unravel their own story in a way which must be something like how they tell it. But as mentioned, they often have very little trouble in reading our stories, as they recognise similarities in behaviour related to different emotional states and they see more subtle changes.

We know this works because we experience it with some of our animals which is often why we keep them. There is nothing new here: throughout history there have been people who have taught animals without any violence willingly to do an enormous range of different, extraordinary things.[2] There are, and have been, of course, also bad, incompetent and violent communicators and trainers or misguided ones, just as there are bad parents or behavioural researchers, but that does not mean that we should dismiss or ignore what inter-species communication can achieve.

After many years of pondering on the hows, whys and wherefores of this, it was when at Berkeley studying with John Searle that I discovered collective intentionality[3] about which he had been philosophising for many years. Collective Intentionality is when two or more individuals willingly do things together which are directed at a common aim. The human example that he used was playing in an orchestra: each individual plays a different instrument, but their common aim is for the sound of the whole orchestra to be as they wish it to be. It is considered, at least by Searle and myself, that animals are capable of this. He quotes

---

[1] F Wemelsfelder, *Animal boredom towards an empirical approach to animal subjectivity*, 1984, Proefschrift, Leiden

[2] MKW, Review: "David Wilson, The welfare of performing animals: A historical perspective. Berlin: Springer, 2015" in *Animals,* 2016, 6:11

[3] John Searle, *The construction of social reality*, 1995, The Free, NY

lionesses hunting as an example: they all want to find and kill their next meal, and they work as a group to achieve this, each doing different things to achieve the aim. This is collective intentionality.

It is possible to have collective intentionality between species too,[1] for example, between a dog and a human. A sheepdog working sheep, goes left and right when asked, but if he is experienced, he knows what the sheep are going to do, and what he is directing the mob to do, and he will do it without these instructions, or he may do something different because he chooses to and/or the instructions are wrong. But both the dog and the human work collectively and willingly to achieve a common aim. This is particularly evident in sheepdog trials: the dog knows what the sheep should do and he may or may not be correctly directed. He may or may not manage to do it very well, either because he has made the wrong decisions or because his handler gave him the wrong commands. It is not possible to achieve as much with a novice dog who may do what he is told, but is not clear what the final aims are. Novice dogs are like the people learning to play their instruments in the orchestra, they are directed by the conductor but do not always get it right! A bullock pulling a cart can be also behave with collective intentionality. There may of course be times when he really does not want to pull it up the hill, when he is too tired, hungry or hot, but he CAN enjoy the doing of it just like the human enjoys driving him and delivering the things in the cart. Riding horses or elephants can, more than perhaps anything else, demonstrate collective intentionality ... but equally things can go very badly wrong when the horse or elephant is forced into going places or doing things he does not want to do, or carrying too heavy a weight and all the other things that can make life unpleasant.

For me, the aim is to try and achieve collective intentionality with the animal with whom you are working. To do this, it is imperative for both the human and the animal to mutually read each other's feelings and intentions. One of the best examples of this is demonstrated in the "art of equitation", where a new and

---

[1] MKW, "Collective intentionality & social ontology," 2008, Berkeley

uniquely subtle language may develop between the rider and the horse which cannot be observed. It looks as if the horse is just doing it all by himself. The aim is to perform more and more elaborate, and different movements in a particular way. But, what if these are achieved by using strong controls of one sort or another such as bits, ropes and pieces of leather, whips or spurs that restrict or restrain the animal? If these movements can only be performed when these restrictions are in place, then it will not be collective intentionality. The horse is restrained and consequently not doing it because he wants to in harmony with the rider, but doing it because he has to.

However, there are times riding a horse when most riders who are really interested, experience this feeling of mutual togetherness, and it is this that some describe as "telepathy". What is certain is that this becomes addictive once it has been experienced, and both the human and the horse get better at it. It may be more obvious when riding because the human and horse are in permanent contact with each other through their bodies and therefore it is easier to read each other's intentions through touch.

Alas, this is neither always achieved nor always encouraged as various restrictions and restraints are usually used in training and even allowed in competitions. If the competing horse and rider can win without these, then they should be highly applauded. We need competitions to begin to reward this.

If, when teaching another species, collective intentionality is the ultimate aim and consequently both are enjoying the experience, then whatever is being done is to be applauded as it brings those animals and humans closer together in a greater understanding of each other.

My life has been one of pursuing freedoms and ideas, and learning to read the stories of my four-footed family and friends, but I know I have only scratched the surface of their knowledge and understanding. First, it is necessary to learn the current dogmas and understandings from every angle, but then it is essential to ponder them, to find where they do not add up. Only then can we perhaps open the doors to different vistas of understanding. I have learnt so much about how to live and what to feel and when not to do something from my four-footed family and friends, but alas this may be something that the next generation of humans will find even more difficult because there will be so few animals around in their world.

## Beyond the Anthropocene

> The idea of the Anthropocene inflates our own importance by promising eternal geological life to our creations. It is of a thread with our species' peculiar, self-styled exceptionalism—from the animal kingdom, from nature, from the systems that govern it, and from time itself. This illusion may, in the long run, get us all killed. We haven't earned an Anthropocene epoch yet. If some day in the distant future we *have*, it will be an astounding testament to a species that, after a colicky, globe-threatening infancy, learned that it was not separate from Earth history, but a contiguous part of the systems that have kept this miraculous marvellous world habitable for billions of years[1]

The anthropocene is today's world, geologically speaking, which is dominated by humans and their activities. Anthropocentrism is the belief that all humans matter more than any other species because they have unique characteristics, particularly mental. Today, this has become more or less a religion world wide. There are different types of anthropocentrism: liberal, social and evolutionary.[2]

Another label often used is "humanist". Strictly speaking, this refers to people who do not believe in God, but it has come to mean those who believe in humans, and is often used instead of anthropocentric because humanists believe that humans are more important than any other species. Some recognise that animals are sentient, and must not suffer "unnecessarily", but gener-

---

[1] https://www.theatlantic.com/science/archive/2019/08/arrogance-anthropocene/595795
[2] YN Harari, *Homo Deus*, 2017, Vintage, London

ally they believe that animals' interests do not really count when in conflict with human interest. Others believe that if human interests are trivial, animal interests might trump human ones.

Anthropocentrism has its origin in almost all the well-known religions: Christianity, Judaism, Buddhism, Hinduism and Islam. In each of these religions one of the central ideas is that humans differ from the rest of the living world, and therefore must be accorded special status. Only animalism, held by traditional believers before other religions arose, recognised that humans were not superior and so they had better make the most of life without threatening others' survival. In fact, some believed it would help to worship the world, including different animals, by showing that you wanted to cooperate with them. "Traditional believers" or "animalists" were stamped out by the other religions. A few sects remain, but they are not well-known religious groups today.

In the Bible it is repeatedly stated that humans must "go forth, multiply and tame the rest of the world" ... and they have certainly done this! Today, the Christian scriptures are interpreted by believers to mean that humans, although in control, must be "good stewards" of the natural world. In other words, the natural world will continue to exist by "grace and favour" of humans, not because without it, humans would not exist: "Behold the birds of the heaven, that they sow not, neither do they reap, nor gather into barns; and your heavenly Father feedeth them. Are not ye of much more value than they?" Matthew 6. verse 27.

The Jews take a very similar stance: "The fundamental presupposition of the rights of the person in Judaism is a belief in the absolute and uncompromising worth of human life. This belief is grounded in the unique value of the individual in the divine scheme of creation."[1]

Muslims believe strongly that humans are superior to other living things, but, also that different groups of humans should be treated differently (for example, males and females), and non-humans treated differently again.

---

[1] https://www.uri.org/kids/world-religions/jewish-beliefs

According to Hinduism, animals are manifestations of God on a lower scale of evolution compared to mankind, although each contains a spark of the divine, and thus is capable of becoming human after numerous reincarnations and achieving salvation. Human life is, therefore, more precious because it comes after many lives of existence in the lower life forms: "In the whole creation only human beings, not even devas (gods), have the opportunity to achieve salvation or ascend to the planes of divinity. Human life is therefore very valuable and unique."[1]

In Buddhism, the doctrine of rebirth holds that any human can be reborn as an animal, and any animal as a human. A dead relative, who was not up to scratch, may be reborn an animal. Because of this, one cannot make a hard distinction between the moral rules applicable to animals and those applicable to humans. The fly might be your granny who therefore must be respected, but, not because he is a fly.

Since the Jain monk wishes to climb the ladder to Nirvana, a super-human state, he must not kill or cause any other being to suffer, even if it is a lower life form, as it might previously have been a human.

Humans are superior in Hindu thinking; that is Hindus believe in anthropocentrism not flyism, even though the fly and the human may end up being partly interconnected:- "the animals were believed to inhabit a distinct 'world', separated from humans not by space but by state of mind ... Rebirth as an animal was considered to be one of the unhappy rebirths, usually involving more than human suffering."

Thus, for all the major religions, humans are superior to any non-human. If this difference is nothing else, it is in their "state of mind". So both religious and non-religious people believe it is humans who hold the moral, emotional, intellectual, and (possibly) physiological high ground. The politics and laws in all human societies reflect this belief, to a point where the interests of a human with disabilities preventing them from developing a fully-operational mind, will trump life-threatening interests of a

---

[1] https://www.hinduwebsite.com/hinduism/essays/animals.asp

healthy adult dog, horse or elephant.

There is a movement in law to recognise that some non-human mammals are "persons", not just "possessions", but complete recognition will be very slow. It has already taken fifteen years to be discussed in jurisprudence.[1] Even then such a change in the law will not recognise that non-human mammals should have equal consideration, or that they are moral agents rather than just beings of moral concern.

But, a consideration in law that non-humans are moral agents is by no means news, the ancient Greeks believed that a "murder whether committed by man, a beast or an inanimate object would arouse the furies". Ironically, in Catholic France in the sixteenth century a French jurist made his reputation at the bar as counsel for some rats who had been put on trial for "feloniously eating up and wantonly destroyed the barley-crop of the province of Autun". The jurist employed various strategies to excuse the rats for not appearing in court. The one that won the day was that they were widely dispersed and had to suffer the attentions of their mortal enemy, cats, if they were to attend, therefore, it was only just that they should be excused! Evans points out how widespread and how long the recognition of the moral agency of animals in the law has been.[2]

Today, almost every philosopher or scientist, every man, woman and even child in the street will say that humans are "superior and special", and not like other species, so they must have special treatment. But, where this distinction is, remains unclear; similarities between us and non-human mammals have long been recognised. Kant in the eighteenth century maintained that we must be nice to non-human sentient beings because being nasty to animals will encourage nastiness to other humans. But if an animal is an inferior being without a mind, this would not matter; after all it

---

[1] Conference on Animal Law, Ethics and Legal Education. Liverpool John Moores University, School of Law in collaboration with the UK Association of Lawyers for Animal Welfare, 2017

[2] EP Evans, *The criminal prosecution and capital punishment of animals.* 2009, MCMVI, Dutton & Co, NY, p9

does not reflect badly on other humans if you bash a stone to pieces ... but it does if you bash a dog. So even Kant recognised that we have something in common with dogs and other non-human mammals.

The Christian ethic of being "good stewards" of the environment and other creatures does not include controlling our own population, or consumption of resources, both of which are threatening life all over the planet. After all, these "good stewards" believe that selfishness, when it comes to humans in conflict with another species, is morally and intellectually respectable because of human superiority. It is therefore "right" that human interests, even trivial ones, trump those of almost anything else.

In the past, when the majority of people believed in a God, the distinction between "us humans" and "them" (non human mammals) was easier; the distinction was made on the grounds that only humans had "souls", although what a soul is has never been defined. It seems to be some non-materialist, not measurable dimension such as having emotions and experiences: a mind. Since we now usually agree that non-human mammals are "sentient" and have a mind with feelings, memories and experiences, they also must have non-measurable "spiritual", soul-like aptitudes. If this is the case, then, surely they should be eligible to join the soul-owning clan?

Scientific functionalism dismisses all this as their followers believe that, eventually, all these mental events will be found in materially-measurable matter, such as electric currents in neurons. But, the problem is that "souls", "personhood" and "individuality" are not neural pathways and brain scans. Emotions, memories, imagination, dreams, thoughts, all of which make up "experiences", exist but they cannot be identified in the mushy white stuff of the brain and the electrical connections of the neurons because they are not there; they are personal, private, feeling stuff which each individual will feel differently.

Although scientific functionalism has been useful to explain aspects of mammalian physiology, it cannot address the un-measurables. We still have a long way to go before we can give up our individuality and consciousness, even if scientific functionalism has become fashionable.[1]

Consciousness or thoughts, ideas, beliefs, choices, decisions, feelings and experiences are the Hard Problem[2] which philosophers have been mulling over for centuries, but neuroscientists only for about thirty years. These are mental, out of body concepts, not materially measurable, or known where or what they are. One way out of this is to deny that they exist,[3] but, this avoidance of the issue (even if you can understand Dennett's peculiar arguments) gets us nowhere because our common sense *knows* that just like my hand or leg, these un-measurable things exist; my dog and I feel, are aware of things when awake and are conscious of being in the world (unless we have been knocked on the head and are unconscious). There have been tomes and tomes written about all this but what "consciousness" is, remains a mystery for science.[4] Gray has a quote from Thomas Hardy (1886) that illustrates this: "How is it that anything so remarkable as a state of consciousness comes about as a result of irritating nervous tissue, is just as unaccountable as the appearance of a Djin when Aladdin rubbed his lamp."

There is non-materialistic, un-measurable sort of "spiritual stuff" out there, which defines each of us mammals, and maybe other living creatures, and there have been people throughout history who have considered that non-humans might also have an un-measurable sort of "spiritual" dimension to their lives.[5] These ideas have tended to be ignored with the rapid spread of anthropocentrism.

Today, it is generally unacceptable to believe that non-human mammals should have equal consideration to humans of any type,

---

[1] eg R Dawkins, *The God delusion*, 2006, Bantam

[2] D Chalmers, "Facing up to the problem of consciousness" in *Journal of Consciousness Studies.* 1995, 2:3, pp200–219

[3] Daniel Dennett: https://curiosity.com/topics/theres-no-such-thing-as-consciousness-according-to-philosopher-daniel-

[4] J Gray, *Creeping up on the hard problem of consciousness*, 2004, OUP

[5] e.g. E. Marais in South Africa 1937. *The soul of the white ant.*

colour, culture, sex or age. Nor is it taken seriously that a non-human may have mental attributes, or a moral creed that we do not have, or that they are better than us in some mental aspects. Recently, one of my students in a fit of rage shouted at me, "you treat me like your dog" ... The person had been with us for over a month in order to learn about a different approach, but their cultural belief was so strong that, for them, it was not possible to entertain another one. Indeed, I hope I did treat the person like one of my dogs: that is with equal consideration and respect, and in return I expect the same, so that we can work together and cooperate. For this person, although interested in horses and dogs, and wanting to learn more about them, it was impossible to escape their deeply held anthropocentrism, the result of the culture in which they had been raised, educated and lived.

I am well aware that my views are likely to get me hung, drawn and quartered, but, whether you agree or not, the reader knows that sometimes, s/he does prefer the company of her/his dog, horse, rabbit or other non-human friend to other humans and considers their needs more carefully. We are also aware that there are, and always have been, people who have lived preferentially with non-humans, not because they have some mental illness, but because they prefer to. As a result, such people may have many common experiences, and a greater depth of understanding of the species or individual they live with. This might, perhaps, be greater than their understanding and knowledge of other humans; but it does not mean they are "mad" or mentally disturbed.

Nearly everyone respects equal consideration when it comes to different human cultures or races, but to genuinely give equal intellectual consideration to another mammal will "just not do", it is fundamentally unacceptable ... Why?

There are cases where individual human interests can trump dogs', and visa versa, but, when this happens, it should be the result of an assessment of the individual case, rather than an assumption that human interests are more important and therefore must win over those of other mammals. Individual case assessment is occasionally happening, for example there are funds being collected for rescuing donkeys from being beaten, when the money could be given to starving children somewhere.

But there remain many questions to ask. For example, the Covid-19 pandemic has made one consider why we accept that humans moving across frontiers without medical assessment is a right, when any animal must be tested for a whole variety of diseases. As a result of this freedom of humans to spread diseases, bovine TB has been reintroduced to the UK by infected humans, when it had been wiped out in the fifties and sixties. This has resulted in both cattle and badgers being put to death for having TB which is transferable to different mammals including humans, even though, although expensive, it is curable with antibiotics (since it is a bacterium).

We are making a poor job of looking after our world; even though it is necessary if humans and other species are going to continue to survive in the world as we know it.

There is also the problem of individual interests being favoured over those of the community. At what point does one make such a decision and why? Animal welfare activities and human medicine generally hold that the individual must be helped at any cost whereas ecologists and wildlife conservationists favour the survival of the species or the community and may sacrifice individual non-humans if this is threatened.

Because both matter, the only way forward is to assess every case on its own merits so that individual suffering is reduced while safeguarding the species, community and environment. This may mean that some individuals will have to be killed to avoid over population of that species which threatens others, but if, when, where and how are the questions that must be asked in each case where this conflict arises, and different decisions may be made in different cases.

For centuries philosophers have used the lifeboat thought experiment to help with such decision making: unless we throw someone out, we will all sink, so who do we throw out of the boat? If we are going to make distinctions between "us" (humans) and "them" (non-human mammals), we must asses why and how we make the choice, just as we do for the individuals in the overcrowded lifeboat. But to do this, first we must address whether we have reasons for believing in the greater importance and superiority of humans. Do humans matter more and, if so, why? Do they have richer mental lives and, if so, where, why, how and when?

Over the last couple of decades almost every mental ability which had previously been thought unique to humans has been found to be present in other species by observational and experimental science, supporting commonsense or folk knowledge. This has led to a slow crumbling of the belief in the all round superiority of humans. Below I will set out how the arguments and supporting science has developed and led to an understanding that each species is unique mentally as well as physically. But, first we must consider what characteristics make humans unique, mentally and physically.

*Humans have opposable thumbs, so they can manipulate and make things.*

In the 1960s there was much discussion of tool-use as a distinguishing feature of humans. But then Jane Goodall, and subsequently others, discovered by carefully observing chimps that they could make and use sticks to fish out termites to eat. Rather than reviewing whether tool-use was unique to humans, chimps were allowed into the wider family of humans: "extremely clever creatures".

Clearly without being able to manipulate things easily, non-human mammals will find making and using particularly elaborate tools difficult, unless they have other body parts that are as handy as hands.

When you think about it, many species use simple tools. They manipulate and use things with their mouths, chins, beaks, tails, feet, toes, not only their hands. It takes a long time for someone who has lost their hands to learn to use their feet to manipulate things, but they often do eventually, so why should this not work the other way around? It will take a long time for a horse to learn to use his nose and teeth like a hand, and hold a paint brush, but he can learn to do this as Shatish has done; it is just not so easy.

Elephants with their hand/nose trunk find it easier to learn to hold a paint brush, a tree trunk, or to break off and fashion a twig to scratch themselves with. Many other animals use tools to manipulate things, sometimes using their paws, their noses or their teeth. However, it may not be as easy, perhaps, as with hands, and they may never be able to perform delicate manipulations, such as threading a needle.

But, when does a tool become a tool? The Oxford dictionary defines a tool as, "Thing used in an occupation or pursuit". Every mammal will use bushes, trees, the earth or any other handy object to scratch themselves: they are using tools. What about birds making nests, or spiders spinning webs? They are also using and making tools; just because humans do not do this, does not disqualify other sentient beings from making tools. Some, may find it difficult to make a particular tool, but this does not mean they are mentally inferior; any more than the person who has no hands or cannot sew!

*Humans have a unique language.*

For a long time scientists and laymen alike have believed that it was the uniqueness of human language which was the cornerstone of what they called humans' "greater cognitive development". Human language is unique (although others may be able to learn to use it). Simply put, human language consists of an elaborate development and use of symbols. The skill for using symbolism has been further elaborated in mathematics, the physical sciences, the arts and of course, modern technology. This may be admirable and unique to humans, but it does not demonstrate that other cognitive (mental) feats that other species have are inferior. We are only beginning to even look for them, and finding them just as extraordinary. Therefore, there are no clear grounds for supposing that human language controls "advanced cognitive development", although it may control advanced symbolic development.

Answering three further questions should clear up any doubt:-

1) *Are humans who do not have language mentally inferior?* Even amongst humans, there is a huge range of different mental at-

tributes that those who have limited language abilities may not have. For example, some autistic people have the ability to draw accurately in three dimensions, even after just a glance, but they do not socialise or talk much. Then there are those who are born blind who develop extraordinary mental skills and sensitivity for smells and feelings which people with sight do not. Children raised with no access to human language, by wolves or chimps, for example, have not been shown to be particularly stupid, even though they do not have human language. It is a pity that no one bothered to study exactly how they see the world as a result of their wolf or chimp experiences, we could have learnt a lot. Deaf people who have never heard human language, develop other different mental skills. Others have perfect pitch and can place every note they hear but may be poor at mathematics. Some do not understand the laws of relativity or philosophical arguments, but they may be much more aware of others' feelings or be extraordinarily good at carpentry.

We are forced to admit that, even in humans, it is not human language which controls the development of mental skills. Because of the different species' different bodies, life styles, habitat, sensory systems and so on, each species has its own mental specialism in addition to individual differences.

2) *Why does the use of symbolism, developed to a high degree in human language, mean we are mentally superior in every way?* Let us consider an alternative view: the different cognitive abilities and resulting world view created by being expert at smells, for example. Humans are not very good at smelling, but we can get a drift of the different way smell causes us to experience the world, because, smells sometimes conjure up memories, experiences, situations, emotions and moments which are not always identifiable in time or place. Unlike auditory or visual experiences, smells may travel in the wind or water or be buried in the earth. Such moments allow us to remember an experience, without identifying its name, time or place. Is it then or now, here, there or anywhere? Normal Newtonian concepts for measured space and time which we learnt at our mother's knee no longer seem to work

when it comes to smells, rather, different experiences take their place and offer different "dimensions" of space and time. I have found some of the ideas emerging in quantum physics mind blowing, but once one begins to get one's head around them, they can be useful with helping to perceive possible alternative world views, mental aptitudes and consciousness of different species. Therefore, humans' specialist view of the world as the only one, or being "superior" must be brought into question.[1]

3) *Can we have similar mental experiences to other mammals?* Nagel, in the 1970s, wrote a paper called: *What Is It Like to Be a Bat?* In this paper, he pointed out that since we do not have sonar, cannot fly, do not sleep upside down and so on, it is not possible to have a complete understanding of what it is like to be a bat. This sparked off further debates about experiences, and how they were, by definition private, that is only experienced by the experiencer, but, we know it is possible to have some understanding of what it is like to be another of a related species at least. We do this by assessing their behaviour and from this, their feelings, and it often works, otherwise we would be unable to teach another species, or have pleasant or unpleasant relationships with them. We have some idea of what it is to be a horse, dog, elephant, or whatever other mammal you choose, because we are also mammals and therefore have much in common.[2] So, it is possible to have some idea of what it is like to be something as different as a bat, because, like us they have sex, babies which they suckle and look after, they squabble and choose their companions, they like and dislike, can be angry or kind, and so on. We will never know exactly what it is like to be a bat, but then I will never know exactly what it is like to be you either. Scientists are beginning to uncover different species' mental aptitudes and their different ways of perception so it is no longer possible to argue that human mental

---

[1] Thomas Nagel. "What 'superior' mental skills exist in other mammals?" in *The Philosophical Review*, 1974, 83:4

[2] MKW, "The mental homologies of mammals. Towards understanding of another mammal's word view" in *Animals* 2017. 7:12 DOI. 10.3390/ani7120087

aptitudes are all superior, although they, like any species, have their mental specialisms.

Human language controls many of our mental abilities, and makes it difficult for us to think "outside the box". As a result, it can be a handicap, making it difficult to come to understand other world views: "Our language affects the way we view the world and how we approach problems."[1] Language takes over our thinking, perceiving, acting, feeling, dreaming and other parts of our mental being.

To conclude this section, there are two areas of expertise that humans have which combine to result in some remarkable technological developments including the engrained symbolism in all the arts and sciences.

1) Manipulating and changing things with opposable thumbs: a physical skill.

2) An elaborate symbolic language: a particular mental skill.

But it is time we considered whether there are other mental attributes that could be as, or more, important in allowing the world to flourish. On the negative side, we are learning fast that human manipulation of technology, among other things, overpopulates and destroys the living world, and humans do not seem to be either able, or willing, to control it.

Having examined particular skills that humans have, I will now examine other mental skills that have been thought to be unique to humans, and point out some non-human mental expertises.

*Sentience, feelings and emotions.*
It is now generally recognised that humans, non-human mammals, reptiles and fish, crayfish and octopuses have feelings. Yet, it is still often maintained that humans have a richer emotional life than non-human mammals.[2] How can we prove this? One of the problems is that in experimental psychology laboratories which have been experimenting with the mental abilities of laboratory-

---
[1] S. Shorrocks 21/09/2013, humanisticsystems.com
[2] SR Hameroff, R Penrose, "Conscious events as orchestrated space-time selections" in *Journal of Consciousness Studies*, 1996

raised animals, any feelings the animals might have, have been disregarded, because they are too difficult to measure. The only feeling that such experimental scientists are often aware of is "motivation" otherwise called "wanting". No animal will do anything voluntarily unless he is motivated, that is he wants to. Wanting (motivation) has been measured by how long an animal will continue to do something, whether he will do it at all, how much "work" he will do to achieve it, and how much enthusiasm he shows i.e. how fast he does it. Usually the "want" (motivation) to eat is used for laboratory animals. As a result, to be effective, experimental subjects are kept at a low body weight (very hungry). Food is used as "reinforcer": that is something they want and will obtain when they have solved the problem. Because they are half starved, they always want to try and obtain food.

Over the last couple of decades, ethologists (those who study animal behaviour outside laboratories) have measured a variety of behaviours that indicate different emotions. The emotion the animal feels is based on what the observer thinks the animal must be feeling: "he is aggressive" when he attacks another, or "he is fearful" when he runs away. This interpretation is based on "folk knowledge", that is a recognition that animals and us behave similarly in similar situations, which is the only way we have of knowing what another being might be feeling. That this is not entirely wrong was confirmed by a researcher in the 1990s who asked some urban humans (who had never had anything to do with pigs) to watch pigs and record what they thought they might be feeling. The results showed that all agreed more often than not, on what they thought the pig was feeling. The interpretation was of course based on common behavioural responses to some of the feelings that the observers have.[1]

Assessing what another mammal may be feeling from his behaviour is old hat for anyone who has had to do with non-human mammals. It is particularly important for those who teach animals because they must know their intentions to encourage them to

---

[1] MKW & J Rendle-Worthington, *Exploding the Myths Large Mammal Welfare, Science & Educational Psychology*, 2012

perform what is required. Folk knowledge (or commonsense) about non-human emotions and feelings has been around for centuries, and often used to good effect.

Folk belief is not folk knowledge, however. Folk belief is unreliable; one of the handicaps of language. If you are told that an animal, or particular breed is very difficult, aggressive, or kind, it becomes a belief and you behave according to this pre-conception when the animal is encountered. Often, the result is that your behaviour may cause the animal to behave in this way because he will have been watching and predicting your intentions. An example of this is the belief that "Rotwielers are dangerous dogs, they are aggressive". While there is no evidence to suggest they are genetically more aggressive, there is plenty of evidence that they can be made so by people.

Folk knowledge about animals is knowledge that has been established over long periods of time by practitioners and established as known fact rather than the result of a particular preconception which might change. For example, "naive horses will snort and run when frightened", although they can be taught not to.

Folk knowledge takes for granted that non-humans show a whole range of feelings:- hunger, thirst, pain, fear, anger, sex, enjoyment and pleasure to name a few. Some people also recognise grief, depression, embarrassment, anxiety and confidence in their animals. In this book, I have shown how important a complex feeling like "confidence", or lack of it is for mammals when they are learning. Having "confidence", by definition, means that the individual *knows* he can do something.[1]

It may also be that some mammals feel emotions that we do not feel.[2] For example, if we know the social norms of horses, we can guess, more or less, what a stallion is feeling when he herds his mares around with a particular "snake face". This is when he puts

---

[1] MKW & J Rendle-Worthington, *Exploding the Myths Large Mammal Welfare, Science & Educational Psychology*, 2012

[2] MKW, "The mental homologies of mammals ..." in *Animals*, 2017, 7:12; MKW, *Right in front of your mind*. 2000

his nose forward, ears back and moves his head up and down, a bit like a snake. He does this when his mares are too close to another male. He may be feeling something like: "please go away, and stay away from that male, or, I will be very angry and might really injure you." But we do not "feel" that emotion, or if we do, we do not show such a characteristic behaviour. There are probably many more emotions that we do not feel, but might be able to have some idea about in many different species, if we do our homework.

The emotions of non-humans have only just begun to be taken seriously so let us not assume that humans have a richer, more complex and deeper emotional life than other mammals before we know. Emotions, as I have previously mentioned, are private and we can never experience exactly what another is experiencing (whether human or non-human), but, if we study them closely, we may come close, and have a better understanding.

So far, the evidence is that emotions are as important to non-human mammals as they are to humans; in fact, it might be that they are more important to some non-human mammals who do not have expertise with symbolism and manipulation. For example, we have seen humans have a context-independent language which can be quite devoid of emotion. But the majority of non-human mammalian communication consists of messages conveying how the communicant is *feeling*. It is up to the observer to look at the context and assess what is going on and why he is feeling that way.[1] This is a different world view, but not an inferior one.

*The Mind.*

If something feels, it must have something to feel with, or out of which some feeling comes; this is what is called the "mind", therefore "feeling", "sentient" beings must have a mind. The mind is something that does mental work, even though it is not always measurable. One of the things a mind must do is put mental events such as emotions, memories, learning, and information together

---

[1] MKW, "The vocalizations of ungulates" in *Zeit fur Tierpsychol*, 1972, 31, pp171-222

to form patterns and have knowledge and experiences so that choices and decisions can be made.

*Memories and Imagination.*
Memories may be short or long term. We now know that non-humans as well as humans have both because memory traces have been found and measured in many mammals. But, there remains a dispute among behavioural scientists as to whether non-humans have "episodic memories". Can they remember particular episodes: the time, place and other information surrounding it? Again folk knowledge knows only too well that this is the case. For example, if a horse, who is being loaded into a trailer, or an elephant when lifting a log, has a bad experience, they will remember that place, that trailer, that person associated with it, in other words, that episode. Non-human mammals therefore have episodic memory just like I do, even if the laboratory scientist who has had no daily experiences of living with his animals does not know this.

Memories consist of actions, situations, occurrences and the feelings associated with the event. When they are replayed, they are sometimes changed, depending on the present feeling and circumstances of the individual. It is not always clear what actually happened and not everyone will report the incident in the same way; it is rather what *you* remember happened.

But, things can happen to that memory as it is replayed, it can become fixed and established or, if the circumstances, feelings, or experiences of the individual change, the memory can be added to, be changed or be used in combination with other memories. This then is the first step of "imagination"; what happened (as only you remember it) may be combined and mixed with other events and experiences, so that it is changed, but it is still related to your experiences. Oberlix, our stallion, would frequently when we came across rocks standing out of the ground on Dartmoor, jump and leap about, and try and run away. There was some association he must have made with these rocks and some scary thing that happened. I knew him well and he had, as far as I know, never had a frightening experience near such rocks. So, was he seeing faces and monsters? Who knows, but there is no reason to doubt that memories when replayed by non-human mammals, will also be subject to change and embellishment and have imagined outcomes just like us. Non-human mammals may not delve into science fiction, or ghost stories, but who knows?

*Beliefs.*
Individuals who have memories, can learn and therefore must have an idea about intentions and outcomes, consequently they have a belief: "this" will follow "that". For example, the horse in essence thinks something like: "This person came with a bucket with food in it before, I believe, therefore, that the bucket will have food in it and I will approach."

Belief must be another facet of all mammals' mental make-up, since they all learn, although they do not say it to themselves in this way.

Many mammals may not develop elaborate ceremonies, or have beliefs attached to religions, but beliefs come in all sorts. A belief in death for example: do they recognise death? They certainly know when another is dead: not moving, not responding. We have had examples of this when one of our old mares decided not to try and get up, and died. She was with her two-year-old son and her nine-month-old foal. The older son smelt her, touched, and then moved away, the young foal found it more difficult to understand perhaps, and stayed near, pawing, and trying to get her to respond. Elephants, and other mammals, show enormous interest, (and sometimes different sorts of ritualised behaviours) when they find the smell, bones or remains of others of their own or different species. Many species appear to recognise death, but there has been little research to date.

*Choices and Decisions.*
If an animal learns, and acquires knowledge, then, inevitably he will exercise choice and make decisions. These choices will be made because he either knows, or believes, certain outcomes and may have some idea about the intentions of others. Consequently,

he makes a decision or chooses to do, or not to do, this or that. Again, different encounters and learning different things, may change his beliefs, choices, decisions and intentions just as in humans.

*Thinking.*
Putting mental aptitudes together is usually called thinking, and it does not require human language to do it. Humans usually "think" in language terms, but not always; after all, people who cannot talk or read, can still think and learn.

Curiously, it is still maintained by some animal cognitive scientists that non-humans cannot think.[1] If "thinking" is defined as "that which only human language users do", then we need another word for the patterns created by all the different mental events all mammals have and do. The exciting question is not, do animals think? It is, what do animals think about?[2]

## Learning, reason and the accumulation of knowledge in all mammals.

*Trial and error & conditioning.*
Over the last century, the study of learning in rats in laboratories has led to the development of human educational psychology, that is, how to teach humans better. This tacitly recognises that learning works the same way in laboratory animals such as rats, as it does in humans; so how does it work?

Infants of praecocial species have to learn a great deal rather quickly after birth if they are to survive. This they do by trial and error, that is learning to get up, to balance and stand, to walk and finally to find the teat of their mother, all of which has to be managed very rapidly after birth. Just getting up is a challenge. A filly foal we watched for the first twenty-four hours of her life started trying within half an hour of her birth, but fell and fell before she had mastered the coordination of her muscles to raise

---

[1] e.g. CDL Wynne, *Do animals think?* 2004, Macmillan

[2] SEG Lea & MKW, "Can animals think?" in V Bruce (ed.), 1997, pp211-240

her hind end and then to balance. Then to walk or wobble over to her mother. Another big challenge was to find her own mother and a teat with milk. Although she had an innate tendency to search under her mother's tummy, and to suck something teat shaped, she had to find it, at the back, not the front of her mother's body, put her mouth around it and suck to obtaining even a very little thick colostrum. All this needed to be learnt within three to six hours of birth, if she was to obtain the immunoglobulins important for her health.

*Reasoning and rationality.*
When the foal, or any other mammal wants to learn, it is called voluntary learning or "conditioning". To learn in this way, the outcome must be reinforced and the performer will reason "if this", "then that". All mammals learn what the outcome of doing or not doing something may be, therefore they reason: whether to do it or not, and then decide and choose: does she try and suck here or there. Which will have the "reinforcer", that is the milk or colostrum? Simple reasoning is a necessary mental ability, and is *not* restricted to humans: "this object is prickly and hard with no milk. This object is warm, pliable and soft, and I sucked some colostrum". The next time she reasons: "go for the soft pliable object between the hind legs," or something of that sort. Reasoning may or may not remain at this level for that individual, whether human or other mammal, but it is a necessary attribute of learning to do things that must or can be learnt.

*The moulding of innate tendencies by learning.*
One innate characteristic of mammals is that the mother takes care of the infant. How she does this will be moulded by learning and her lifetime experiences, as we found with cows who were asked to adopt a *second* calf. If the calf was mothered with an adoptee, she would accept an adoptee much more willingly when she calved. Her innate tendencies had been moulded by her lifetime experiences; she had learnt to accept a second calf to suckle although it is not characteristic of most suckler cows. It was the result of her lifetime experiences when she was a calf: observing

her mother and suckling with another calf.

The individual has innate tendencies to do this or that, or associate in this or that type of group, but whether, how, where, when, or if she does, depends on her experiences (what she has learnt and felt) during her life.

*Observational or social learning.*
To learn useful things it is necessary to watch and learn from the right individual. It would be a mistake to follow a youngster into the river, a better idea would be to follow an older individual who knows the river. Therefore, to avoid making what could be costly mistakes, it is necessary to know what knowledge another has, that is who to follow into the river. Social learning is not just learning who is who, and watching what they do, but also being able to assess his/her knowledge and whether it would be wise to follow or imitate him or her. If a member of the group, s/he must learn and know the roles of the others, and in this way, something about what the other knows.[1]

It is safer and quicker to acquire necessary survival knowledge from others, by watching, observing and imitating them, than by trial and error. It is not always necessary to learn how and what to do by trial and error, especially when there can be a risk of making a big mistake, for example eating a poisonous plant. Watching another and learning from his mistakes and achievements, and imitating one whom you know and "trust" (that is, I can predict to a degree how he will behave, therefore I know something about what he knows) is worthwhile. That most mammals do this is clear from the fact that adults often do not follow or imitate a juvenile because they know the youngster does not know much.

*Imitation.*
Imitation remains a sensitive issue when it comes to non-humans. Do they imitate novel acts (which is how it is defined) or are they doing it because it is a normal action in their repertoire, and

---

[1] e.g. A Dickinson, *Contemporary learning theory*, 1980, CUP

others are doing it? The second case is not really imitating, but rather "social facilitation". It is the same as when you enter a room and everyone is sitting down, you will usually look for a chair to sit on too.

We have found that equines, elephants, guanacos, puppies and cows imitate another to learn to do something novel. For example, we had a foal who did the Spanish walk (lifting each front leg higher than normal) by imitating his mother. He did what his mother was doing, but it was not a normal act in his behavioural repertoire. We also discovered that these mammals learn to imitate new actions performed by another species: humans. Thus, to encourage imitating the teacher can be a useful teaching aid, just as when teaching children to do novel acts. Mirror neurons are said to control imitation, and initially, they were only found in humans. But when neurophysiologists looked, low and behold, they found them in other mammals too!

*Silent Learning.*
All mammals also learn "silently" or "cognitively".[1] This is another way of saying that we acquire knowledge or information about the world or things in it, without particularly "wanting" to. We pick up information as we wander around and it does not need "reinforcement", wanting or reasoning. It includes things like finding our way around an unfamiliar place, or recognising when something has been changed in the environment.

All mammals, as mentioned earlier, must acquire a great deal of ecological knowledge: they must become good natural botanists, zoologists, geologists, ethologists, geographers; in other words good all round "natural ecologists" in order to survive and reproduce. Much of this is learnt by silent learning. For example, most horses, who are familiar with a path will have taken note of any log and the angle at which it lies to a hedge. If it is moved, they will look twice, snort or jump when they see its changed position, even though there is no evidence of disturbance. But many humans will

---

[1] MKW & H Randle. "First steps in comparative animal educational psychology," 1996

not notice this, unless specifically instructed; silent learning is not unique to humans.[1]

*Can humans learn more than other mammals?*
Because of their context independent language, humans communicate symbolically and can also write things down, thereby using more symbols. Thus, they can find information, even if they do not always remember it. There is no doubt that the development of the "science of symbolism" is their speciality, but that does not mean they can learn more about *every* aspect of the world than a dog, horse or elephant. They learn about different things and each species will find some things easier than others because of his size, lifestyle, sense organs, brain, and resulting body/mind "being"; each will consequently have its own different mental expertise. Super-imposed on species differences are individual special aptitudes. For example, I can learn more or less how to use a computer, and so can a chimpanzee, an orangutan or even a dog, although they may not be as good as I am (although if they were taught enough, they might be). But I know nothing about how to programme a computer. I could learn a bit about this, but it is very unlikely I would be any good at it, or want to learn more, but there are plenty of others who are and do want to learn more.

The point is that we can all begin to learn each others' mental aptitudes, although we may never be as good as they are. I can learn to read very slight visual or tactile messages conveyed by horses better than some other humans perhaps, and the horse can learn the meaning of many words in my verbal human language, if I bother to teach her. But, because of the species, life style and her particular body, there are some things that are much more difficult for that species to learn than others. But, to believe they cannot learn to do something is premature, when we have not even tried to teach them.

I have had a long-standing debate with various philosophers who maintain that my dog could never understand the concept of "bank manager". However, if we break the concept into bits, it

---
[1] A Dickinson, *Contemporary learning theory*, 1980, CUP

does not seem improbable. To start with, my dog can certainly learn concepts; that is that things are classed as similar if they have a common function.

So what is the concept of "bank"? It is a place where a lot of precious stuff is kept. The fact that money is symbolic of, for example, bones for a dog would not be an insurmountable difficult belief for the dog to grasp (although he may have problems with adding and subtracting large sums, but so do I). He can learn some idea of numbers and do simple adding and subtracting and anyway, the bank manager probably does not do much of this. Then, what is a "manager"? It is someone who is in charge of a bank, at least that is what I think a bank manager is. He makes sure that others do not steal the money, or squander it, or give it away and he tries to get as much of it as he can. This is certainly not difficult for Moto, my collie dog, who is only too keen to get whatever she can and as much as possible, then protect it and hide it from others.

The manager also directs others to do things with the money; can dogs direct others? They can certainly go where directed and do what is asked, and occasionally, they direct me to do something by, for example, whining because they want to go for a walk, and sheep dogs learn how to direct sheep. So, I do not think this would be impossible, but it would need some teaching.

Moto, my canine friend, understands many of these concepts. She certainly understands possession: mine and yours, and do not touch this which I am collecting and hiding from you, but you can take that. She can direct others, including humans to go places and do things. So, on balance with step-by-step teaching, I think Moto could learn the concept of a "bank manager", just like my grandchild Ziggy can, if someone teaches him in the same way ... but whether at the end of the day Moto would be interested in what a "bank manager" does is an open question.

This does not mean that dogs would ever be able to fly to the moon in a space craft they have designed. We humans can, with the technology we have designed. But, we cannot learn, without technical aids, to follow a scent and find mines that others have left lying around to blow up other humans. A dog can without any technical aids. Yes, we have invented technology, but non-humans can at least begin to learn something about our lives and language, even the idea of a "bank manager" perhaps, should we not begin to learn something about theirs?

Do non-humans learn slower than humans and can they learn as much? This depends on what is being taught. Teach them to do something within their area of expertise, and they learn very fast, for example, when horses are frightened they learn in one trial not to go there or to avoid that. Dogs learn to sit down very fast because it is easy for them, but a horse or a cow takes longer to learn this as it is much more difficult for them and not something they do many times a day.

The amount of information non-humans can acquire remains a mystery, but we need to be cautious if it is maintained that they cannot learn as much. For example, recently a sheep dog demonstrated that he could learn to identify three hundred objects, even though these were human-designed toys. I wonder if the experimenter knew them all with out looking at his list? Several chimpanzees and now dogs have been taught to use a computer to ask for things and it may be just a matter of time before they are able to do more. Shatish remembered the fifty-one gates and all the hundreds of possible changes of route on a recent ride we had ridden together once in the other direction when I could not.

Non-humans are usually slower to learn symbolic structures than humans. But, remember, humans have had a lifetime experience with this. When horses, dogs and various other mammals have been tested to see if they could distinguish "same" from "different", human geometric shapes have usually been used, such as circles, squares and triangles. Many species have a hard time doing this, but they are very unfamiliar with these shapes, as well as the laboratory situation, whereas the infants, to whom they are compared, are not. We found that our horses learnt quite rapidly the name of a colour and also to pick out that colour from others when asked because, presumably, they are familiar with colours in their lives and they just had to learn its human name.[1]

*Are non-human mammals moral agents?*
An agent is a living, feeling being who is "active and operational". A moral being is one who knows the difference between "right" and "wrong". Therefore, a moral agent is a being who actively makes moral choices. Non-humans are not usually considered to be "moral agents", but rather of "moral concern" which means that humans have "rights" and "wrongs" in the ways we treat and have to do with animals.

However, all mammals live in groups where they have a social contract and rules of that society,[2] and they have a choice between obeying or not these rules. Such rules include "do not kill youngsters" or "do not go on attacking an individual who has given up". Although individuals do not always obey these rules, if they disobey frequently enough, then they may be injured or kicked out of the society, so to continue to live in the society, it is worth knowing and obeying the rules.

Moral behaviour of a moral agent is knowing the difference between "right" (obeying the rules) and "wrong" (not obeying the rules), whether the individual thinks the rules are sensible or not. Anyone who knows the rules, makes choices and acts on them, knows the difference between "right" and "wrong".

This means that humans can teach other mammals human moral rules and they can teach us theirs. For an elephant this may be "do not grab, lift or stamp on a person or an infant elephant", or for a dog "go towards someone who is calling you". The more thoughtful also respect their rules, such as "let a mother look after her baby". Although neither animals nor humans always obey the rules, both are moral agents.

---

[1] E Hartmann & MKW, "Enhancing learning of verbal cues in horses (*Equus caballus*) through cooperative teaching", Poster ISES and *Eco Research* pub 31. 2005

[2] MKW, "The mental homologies of mammals. Towards understanding of another mammal's world view" in *Animals* 2017, 7:12, DOI 10.3390/ani7120087

*Do non-humans have cultures?*
The social contract may vary depending on the species, the environment in which they live, the population density and so on. There may be different rules in different areas which results in different "cultures".[1] Cultures are ways in which individuals of a species or group behave. The cultures in human and non-human species depend on variations in the physical and social environment which controls the different expected ways of behaving. Today, we are beginning to learn how varied these can be. As a young research student in Uganda, I went to learn and help American researcher Dr Bruchner with his early studies on Uganda cob (an antelope) in the Semliki valley. He had found that the males in that area took up very small territories next to each other (around ten to fifteen metres in diameter); "territorial grounds", he called them. The females would arrive and wander through these territories choosing their male to copulate, while all the males display to them as best they could. But Uganda cob in other areas do not have these territorial grounds; the males have much larger territories, and can hardly see each other, so they do not fight each other or display to females as much. What the consequences of these differences were in terms of breeding and so on is not known, but what is known is that they had different "cultures", or different ways of behaving in the different areas. This may have been because of population pressure, availability of food or for other reasons. Horses in different stables also show different "cultures". In some stables all are very nervous and leap about a lot, in another, the same race or breed will be quite calm and relaxed due, of course, to their lifetime experiences and learning how to behave with each other.

*Communication, "theory of mind" and attention.*
Clearly, in a society with rules and a social contract, there is a need for communication. To receive messages, it is necessary to understand what the other is communicating, and therefore his inten-

---

[1] Kevin N Laland, Bennett, Galef (eds) *The question of animal culture*, 2009 Harvard University Press, p351

tions: is he friendly or not? is he interested or not? is there something scary over there? and many others. These messages are usually concerned with how the other is feeling but their understanding is central for the continuation of society. To recognise the other has feelings and intentions means that the other must have a mind. Recognising this has been called having a "theory of mind": that is, one knows the other has a mind. This, many argue, was a breakthrough in the understanding of where the differences between human and non-human mental abilities lay. For a few decades, it was maintained by the elite scientists, that it was only humans who had any idea that the one they were communicating with also had a mind and feelings. This, even though, in the same period (1970s to 1990s), it was becoming universally recognised that non-humans had emotions and feelings and communicated them by showing "intentions", and this was even recognised in law in some countries. More recently, other species of mammals, such as chimpanzees, cetaceans and elephants, have been allowed to have a "theory of mind" and therefore to enter this field of glory.[1]

But the "lower" mammals by our fireside, in our stable, or leaping about on the plains or in the forests, are often excluded from this exclusive band. This belief about "theory of mind" which is still held by many cognitive ethologists and psychologists has always struck me as daft. This is because, to retain a social group and live together, it is necessary to have some communication. Communication by definition transfers information and that information concerns feelings: how that individual is feeling and consequently his likely intentions; is s/he aggressive? is s/he friendly? has s/he seen something interesting and is frightened or curious? If he has intentions and others can see them, then, even according to the "theory of mind" theorists themselves, he *must* have a "theory of mind" because, as we have seen, he cannot have

---

[1] HS Terrace & J Mecalfe (eds) *The missing link in cognition. Origins of self-self-reflective consciousness*, 2005, OUP; G Clement, "Animals and moral agency: The recent debate and its implications" in *Journal of Animal Ethics*, 2013, 3:1, pp1-14

feelings and show intentions without having a mind.

Everyday folk knowledge on how we communicate with other species is based on knowing that the other has a theory of mind. If you are feeling nervous, he reads this and reacts to it. For example, if you wish to approach another who is unfamiliar with you, whether human or non-human, if you rush at him, shout and wave your arms about, *you know* that s/he will interpret this as threatening or frightening and will run away. When one thinks about it, it is clear that without knowing that the other has intentions, feelings and a mind, it would be impossible to associate together for any length of time, to learn from or teach each other in the every day world where most of us humans and non-human mammals live. Non-feeling robots do not live there ... yet.

There are of course, different ways of communicating. Humans mainly use verbal structured symbolic language, horses are expert visual communicators, dogs are good on smells, and elephants who use their trunk to communicate a great deal, use all their senses at once! Curiously, humans (particularly writers), fancy themselves as very good visual communicators. In almost every novel the author mentions that the human's story is in "the expression of the eyes". As a rule, all of us are rotten visual communicators when compared to an expert visual communicator like an equine. We concentrate on verbal messages so only an exaggerated, obvious facial expression, such as a smile or a frown, may tell another human how we feel. One rapid blink, or eye movement, a very slight twitch or relaxation of a muscle will usually be missed by another human, although it may be picked up by dogs or horses, and tell them about your feelings and intentions.

We humans pay attention to words and phrases, and understand the context-independent message that is being conveyed in any language, once we have learnt the meaning of the words. However, we are not very good at reading the expression and the mood of the communicator when compared to non-humans who have learnt to read humans. When the author "reads the expression in the eyes", this usually reflects a preconception of the observer's belief concerning what that individual is feeling, but not of the subject.

Of course, politeness and good manners in interactions with others is necessary for group stability and pleasure, but trying to please others by lying about one's emotions is not always the best policy when dealing with non-humans because they can read your insincerity, while humans may be easier to hoodwink. Therefore, when trying to learn another mammal's language, learning how to communicate one's emotions more honestly helps, but also one needs to be more subtle about it. Often, this is a better starting point to developing a pleasant human-animal relationship; but it can also be used to help with human relationships. Some who used "therapy animals" to help human misfits have, perhaps, recognised this, tacitly at least.

Some years ago, there was an extraordinary idea put around by the cognisate (respected comparative psychologists) that non-humans did not know what they were looking at. However, it is not possible for mammals to live in the world or in a group where they have to make decisions and choices, unless they see things and know what they are. They do not name these in human language, but they certainly know what they are looking at, and go one step further and know what *you* are looking at. We did some experiments teaching Shemal and Oberlix to "touch the paper that I am looking at". The pieces of A4 paper were fastened fifty centimetres apart from each other, and the horse had to walk seven metres from where I was standing and staring at the chosen paper to touch it. They touched the right piece ninety-five per cent of the time. This was rather better than I could do when watching another human. Reading a human's direction of visual attention from the eyes is relatively easy, at least for a horse, because the pupil is surrounded by the white of the eye. Detecting where a horse, or even a chimpanzee may be looking is much more difficult for us as they have no whites of the eyes, yet horses and many other species can do it. Maybe it is humans who often do not know what they are looking at?

*Group belief and "brain-washing".*
Because humans, like most other mammals, like to do what others do: to follow, imitate, accept and believe in something that they have learnt from others, it is relatively easy to change the motivation of a group, and it can go as far as self-sacrifice. People, horses, dogs or elephants, can be quite easily encouraged to sacrifice their lives. This is because the majority tend to believe that whatever is put across by the group they are part of allows them to remain a part of that group. Even an individual who does not want to follow may know that walking out of step may lead to being thrown out of the group, a worse consequence. When pacifists were presented with a white feather in the First World War, this was a symbol that they were no longer part of the group. In the same way a cavalry horse or an elephant who will not join the "charge of the light brigade" will be excluded from the group by being sent away or shot.

Some of the general beliefs and behaviours performed by the group may be to the species' and/or individual's advantage, but, some may lead towards the ultimate demise of that group or species.

*Self-Awareness.*
A belief, still strongly held by many, is that only humans are "self-aware". It has long been considered that if an individual did not recognise himself in a mirror, then he could not be self-aware. However, there is a great deal more to "self-awareness" than recognising oneself in a mirror, for example, if you feel pain, you know it is YOU feeling pain, and you know where ... so does a horse or a dog.

In addition, the elephant, dog, horse, rat and any other mammal, does not only feel, but also sees at least parts of his body, and knows it is his. He learns to direct parts of his body in a whole variety of ways, because he can see and feel it and therefore knows it is his.

This learning to do certain things can be divided into learning and knowing "that", that is what to do. For example seeing a bicycle and knowing that it is for riding, a jumping horse sees a jump and knows it is to be jumped, or an elephant seeing a particular log knows it is to be lifted and placed on a pile. But knowing "that" does not mean that s/he can actually do it; he has to coordinate his nerves, muscles, limbs and so on to do it. In other words, he has to learn "how". The philosopher Ryle in 1949,[1] was the first to point out how learning is divided into these two parts. Take for example our horses learning to do the Spanish walk, they learn "that" they must raise their front legs one at a time higher than normal and at the same time move forward; but doing it is another matter. They can do the lifting but when they put their leg down, their body is so stretched that they are often stuck because they have not learnt to move their back legs forward.

*Lifetime experiences, and group or species conservation.*
One thing that is becoming increasingly realised is that the lifetime experiences of any mammal, human or not, are crucial to an understanding of the particular individual's world view. This is amassed from what he knows, what he has learnt, what he feels, and from other past experiences. Lifetime experiences therefore are important to all mammals, not just humans, and they matter.[2]

It also means, that if you and an individual of a different species have had common experiences, perhaps by living together for much of your lives, you may have more in common in how you see the world with the non-humans that you live with, than with a human from a different environment or culture. This is the experience of many, and myself. For example, we have had many motivated, interested humans come to stay or learn with us, but, although they are of the same species, they often have such different lifetime experiences from ours that how we live our lives and the experiences we have can be more similar to those of our dogs

---

[1] Gilbert Ryle, *The concept of mind*, 2002, University of Chicago Press
[2] JT Bonner, *Evolution of culture in animals*, 1986, Princetown UP, Eytan Avital & Eva Jablonka, *Animal traditions: Behavioural inheritance in evolution*, 2000, Cambridge UP; Philippa Brakes, Sasha RX Dall, Lucy M Aplin, et al, "Animal cultures matter for conservation" in *Science*, 2019, 363:6431, 1032-1034. DOI: 10.1126/science.aaw3557

or horses, than to them and their anthropocentric view.

Each of us mammals remain the species we were born, but our individual epistemology (world view or knowledge of the world) changes as a result of our experiences. We have innate tendencies to behave and learn certain things, but the experiences of our lives mould them, and may even trump the genetic tendencies. For example, although humans and dogs have genetic tendencies to live in groups and eat particular types of foods, there are, and always have been, people and dogs, who have lived preferentially alone, with a different species and eat different things.

*Are humans more cooperative than other mammals?*
It is fiercely argued today by the cognisate (e.g. 1,2,3,) that humans cooperate more and are better at it than other species, as well as being more compassionate and empathetic. This leads one to ask why, if this were the case, they do not control their population and stop galloping consumerism, both of which result in outstripping the planet's renewable and non-renewable resources. It is only humans and their activities which are causing a species a day to become extinct, unbalancing the ecosystem. Only humans manipulate the genes of species reducing their adaptability. Only humans spray poisons of one sort or another all over the land to try to grow more food for themselves, thereby killing trillions of other individuals and species.[1] Finally, it is only humans who kill each other by the hundred thousand in wars, and rape and murder each other daily, all over the world. These are just a few of the present human game plans. How can humans be said to be more far sighted, more cooperative, more tolerant and more compassionate than any other species since no other species does these things?

Perhaps it is something to do with the evolutionary background of humans. The primate social contract favours anything that will lead to their own individual short-term advantage and anthropomorphism has developed "selfishness" as a specialism. However, it is now emerging that the rest of the living world, from bacteria

---

[1] M McCarthey, *The moth snowstorm*, 2015, New York Review

onwards, live symbiotically with each other, and it appears that it is this symbiosis which is the end game of evolution.[1]

In primates among the mammals, food is not readily available everywhere (such as it is for most herbivores), but is found packaged. So there is likely to be competition for it and this may cause injury to individuals. The evolutionary story is that, in such a group, there is a known distribution of power among the individuals which is called a "dominance hierarchy" to control access to restricted resources and reduce fights and injury to individuals. This power distribution or "dominance hierarchy" is where each individual in the group knows his place in the frequent competitive social world. Most members of the group go along with their "dominance status", until another opportunity arises. But, obtaining a higher dominance status, or, obtaining power over others, is desired. By getting people to believe in your "dominance", more members of the group are likely to believe in the power that you as an individual have. This is not the result of mutual cooperation, it is the result of competition and a specialism of primates and other species where food is unequally distributed.

But, all mammal societies do not run along the same lines, even though the idea of constant competition and the resulting "dominance hierarchies" have become a dogma for ethologists, popular science and television wildlife programmes. In some species, this is not how their society operates. Large herbivores, although they can be forced to develop power struggles and dominance hierarchies (such as when kept with very limited resources), evolved to live in groups in the great outdoors. In their habitats, either all have access to resources (food, water, shelter) or there is none available for any. When studied carefully, large herbivores resist power struggles in favour of demonstrating how to stick together and cooperate.[2] For example, we measured ninety-six behaviours in a group of elephants, including who did what to whom, and

---

[1] Marguis & Sagan, *Microcosmos*, 1987; Tsing, Swanson, Bubandt, *Arts of living on a damaged planet*, 2017
[2] MKW, "Communication in a small herd of semi-domestic elephants," 2019

what the responder did. We found that each individual had a particular role, or personality; some were more socially involved than others, some more aggressive, some more affiliative than others, and some were more often receivers of messages than others. There were hierarchies of affiliation, aggression, showing interest in others and so on, but all performed significantly more behaviour related to cooperation or "sticking together", that is, being nice to each other, rather than having conflict which could result in splitting the group. Even though we correlated the ranking of all the individuals across the various behaviours with each other, we could not come up with a competitive or power structured parameter such as a dominance hierarchy.

There were some other interesting results. For example, many directed messages were ignored by the recipient. Also, significantly more behaviours were reciprocated with the same behaviour than would be expected by chance. This looks as if "do as you would be done by", or "be done by as you did"[1] are important in elephants and other large herbivores. Another surprise was that behaviours that indicate that the individual was uncertain how to behave or respond to another individual were very frequent. One is tempted to consider that this is another method of encouraging sticking together, a sort of demonstration of "good manners".

Instead of individuals competing in the group, they demonstrated cooperation and obeying the social contract.

But why do they need to stay together? The most obvious reason is that they need to acquire an enormous amount of information to survive. They have to become good natural ecologists; good botanists to know what to eat (there are many plants, which ones are good to eat and which are not) and where to find them. They must be good natural geographers, know where they are and find their way to other places, good natural zoologists to know which other animals are around, whether predators or not. They need to become natural ethologists to know how other animals are behaving, are the lions hunting or relaxing? That this is the case is

---

[1] Moral laws to help educate children in *The Water Babies* by Charles Kingsley, 1885, MacMillan, London

shown by many antelope (springbok, Grants and Thomson's gazelles) who will approach close to sleeping replete lions; but when the lions wake up and begin to look hungry they spronk off as fast as they can. All mammals must become good natural physicists too; they know about gravity, for example, when to get out of the way of things falling. They must become good natural geologists to know where they can walk and where they will slip on different rocks. The chamois who live in the Alps can negotiate some almost perpendicular cliffs without falling, but certain cliffs with different rocks they do not try to climb.

In effect, all mammals must become *good natural historians to survive and breed.* They must also of course become *good sociologists* to get on and live in their society so they can acquire this information.

The easiest way of accumulating knowledge is to live in a stable society where individuals get along and learn from each other. The elephant, rhinoceros and horse groups we studied, were primarily being cooperative and nice to each other. But, if someone broke the social rules by, for example, being aggressive without justification, he was aggressed back. These societies were not about competition between individuals and obtaining power; they fostered cooperation and sticking together where every individual had their own role.

Maybe "dominance hierarchies" and power struggles are normal for humans and some other primates, but we could learn a thing or two from non-primates. Perhaps, we need to learn that it is more cooperation that we need if we are going to keep our planet alive.

There *must* be multi-national agreement for a cooperative effort which will have to take the place of competition and power struggles among the not-very-cooperative human species.

We now know that there are many examples of extraordinary types of cooperation in non-humans, both between and with other species, including with humans. There is a also a growing understanding of the enormous importance of symbiosis (cooperation for mutual benefit) to make life work.[1]

By looking at things differently, we find that most mammals' social ontology,[2] the things they do as a society, are principally cooperating and acting together. I am constantly amazed by the extraordinary ability of our four-legged friends to cooperate with each other and particularly with humans. If you learn to listen to their stories, they surprise you, and make you realise how much we can learn from them in their willingness to cooperate.

Each species has their own physical and mental specialism; no one is overall superior to another, we are all different. The more we study and look at this in the different species, the more this is confirmed. The day before writing, Shatish and I rode over the moors, through the villages, across the roads, to visit a friend fifty miles away. Going there I was constantly reading the map to stay on the bridle paths and to go in the right direction. On the way back, I had lost the map. I left it to Shatish, and although I often had no idea which way we had come, he choose the correct way every time almost without pause, and there must have been around three hundred choices at least ... once travelled, well remembered. This was something I could not do, but he had no difficulty.

In the future there will be more ideas floated concerning how humans are better, superior, more cooperative, sympathetic, intelligent, and mentally able than any other species, but the belief in the all time "superiority" of humans, rather than their "species differences, and special mental skills" is more difficult to maintain, other than by worshipping technological genius.

Mental abilities do not arrange themselves like a tree with higher and lower branches, they are like an ever-branching, growing, expanding bush where some branches intertwine or graft onto each other, as we learn something about others and their world views. Now is the time to look in depth at all aspects of the body, behaviour, brain and mind of each species and begin to outline the

---

[1] Tsing, Swanson, Bubandt, *Arts of living on a damaged planet*, 2017
[2] J Searle, "Social Ontology some basic principles", 1 March 2006 Research Article https://doi.org/10.1177/1463499606061731.

species' world view: their epistemology. A few scientists and philosophers are beginning to establish a methodology for doing this,[1] but there are an awful lot of mammals to consider. Consider them we must before they become extinct, lost forever without the opportunity of learning from them. If only we would pause to think and learn from our close relatives, we might also find many answers to our current environmental and social problems.

It is a windy day, I have been lying on the lawn of our mountain home and listening to the wind as the sun dodges in and out of the clouds ... each tree makes a different noise as it is blown: the conifers whistle a little and bend, the ash are covered in their finger-like seeds with large leaves, make the most noise, rattling and waffling, while the apricot next to my window sometimes feels the blast and ripples with it; sometimes it is still, almost listening to the others. Thinking about this made me realise that most people raised in cities, have never heard the wind in the trees, never mind know that different trees make different noises. When the wind blows, there is a whole orchestra out there. They have never felt the silence of a hot still afternoon. Silence is unknown to them, there is always traffic noise, noise of machines, people walking, running, shouting and laughing, the fridge, the washing machine, even the lights making noises. Many people when they come here find the lack of human-made noises disturbing, some finally relax, but all remark on it. In the early morning there is regularly a bird chorus, chattering and singing, whooping and calling to each other, loud or soft, near and far. In the evenings in the barn, the munching, gurking, farting, sighing of animals eating, sleeping, and relaxing. During the day the chickens clucking and scraping, the cock proclaims his finds to call his girls, eggs being laid and loudly announced. During the night, rustling and cracking of animals moving around, the odd dog barking every now and then if the wild boars are too close, or the still quietness of the moonlight. Many people, it turns out, have never seen the stars, but we watch them almost every night coming and going,

---

[1] De Waal, *Are we smart enough to know how smart animals are?*; MKW, *Right in front of your mind ...* 2000

twinkling and establishing us within the universe. I do not know one from the other, and I do not need to. It is the reassuring presence of them in their millions, their being there that matters; yet if you have never seen them, you would never know.

Humans in cities have put themselves into the equivalent of an animal-intensive farm. They are well fed, and reproduce themselves; in fact, their reproduction has become one of the great highlights of their lives, even though the result is bound to destroy the world with too many of them. They accept the lack of freedoms to do so many things that to us are just normal. They think they have improved their lives because they are always warm, physically comfortable and have each other to argue, laugh and shout with. But they do not know the feeling of the sun on their backs or the sight of the stars, the silence of the black nights or the still, bright moonlight, the noise of the wind or rain, the rumbling, blasting of the thunder and lightening, the dreadful coldness of walking home through the mountains in the dark in the snow, the necessary self-reliance if you fall in a ditch and strain your ankle or break your leg, with no medical back-up, the crispness of a frosty morning, the intense pain of being bitten by a wasp.

They may have a dog or a cat they like very much, and have some emotional exchanges with, but they never experience the joy of a group of young horses or heifers as they rush out into the crisp autumn morning leaping and cavorting, or the overpowering sensation of the silliness of your worries as you bury you face in the neck of your horse, who is tolerant but not overtly sympathetic. Yes, as rural dwellers with our animals, we suffer daily from cold, heat, flies, hunger, thirst, aches and pains, exhaustion, stiff backs, shoulders, thorns in the feet and many more difficulties, but we can make choices and have freedoms, and we have a whole world of experiences that those living in cities never have in their comfortable homocentric world. They have their parties, drinking bouts, drugs, social get-togethers, theatres, operas, art, music of many kinds, television and talking. We can, and do, have these as well ... if we wish. The everyday living of an urban dweller is what we would call a half life. Some fall through the net in the city and are hungry or cold, but there is a net.

Humans can become aware of the world and perhaps begin to value and worship it again like the traditional believers did. One way of beginning this journey is to have contact with other species and to get to know a whole other world, both physically and mentally. This does not come from taking your dog for a walk, (although this is a start), it comes from throwing away preconceptions about dogs and then reading their stories: what is it they are thinking, experiencing and believing about now, then, or all the time? Can I understand and if I can, how is it different or similar to my world view? Can it tell me things I have never known before? Can I enrich my own life by this contact?

There are thousands of different species out there to learn from and they all have different species' stories, and of course, each individual has a different story to tell too.

If, to start with, we restrict our story-reading to other mammals, because we are one, we will gradually recognise our similarities, and then our differences. This will add layers of enjoyment and experiences to our lives, and, perhaps change the way we live so that we all continue to exist.

My life has been one of trying to learn to read the stories of my four-footed family and friends, but so far we have only scratched the surface of their knowledge and understanding. I have learnt so much about how to live and what to feel, and when not to from them. It may be that the next generation of humans will find it even more difficult to do this because of the lack of animals, and their isolation from the living world. But one thing is certain, if anthropocentrism does not lose favour, humans will destroy the living world, and all its different aspects. They will destroy not only their remaining liberties, joys and delights, but eventually, their lives, even if along the way they become robots who do not know what they are missing.

We have to act now and fast as many humans continue to believe they are superior beings, increase their numbers and destroy the living world, because they can. We can change things, if we are serious. This will require some radical lifestyle changes which may be difficult, and there is no knowing if these will make you happier. But, there is no option, it is only such changes that will keep us alive, living and smiling in the real, living world.

# Publications by Marthe Kiley-Worthington referred to in text - in year order

Spacial distribution and sexual behaviour of the waterbuck (*Kobus defassa & K.ellipsipirmnus*) in East Africa, in *Mammalia*, 1965, 29, pp199-204

A preliminary investigation into the feeding habits of the waterbuck in East Africa by faecal analysis, in *East African Wildlife Journal*, 1966, 4, pp153-157

*Some displays of canids, felids and ungulates with particular reference to their causation*, 1969, Dphil thesis, University of Sussex

The vocalisation of ungulates, in *Zeit fur Tierpsychol*, 1972, 31 pp171-222

The tail movements of ungulates, canids and felids with particular reference to their causation and use of displays, in *Behaviour*, 1976, LVI, pp69-115

*The behavioural problems of farm animals*, 1977, Oriel Press

The visual displays of eland, in *Behaviour*, 1978, LXVI, pp179-222

The social organisation of a small herd of captive eland, oryx and roan antelope, in *Behaviour*, 1978, LXVI, pp32-40

Individual differences in a small herd of captive eland, oryx and roan antelope with particular reference to personality profiles, in *Behaviour*, 1978, LXVI, pp44-55

The problems of modern agriculture, in *Food Policy*, 1980, August, pp208-215

*The behaviour of beef suckler cattle* (Bos taurus), 1983, Verlag Birkhauser, Basel [with S de la Pain]

The behaviour of confined calves raised for veal. Are these animals distressed, in *International Journal of the Study of Animal Problems*, 1983, 4, pp198-213

Animal language? Vocal communication of some ungulates, canids and felids, plenary talk, in E Erkinars (ed.) *Proceedings of International Thereological Congress*, 1984, Helsinki

Ecological agriculture. A case study of an ecological farm in the South of England, in *Biology, Agriculture and Horticulture*, 1986, 2, pp 101-133 [with CC Rendle]

Cattle husbandry in organic agriculture, in *International Federation of Organic Agricultural Movements Colloquium*, 1986, Kassel, Germany

*The behaviour of horses in relation to management & training*, 1987, JA Allen, London, reprint 1998 and translations

Ethological, ecologically and ethically sound environments for animals, in *Journal of Agricultural ethics*, 1989, 2, pp323-247

*Animals in Circuses and Zoos, Chiron's world?*, 1990, Eco Farm, Basildon

Animal welfare. Towards symbiosis for the 21st century?, *International Ethological Congress*, 1991, Kyoto

*Ecological agriculture. Food first farming*, 1992, Souvenir, London

Implications of semi-intensive management on the breeding of Black Rhino (*Diceros bicornis*), in *International Society of Applied Ethology*, 1996, Eco Research Centre publication 29 [with CC Rendle]

First steps in comparative animal educational psychology, in *International Social Comparative Psychology*, 1996, Eco Research Centre publication 30 [with H Randle]

*Equine watch*, 1997, JA Allen, London

Can animals think?, in V Bruce (ed.) *Unsolved Mysteries of the Mind*, 1997, Erlbaum

Communication in horses. Competition or cooperation?, 1997, Eco Research Centre publication 21

Cooperation & Competition – A Detailed Study of Communication and Social Organisation in a Small Group of Horses at Pasture. *Eco Research Centre.* 21, 1997

A comparison of behaviour of group and individually caged male Angora rabbits, *Talk University Federation for Animal Welfare*, India, 1998, Eco Research Centre publication 33 [with C Daley]

Ecological agriculture and improved animal welfare for development in women's agriculture, *Address for Women's World Summit Foundation prize for creativity in rural life*, 1998

*Right in front of your mind. Equine and elephantine epistemology*, 2000, MPhil thesis, University of Lancaster

Integrating food production and wildlife conservation on ecological farms. Wildlife ambassadors helping rural women, *Invited talk for Commonwealth Heads of State Environmental Committee*, 2003, Abujan, Nigeria, Eco Research Centre publication 22

*Exploding the myths. Large mammal welfare science and educational psychology*, 2012, Exlibris [with Jake Rendle-Worthington]

Ecological agriculture, integrating low input, high productive farming with wildlife conservation, results from the experimental farm La Combe, Drome, France, in *Open Journal of Ecology*, 2014, 4, pp744-763

The roles of individual and social networking in a small herd of domestic horses, in *Journal of Animal Health and Behavioural Science*, 2017, 1:1 [with C Ricci-Bonot]

The mental homologies of mammals. Towards understanding of another mammal's world view, in *Animals*, 2017, 7:12, DOI. 10.3390/ani7120087

Communication in a small herd of semi-domestic elephants. Another interpretation of their social organization, social contract and world view, in *Journal of Animal Research and Veterinary Science Research Article*, 2019, 3, DOI. 10.2466/ARVS-3751/100012

How to measure quality of life in equines. *Revista General de Derecho Animal y Estudios Interdisciplinares de Bienestar Animal / Journal of Animal Law & Interdisciplinary Animal Welfare Studies*, 2019, 3

Proceedings of a Workshop on establishing welfare standards for captive or semi-captive elephants. ZEWACT, Victoria Falls. 2019

# Select Index

Agricultural Research Council ...128
Andrew Fraser ... 129
Antelope ... 16, 25, 27, 51, 86, 94. 97, 101, 137ff, , 137, 146, 158, 233, 317, 354, 385, 404
Barak Obama ... 32
Bilharzia ... 30, 50, 130
British Council ... 246
British Horse Society ... 56, 173
Cape Buffalo ... 267, 362ff
Carol Jones ... 30
Churchill Memorial Travel Fellowship ...165, 324
Circus ... 60, 104, 169, 207, 248, 250
Club of Rome ... 127
Collective intentionality ... 60-61, 146, 342, 366ff
Compassion in World Farming ... 163, 270ff
Consciousness ... 60, 363ff, 374, 380
Crib bite ... 57, 66
Cynthia Moss ... 130
David Woodgush ... 129
Donald MacNellish ... 190
Donkey/Donkey Sanctuary ... 82, 150, 244-246, 248, 250, 259, 266, 376
Double-sucklers ... 163-164, 168, 231, 270
EF Shumaker ... 126
Elephant Bill ... 259, 280
Emotions/feelings ... 22, 68, 98, 119ff, 133, 154ff, 164, 169, 179ff, 204, 206, 213, 219ff, 224, 235, 239, 255, 272, 292, 319, 327. 329ff, 334ff, 344, 346, 355ff, 364, 366, 371, 373,
379, 396, 398, 407
Eve Balfour ... 126
Fish ... 6, 14, 18, 321
French Equitation ... 288
Fresh Water Research Centre ...14
Game/Nature Reserves ... 25, 36, 61, 114, 125, 132, 137, 145, 169, 176ff, 226ff, 247, 251, 253ff, 259, 261ff, 266, 277, 294, 296, 303, 310, 313, 316ff, 320, 351, 353
George Carter ... 13
George Clay ... 102, 112
George Iwanowski ... 138ff, 154
Guide Dogs for the Blind ... 244, 360
Horse Whisperer ... 150
Human therapy ... 183, 306, 347, 358
Hunting ... 28ff, 46, 56, 60, 69, 173, 177, 233, 260, 301ff
Idi Amin ... 45
International Biological Programme ... 44
International Union for Conservation of Nature ... 94, 102, 111
Jack Dyer ... 19
Jane Goodall ... 25, 99, 377
Jenny Car ... 30
Johanna – syce ... 22, 31, 45, 48ff
John Corner ... 13
John Maynard Smith ... 114
John Searle ... 366
Jomo Kenyatta ... 42
Jonathon Porritt ... 127
Konrad Lorenz ... 23
Language/communication ... 48, 56, 59, 81, 87, 99, 113ff, 119, 121, 123ff, 137ff, 140, 147ff, 187, 231, 241, 246, 253, 258,

299, 304, 327, 338, 346, 354, 358, 360, 366, 368, 378ff, 381, 383ff, 387, 391, 394ff, 397ff
Lesley Rogers ... 116
Louis Leakey ... 26, 99
Lippizana ... 138ff
Lord Hailey ... 13
Mammal Research Institute ... 134
Mary Midgley ... 205
Monty Roberts ... 149
Mulesing ... 225
Multi-species home ... 230, 338
Nikolaas Tinbergen ... 23
Norman Travers ... 246ff, 254, 264
Nuffield Project ... 144ff
Peter Scott ... 324
Peter Singer ... 205
Rabbits ... 213, 223, 232
Rachel Carson ... 126
Richard Laws ... 88
Robert Hall ... 55, 59, 68, 139, 154
Roger Ewbank ... 129
Rori Hensman ... 248
Rudi Shenckel ... 95, 98
Ruth Harrison ... 126
Sheelagh Wakely ... 32
Sleeping sickness/Trypanosomiasis ... 46, 188, 267
Social contract/moral code ... 146, 153ff, 170ff, 186, 199, 207, 222, 276, 300, 332, 338ff, 344, 349, 358, 394ff, 401, 403

Society for the Prevention of Cruelty to Animals ... 274, 277
RSPCA ... 58, 146, 157, 180, 207ff, 218, 220ff, 231
Society of Veterinary Ethology ... 129
Soil Association ... 166
SPANA ... 151
Spanish Riding School ... 55, 138ff
Thinking cow ... 159, 193ff, 303, 347
University Federation for Animal Welfare ... 279
Veal calf/meat study ... 163ff
Vegan implications ...176ff, 272ff
Waterbuck ... 16, 28, 86ff, 93, 95ff, 108, 110, 138
Weaving ... 58ff 119, 147ff, 151, 340
WWOOFers ... 164, 168, 173, 182ff, 190, 204, 242ff, 308
Zoo ... 17, 114, 138, 145, 208, 210ff, 233, 254, 264

www.ingramcontent.com/pod-product-compliance
Lightning Source LLC
Chambersburg PA
CBHW032008220426
43664CB00006B/180